Methods in Enzymology

Volume 357
CYTOCHROME P450
Part C

METHODS IN ENZYMOLOGY

EDITORS-IN-CHIEF

John N. Abelson Melvin I. Simon

DIVISION OF BIOLOGY
CALIFORNIA INSTITUTE OF TECHNOLOGY
PASADENA, CALIFORNIA

FOUNDING EDITORS

Sidney P. Colowick and Nathan O. Kaplan

Methods in Enzymology

Volume 357

Cytochrome P450

Part C

EDITED BY

Eric F. Johnson

DIVISION OF BIOCHEMISTRY
DEPARTMENT OF MOLECULAR AND EXPERIMENTAL MEDICINE
THE SCRIPPS RESEARCH INSTITUTE
LA JOLLA, CALIFORNIA

Michael R. Waterman

DEPARTMENT OF BIOCHEMISTRY
VANDERBILT UNIVERSITY SCHOOL OF MEDICINE
NASHVILLE, TENNESSEE

EDITORIAL ADVISORY BOARD

Francis Durst
Cossette J. Serabjit-Singh

ACADEMIC PRESS

An imprint of Elsevier Science

Amsterdam Boston London New York Oxford Paris
San Diego San Francisco Singapore Sydney Tokyo

Academic Press
An imprint of Elsevier Science.
525 B Street, Suite 1900, San Diego, California 92101-4495, USA
http://www.academicpress.com

Academic Press
84 Theobalds Road, London WC1X 8RR, UK
http://www.academicpress.com

International Standard Book Number: 0-12-182260-5

PRINTED IN THE UNITED STATES OF AMERICA
02   03   04   05   06   07   SB   9   8   7   6   5   4   3   2   1

Table of Contents

Section III. Regulation of Expression

Section IV. Metabolism

Section V. Invertebrate P450s

Contributors to Volume 357

Article numbers are in parentheses following the names of contributors.
Affiliations listed are current.

DOUGLAS J. A. ADAMSON (19), *ICRF Molecular Pharmacology Unit and Biomedical Research Center, Ninewells Hospital and Medical School, Dundee DD1 9SY, Scotland, United Kingdom*

TOMOYOSHI AKASHI (35), *Department of Applied Biological Sciences, Nihon University, Fujisawa, Kanagawa 252-8510, Japan*

METTE DAHL ANDERSEN (32), *Protein Design, Novo Nordisk A/S, DK-2880 Bagsværd, Denmark*

TOMMY B. ANDERSSON (14), *Drug Metabolism and Pharmacokinetics and Bioanalytical Chemistry, AstraZeneca R&D Mölndal, S-431 83 Mölndal, Sweden*

PAVEL ANZENBACHER (15), *Department of Pharmacology, Faculty of Medicine, Palacky University, 775 15 Olomouc, Czech Republic*

TOSHIO AOKI (35), *Department of Applied Biological Sciences, Nihon University, Fujisawa, Kanagawa 252-8510, Japan*

ALEXANDER I. ARCHAKOV (10), *Institute of Biomedical Chemistry, Russian Academy of Medical Sciences, Moscow 119992, Russia*

SHIN-ICHI AYABE (35), *Department of Applied Biological Sciences, Nihon University, Fujisawa, Kanagawa 252-8510, Japan*

CLAUDE BALNY (15), *French National Institute for Health and Medical Research U 128, IFR 24, F-34293 Montpellier Cedex 5, France*

FRÉDÉRIC BANCEL (15), *French National Institute for Health and Medical Research U 128, IFR 24, F-34293 Montpellier Cedex 5, France*

IRÈNE BENVENISTE (34), *Department of Cellular and Molecular Enzymology, Institute of Plant Molecular Biology/CNRS, F-67083 Strasbourg Cedex, France*

CHRISTOPHER A. BRADFIELD (20), *McArdle Laboratory for Cancer Research, University of Wisconsin Medical School, Madison, Wisconsin 53706*

CYNTHIA BRIMER-CLINE (31), *Department of Pharmaceutical Sciences, St. Jude Children's Research Hospital, Memphis, Tennessee 38105*

KATHY CARROLL (17), *Department of Molecular Biology, XenoTech LLC, Lenexa, Kansas 66219*

WING-TAI CHEUNG (22), *Department of Biochemistry, The Chinese University of Hong Kong, Shatin, New Territories, Hong Kong, China*

JOSÉ COSME (12), *Astex-Technology, Cambridge, England CB4 OWE, United Kingdom*

BRIAN R. CRANE (13), *Department of Chemistry and Chemical Biology, Cornell University, Ithaca, New York 14853*

MARK W. CRAVEN (20), *Department of Biostatistics and Medical Informatics, University of Wisconsin Medical School, Madison, Wisconsin 53706*

CHARLES L. CRESPI (26), *BD Biosciences, Woburn, Massachusetts 01801*

MACIEJ CZERWINSKI (17), *Department of Molecular Biology, XenoTech LLC, Lenexa, Kansas 66219*

JOHN H. DAWSON (13), *Department of Chemistry and Biochemistry, University of South Carolina, Columbia, South Carolina 29208*

ILIA. G. DENISOV (11), *Department of Biochemistry, University of Illinois, Urbana, Illinois 61801*

ELIZABETH S. DENNIS (37), *Division of Plant Industry, Commonwealth Science and Industrial Research Organization, Canberra, ACT 2601, Australia*

IVAN J. DMOCHOWSKI (13), *Beckman Institute, California Institute of Technology, Pasadena, California 91125*

NORMAN R. DRINKWATER (20), *McArdle Laboratory for Cancer Research, University of Wisconsin Medical School, Madison, Wisconsin 53706*

ALEXANDER R. DUNN (13), *Beckman Institute, California Institute of Technology, Pasadena, California 91125*

FRANCIS DURST (34), *Department of Cellular and Molecular Enzymology, Institute of Plant Molecular Biology/CNRS, F-67083 Strasbourg Cedex, France*

JEAN-MICHEL FABRE (30), *French National Institute for Health and Medical Research U 128, F-34293 Montpellier, France*

JEAN-BERNARD FERRINI (30), *French National Institute for Health and Medical Research U 128, F-34293 Montpellier, France*

JOACHIM FISCHER (5), *Transgenomic, Ltd., Crewe, Cheshire CW1 6UZ, United Kingdom*

KEN-ICHI FUJITA (29), *Department of Pharmacy, Miyazaki Medical College Hospital, Miyazaki-gun, Miyazaki 889-1692, Japan*

LAWRENCE L. GAN (17), *Bristol-Myers Squibb Company, Newark, Delaware 19714*

SABINE GERBAL-CHALOIN (30), *French National Institute for Health and Medical Research U 128, F-34293 Montpellier, France*

SHARI D. GOODZ (7), *Center for Addiction and Mental Health, Department of Pharmacology, University of Toronto, Toronto M5S 1A8, Canada*

SANDRA E. GRAHAM (2), *Department of Biochemistry, The University of Texas Southwestern Medical Center, Dallas, Texas 75390*

HARRY B. GRAY (13), *Beckman Institute, California Institute of Technology, Pasadena, California 91125*

MICHAEL T. GREEN (13), *Beckman Institute, California Institute of Technology, Pasadena, California 91125*

ALAIN HEHN (33), *Department of Plant Stress Response, Institute of Plant Molecular Biology, CNRS UPR 2357, F-67000 Strasbourg, France*

CHRIS A. HELLIWELL (37), *Division of Plant Industry, Commonwealth Science and Industrial Research Organization, Canberra, ACT 2601, Australia*

CHRISTIAN HELVIG (34), *Department of Cellular and Molecular Enzymology, Institute of Plant Molecular Biology/CNRS, F-67083 Strasbourg Cedex, France*

SUSAN M. G. HOFFMAN (4), *Department of Zoology, Miami University, Oxford, Ohio 45056*

YASUSHI HOJO (23), *Department of Biophysics and Life Sciences, University of Tokyo at Komaba, Meguro, Tokyo 153, Japan*

YEON-PYO HONG (6), *W. Harry Feinstone Center for Genomic Research, University of Memphis, Memphis, Tennessee 38152*

GASTON HUI BON HOA (15), *French National Institute for Health and Medical Research U 473, F-94276 Le Kremlin Bicetre, France*

MAGNUS INGELMAN-SUNDBERG (3), *Division of Molecular Toxicology, IMM, Karolinska Institutet, SE-171 77 Stockholm, Sweden*

YURI D. IVANOV (10), *Institute of Biomedical Chemistry, Russian Academy of Medical Sciences, Moscow 119992, Russia*

ERIC F. JOHNSON (8, 12), *Division of Biochemistry, Department of Molecular and Experimental Medicine, The Scripps Research Institute, La Jolla, California 92037*

STACEY A. JONES (16), *Department of Nuclear Receptor Functional Analysis, High Throughput Biology, GlaxoSmithKline, Inc., Research Triangle Park, North Carolina 27709*

TETSUYA KAMATAKI (29), *Laboratory of Drug Metabolism, Hokkaido University, Kita-ku, Sapporo 060-0812, Japan*

SUGURU KAWATO (23), *Department of Biophysics and Life Sciences, University of Tokyo at Komaba, Meguro, Tokyo 153, Japan*

DIANE S. KEENEY (4), *Medical Research Service, Department of Veterans Affairs Medical Center and Department of Medicine/Dermatology, Vanderbilt University School of Medicine, Nashville, Tennessee 37232*

TETSUYA KIMOTO (23), *Department of Biophysics and Life Sciences, University of Tokyo at Komaba, Meguro, Tokyo 153, Japan*

KATHRIN KLEIN (5), *Dr. Margarete Fischer-Bosch Institute of Clinical Pharmacology, D-70 376 Stuttgart, Germany*

STEVEN A. KLIEWER (16), *Department of Molecular Biology, University of Texas Southwestern Medical Center, Dallas, Texas 75390*

TONI M. KUTCHAN (36), *Department of Natural Product Biotechnology, Leibniz-Institut für Pflanzenbiochemie, D-06120 Halle/Saale, Germany*

THOMAS LANG (5), *Epidauros Biotechnologie AG, D-82347 Bernried, Germany*

REINHARD LANGE (15), *French National Institute for Health and Medical Research U 128, IFR 24, F-34293 Montpellier Cedex 5, France*

SUSANNA SAU-TUEN LEE (22), *Department of Biochemistry and Environmental Science Program, The Chinese University of Hong Kong, Shatin, New Territories, Hong Kong, China*

WING-SUM LEE (22), *Department of Biochemistry, The Chinese University of Hong Kong, Shatin, New Territories, Hong Kong, China*

HUIYING LI (9), *Departments of Molecular Biology and Biochemistry and Physiology and Biophysics, University of California, Irvine, California 92697*

LUKE K. LIGHTNING (28), *Department of Medicinal Chemistry and Molecular Pharmacology, School of Pharmacy, Purdue University, West Lafayette, Indiana 47907*

GANG LUO (17), *Bristol-Myers Squibb Company, Newark, Delaware 19114*

NEIL MACDONALD (24), *Central Toxicology Laboratory, Syngenta, Macclesfield SK10 4TJ, United Kingdom*

AJAY MADAN (17), *Department of Molecular Biology, XenoTech LLC, Lenexa, Kansas 66219*

THOMAS M. MAKRIS (11), *Department of Biophysics and Computational Biology, University of Illinois, Urbana, Illinois 61801*

JENNA S. MAMMEN (6), *W. Harry Feinstone Center for Genomic Research, University of Memphis, Memphis, Tennessee 38152*

COLLEN M. MASIMIREMBWA (14), *Drug Metabolism and Pharmacokinetics and Bioanalytical Chemistry, AstraZeneca R&D Mölndal, S-431 83 Mölndal, Sweden*

PATRICK MAUREL (30), *French National Institute for Health and Medical Research U 128, F-34293 Montpellier, France*

BOWMAN MIAO (18), *Bristol-Myers Squibb Company, Experimental Station, Wilmington, Delaware 19880*

VAUGHN P. MILLER (26), *BD Biosciences, Woburn, Massachusetts 01801*

BIRGER LINDBERG MØLLER (32), *Plant Biochemistry Laboratory, Department of Plant Biology, and Center for Molecular Plant Physiology, Royal Veterinary and Agriculture University, DK-1871 Frederiksberg C., Copenhagen, Denmark*

LINDA B. MOORE (16), *Systems Research, GlaxoSmithKline, Inc., Research Triangle Park, North Carolina 27709*

RICK MOORE (21), *Laboratory of Reproductive and Developmental Toxicology, National Institute of Environmental Health Sciences, National Institutes of Health, Research Triangle Park, North Carolina 27709*

MARC MORANT (33), *Department of Plant Stress Response, Institute of Plant Molecular Biology, CNRS UPR 2357, F-67000 Strasbourg, France*

DANIEL R. MUDRA (17), *Department of Molecular Biology, XenoTech LLC, Lenexa, Kansas 66219*

RANJAN MUKHERJEE (18), *Bristol-Myers Squibb Company, Experimental Station, Wilmington, Delaware 19880*

MASAHIKO NEGISHI (21), *Laboratory of Reproductive and Developmental Toxicology, National Institute of Environmental Health Sciences, National Institutes of Health, Research Triangle Park, North Carolina 27709*

DAVID R. NELSON (1), *Department of Molecular Sciences, University of Tennessee, Memphis, Tennessee 38163*

PETER OPDAM (17), *Department of Pharmacology, Toxicology, and Therapeutics, University of Kansas Medical Center, Kansas City, Kansas 66100*

MIKAEL OSCARSON (3), *Division of Molecular Toxicology, IMM, Karolinska Institutet, SE-171 77 Stockholm, Sweden*

COLIN N. A. PALMER (19), *ICRF Molecular Pharmacology Unit and Biomedical Research Center, Ninewells Hospital and Medical School, Dundee DD1 9SY, Scotland, United Kingdom*

ANDREW PARKINSON (17), *Department of Molecular Biology, XenoTech LLC, Lenexa, Kansas 66219*

JEAN-MARC PASCUSSI (21), *Laboratory of Reproductive and Developmental Toxicology, National Institute of Environmental Health Sciences, National Institutes of Health, Research Triangle Park, North Carolina 27709*

W. JAMES PEACOCK (37), *Division of Plant Industry, Commonwealth Science and Industrial Research Organization, Canberra, ACT 2601, Australia*

SHARRON G. PENN (20), *Amersham Biosciences, Sunnyvale, California 94085*

JULIAN A. PETERSON (2), *Department of Biochemistry, The University of Texas Southwestern Medical Center, Dallas, Texas 75390*

LYDIANE PICHARD-GARCIA (30), *French National Institute for Health and Medical Research U 128, F-34293 Montpellier, France*

FRANCK PINOT (34), *Department of Cellular and Molecular Enzymology, Institute of Plant Molecular Biology/CNRS, F-67083 Strasbourg Cedex, France*

THOMAS L. POULOS (9), *Departments of Molecular Biology and Biochemistry and Physiology and Biophysics, University of California, Irvine, California 92697*

ZHENG QIAN (6), W. Harry Feinstone Center for Genomic Research, University of Memphis, Memphis, Tennessee 38152

DAVID R. RANK (20), Amersham Biosciences, Sunnyvale, California 94085

MARIANNE RIDDERSTRÖM (14), Drug Metabolism and Pharmacokinetics and Bioanalytical Chemistry, AstraZeneca R&D Mölndal, S-431 83 Mölndal, Sweden

RUTH ROBERTS (24), Centre de Recherche de Paris, F-94400 Vitry-sur-Seine, France

JEAN-PIERRE SALAÜN (34), Department of Cellular and Molecular Enzymology, Institute of Plant Molecular Biology/CNRS, F-67083 Strasbourg Cedex, France

LINDA D. SANTOMENNA (18), Chesapeake City, Maryland 21915

JOACHIM SCHRÖDER (36), Institute for Biology II, Universität Freiburg, D-79104 Freiburg, Germany

ERIN G. SCHUETZ (31), Department of Pharmaceutical Sciences, St. Jude Children's Research Hospital, Memphis, Tennessee 38105

COSETTE J. SERABJIT-SINGH (27), Department of Drug Metabolism and Pharmacokinetics, GlaxoSmithKline, Inc., Research Triangle Park, North Carolina 27709

MAGANG SHOU (25), Department of Drug Metabolism, Merck Research Laboratories, West Point, Pennsylvania 19486

IVIN S. SILVER (27), Department of Drug Metabolism and Pharmacokinetics, GlaxoSmithKline, Inc., Research Triangle Park, North Carolina 27709

STEPHEN G. SLIGAR (11, 13), Departments of Biochemistry, Chemistry, Biophysics, and Beckman Institute for Advanced Science and Technology, University of Illinois, Urbana, Illinois 61801

C. DAVID STOUT (8), Department of Molecular Biology, MB-8, The Scripps Research Institute, La Jolla, California 92037

DAVID M. STRESSER (26), BD Biosciences, Woburn, Massachusetts 01801

PAUL T. STRICKLAND (6), Department of Environmental Health Science, Johns Hopkins University, Bloomberg School of Public Health, Baltimore, Maryland 21205

TATSUYA SUEYOSHI (21), Laboratory of Reproductive and Developmental Toxicology, National Institute of Environmental Health Sciences, National Institutes of Health, Research Triangle Park, North Carolina 27709

SHAOXIAN SUN (18), Pfizer, San Diego, California 92121

CARRIE HAYES SUTTER (6), W. Harry Feinstone Center for Genomic Research, University of Memphis, Memphis, Tennessee 38152

THOMAS R. SUTTER (6), W. Harry Feinstone Center for Genomic Research, University of Memphis, Memphis, Tennessee 38152

RUSSELL S. THOMAS (20), McArdle Laboratory for Cancer Research, University of Wisconsin Medical School, Madison, Wisconsin 53706

LI TIAN (22), Department of Biochemistry, The Chinese University of Hong Kong, Shatin, New Territories, Hong Kong, China

NATHALIE TIJET (34), Department of Cellular and Molecular Enzymology, Institute of Plant Molecular Biology/CNRS, F-67083 Strasbourg Cedex, France

WILLIAM F. TRAGER (28), Department of Medicinal Chemistry, School of Pharmacy, University of Washington, Seattle, Washington 98195

RACHEL F. TYNDALE (7), *Center for Addiction and Mental Health, Department of Pharmacology, University of Toronto, Toronto MSS 1A8, Canada*

DANIELE WERCK-REICHHART (33), *Department of Plant Stress Response, Institute of Plant Molecular Biology, CNRS UPR 2357, F-67000 Strasbourg, France*

MICHAEL R. WESTER (8), *Division of Biochemistry, MEM-255, The Scripps Research Institute, La Jolla, California 92037*

JONATHAN J. WILKER (13), *Department of Chemistry, Purdue University, West Lafayette, Indiana 47907*

JAY R. WINKLER (13), *Beckman Institute, California Institute of Technology, Pasadena, California 91125*

G. BRUCE WISELY (16), *Department of Gene Expression, Protein Biochemistry, GlaxoSmithKline, Inc., Research Triangle Park, North Carolina 27709*

STEPHEN A. WRING (27), *Department of Drug Metabolism and Pharmacokinetics, Trimeris, Inc., Durham, North Carolina 27707*

JASON K. YANO (9), *Department of Molecular Biology and Biochemistry and Physiology and Biophysics, University of California, Irvine, California 92697*

PETER R. YOUNG (18), *SUGEN, Inc., San Francisco, California 94080*

ISMAEL ZAMORA (14), *Drug Metabolism and Pharmacokinetics and Bioanalytical Chemistry, AstraZeneca R&D Mölndal, S-431 83 Mölndal, Sweden*

ULRICH M. ZANGER (5), *Dr. Margarete Fischer-Bosch Institute of Clinical Pharmacology, D-70 376 Stuttgart, Germany*

JI HU ZHANG (18), *Novartis Pharmaceuticals, Summit, New Jersey 07901*

Preface

In the Preface to the first of the P450 volumes of *Methods in Enzymology* (Volume 206 in 1991) the number of CYP genes was noted to be 160. In the Preface to the second of these volumes (Volume 272 in 1996) this number was 477. From the Nelson home page on cytochromes P450 (http://drnelson.utmem.edu/cytochromeP450.html) we learn that in June 2002 this number exceeds 2000 with an expectation of 2500 by the beginning of 2003. This exponential growth of known P450s over the past eleven years is a consequence of accelerated activity in genomics involving both eukaryotes and prokaryotes. We are now facing a new challenge in this very active area of research, "so many P450s, so little functional information." The goal for the future is not only the continued identification of P450s but greater effort in establishing their functions. For example, we now know that high resolution structures of P450s with unknown function are available. Many of the articles in this volume provide insights into how function of P450s can be examined. Genomics captures a complete section for the first time, and extensive details on determination of structure/mechanism and transcriptional regulation of P450 gene expression by nuclear proteins are found.

The field is becoming so large that we sought help to identify all the emerging excitement within it, and are grateful to our editorial advisors, Cossette J. Serabjit-Singh and Francis Durst, for their insights in regarding new areas of research. We also appreciate the efforts of each of the authors in assuring that the most up-to-date and detailed experimental approaches are concisely described. Through their efforts, a very useful and important collection of experimental approaches used to address the most important problems in P450 research is found within.

ERIC F. JOHNSON
MICHAEL R. WATERMAN

METHODS IN ENZYMOLOGY

Section I

Human Genomics and Genetics

[1] Mining Databases for Cytochrome P450 Genes

By DAVID R. NELSON

Introduction

The growth of nucleotide databases is currently exponential. GenBank release 122 (Feb. 15, 2001) had 11.7 billion base pairs (bp) of sequence. The counter on June 6, 2001 was 15.8 billion[1] for an increase of 4.1 billion bp in 111 days, or 37 million bp per day. The rate since GenBank release 123 (April 15, 2001) is 65 million bp per day. The species represented include not only human but a wide variety of prokaryotes and eukaryotes. There are now five completed eukaryote genomes,[2] not including human, and 53 complete prokaryote genomes in GenBank. More than 300 complete prokaryotic genomes are known in private databases.

This data trove offers many opportunities to find genes. The cytochrome P450s are richly represented. These genes make up 273 of about 25,000 *Arabidopsis* genes or 1% of the whole genome.[3–5] In *Drosophila* there are 90 P450 genes[6,7] in about 13,600 or 0.7% of the genome. Mice have 82 P450 genes[8] now in approximately 30,000 total genes for about 0.3%. As more and more data come in from species like zebrafish, *Xenopus*, sea urchin, and *Dictyostelium*, it becomes necessary to systematically search these sequences to find and assemble the P450 genes. The same is true for any other gene family. This chapter outlines how I have done this for some species, giving summaries of the results with emphasis on recent findings.

GenBank Substructure

GenBank is not just one database. There are multiple parts to it and when a BLAST search is done, only one section is searched at a time. Table I lists the main sections of GenBank. The default section is nr (for nonredundant), where you will search when you do a BLAST unless you actively select another section. It contains the cDNAs and genes that have been sequenced by individual laboratories over

---

[1] www.ncbi.nlm.nih.gov/entrez/query.fcgi?db=Nucleotide
[2] *Saccharomyces cerevisiae, Caenorhabditis elegans, Drosophila melanogaster, Arabidopsis thaliana, Schizosaccharomyces pombe.*
[3] drnelson.utmem.edu/Arablinks.html
[4] www.biobase.dk/P450
[5] S. M. Paquette, S. Bak, and R. Feyereisen, *DNA Cell Biol.* **19,** 307 (2000).
[6] drnelson.utmem.edu/dros.html
[7] p450.antibes.inra.fr
[8] drnelson.utmem.edu/fasta.mouse.P450s.html

TABLE I
SIZES OF GenBank SUBDIVISIONS

Section of GenBank	Bases (millions)
nr	3214
EST	3502
Human	1631 (46%)
Mouse	734 (21%)
Others	1140 (33%)
HTGS	4356
GSS	1336
STS	89

the years. Unless a submitter is sending in a batch submission of large numbers of sequences for the other sections of the database their sequence goes here. Note that nr has 3.2 billion bases or about one-fifth of the total GenBank bases. If you only search here for a match to your sequence, you are neglecting almost 80% of GenBank sequence data.

ESTdb[9] is the division for sequencing projects generating cDNAs. Usually, large numbers of ESTs are deposited from these projects. The ESTs are broken down into human, mouse, and other ESTs to show that almost one-half are human, one-fifth are mouse, and one-third are other. Note that there are more bases in the EST section than in the nr section. The HTGS (high throughput genomic sequence[10]) section is for genomic sequencing. This is where the genome projects send their data. It is the largest section of GenBank. It has a large amount of human and mouse sequence. GSS stands for genome survey sequence.[11] Most genome projects are based on sequencing clones of differing sizes. Some are cosmids and some are bacterial artificial chromosomes (BACs). One way to tag BACs is to end sequence them to get two ~500-bp sequence markers for mapping and later assembly. The BAC end sequences go in the GSS section of GenBank. They are similar in size to ESTs but they are from genomic DNA. There are over a billion bases of the GSS sequence. Search all four sections for matches to any sequence you have, if you do not want to miss anything. There is a smaller section called STS for sequence-tagged sites[12] (only 89 million bases). These are used for mapping. Search there too, even though there is not as much data because missing exons can also be found in the STS database that are not in any other section of GenBank.

[9] www.ncbi.nlm.nih.gov/dbEST/index.html
[10] www.ncbi.nlm.nih.gov/HTGS
[11] www.ncbi.nlm.nih.gov/dbGSS/index.html
[12] www.ncbi.nlm.nih.gov/dbSTS/index.html

TABLE II
MAJOR DIVISIONS OF nr DATABASE[a]

nr Section	Entries	Bases (millions)
PRI (primate)	162,557	1512
ROD (rodent)	65,765	124
MAM (other mammal)	30,048	26
VRT (other vertebrate)	54,414	49
INV (invertebrate)	87,610	431
PLN (not animal, bacterial, or viral)	153,924	381
BCT (bacteria, archaea)	104,940	283
VRL (viral)	118,665	104
Total	777,723	2910

[a] GenBank Release 123, April 15, 2001, nr Database Sections.

Looking more closely at the nr section, release notes for the April 15 release 123 show the following.[13] The GenBank release is broken into several hundred files. If the EST, GSS, STS, and HTGS files are ignored, nr is what is left. The nr sequences are categorized into eight groups listed in Table II. These designations appear in the top line of any sequence from the nr section of GenBank. Primate is the largest section with the human sequence in it. Rodent is about 10 times smaller, with other mammal and other vertebrate (zebrafish, *Xenopus,* etc.) tiny by comparison. Invertebrate is fairly large, as it has both fly and *Caenorhabditis elegans* sequences in it. Plant is a catch-all section that has every species that is not animal, bacterial, or viral. This includes fungi and protists, even though they are not plants. It is fairly large because *Arabidopsis* is included here. Bacteria is also fairly large, as all 53 public prokaryotic genomes are here.[14] Viral rounds out the nr section. There are 607 complete viral genomes in GenBank.

When doing a BLAST search,[15] you can specify any of these categories except invertebrate. You can also select all animals (metazoa) or all arthropods, which gives you a subset of invertebrates, but not all. There are also 27 common species in the BLAST pull-down menu; however, you can type in any taxon name in another window to limit a search. For instance, if you wanted to search only soybean sequences, you could enter *Glycine max*. It is often important to do this to reduce noise. The BLAST output can limit your results to 100 alignments. In a sample search of the db EST with a known *Arabidopsis* sequence with the filter set for Viridiplantae (green plants), 100 alignments were recovered but only 26 were

[13] ftp.ncbi.nih.gov/genbank/gbrel.txt
[14] www.ncbi.nlm.nih.gov/PMGifs/Genomes/a_g.html   and   http://www.ncbi.nlm.nih.gov/PMGifs/Genomes/eub_g.html
[15] www.ncbi.nlm.nih.gov/blast

TABLE III
dbEST RELEASE 051101 SUMMARY BY ORGANISM[a]

Organism	Sequences
Homo sapiens (human)	3,510,107
Mus musculus + domesticus (mouse)	1,961,530
Rattus sp. (rat)	285,326
Bos taurus (cattle)	168,677
Glycine max (soybean)	168,157
Drosophila melanogaster (fruit fly)	125,727
Medicago truncatula (barrel medic)	122,365
Lycopersicon esculentum (tomato)	115,070
Arabidopsis thaliana (thale cress)	113,000
Caenorhabditis elegans (nematode)	109,215
Xenopus laevis (African clawed frog)	91,558
Zea mays (maize)	89,125
Danio rerio (zebrafish)	88,276
Oryza sativa (rice)	75,057
Hordeum vulgare (barley)	68,480
Sorghum bicolor (sorghum)	65,092
Chlamydomonas reinhardtii (green algae)	64,973
Sus scrofa (pig)	64,709
Triticum aestivum (wheat)	60,022

[a] Number of public entries: 7,965,906. May 11, 2001.

from soybean. When the filter was set for soybean, 100 hits recovered were all soybean. The filter eliminated the noise from all the other plant sequences.

The db EST has a summary by organism that shows the most sequenced species.[16] The top 19 organisms are shown in Table III. Human and mouse are at the top with more than 5 million EST sequences between them. The top 4 are mammals, but 9 out of the top 19 are plants. Surprisingly, cattle have been sampled pretty heavily. The human CYP26C1 gene was missing exons 2 and 3 from all human sources I could find, including Celera Genomics public data. An exon 1 and 2 fragment is present in cattle ESTs. I know this because cattle exon 1 is 91% identical to the human sequence. In May 2001, the two missing human exons were deposited in a new version of a human genomic sequence so it pays to look often for new hits. The chance of finding a gene or a close homolog in the EST database is good, even rare transcripts may be represented. For humans, with approximately 30,000 genes, there are on average over 100 ESTs per gene. This does not mean every gene is represented. There are five human P450s that are not pseudogenes yet they have no ESTs associated with them. These are CYP2A13, 2C19, 4A22, 26C1, and 27C1. This shows that the EST database does not have every gene represented

---

[16] www.ncbi.nlm.nih.gov/dbEST/dbEST_summary.html

TABLE IV
dbGSS RELEASE 051101 SUMMARY BY ORGANISM[a]

Organism	Sequences
Homo sapiens (human)	869,266
Mus musculus (mouse)	835,820
Tetraodon nigroviridis (freshwater pufferfish)	188,963
Oryza sativa (rice)	93,107
Trypanosoma brucei (African sleeping sickness)	90,540
Strongylocentrotus purpuratus (sea urchin)	76,019
Arabidopsis thaliana (thale cress)	61,265
Entamoeba histolytica (human parasite)	49,129
Drosophila melanogaster (fruit fly)	45,143
Fugu rubripes (Japanese pufferfish)	42,896
Magnaporthe grisea (rice blast fungus)	12,674
Trypanosoma cruzi (Chagas' disease)	12,232
Leishmania major (Leishmaniasis)	11,929
Plasmodium vivax (malaria)	10,682
Saccharomyces cerevisiae (Baker's yeast)	7,637
Anopheles gambiae (malaria mosquito)	7,514
Cryptosporidium parvum (human parasite)	7,082
Plasmodium berghei (rodent malaria)	5,476
Glycine max (soybean)	4,818

[a] Number of public entries: 2,493,887. May 11, 2001.

even though there are 3.5 million sequences. These five P450s represent 9% of the known human P450s so gene coverage in the human EST database is about 91% based on known P450 genes. There still may be ESTs from the 3 prime (3′) end of the message, but it is difficult to recognize them without a 3′-untranslated sequence for comparison.

The GSSdb has over 1.3 billion bases of genomic sequences from many different organisms.[17] The top 19 are shown in Table IV. This section of the database is especially interesting because it has a large number of fish sequences from both freshwater and salt water pufferfish. There are even 76,000 GSSs from sea urchin. There are also many parasite ESTs here, including *Plasmodium* (malaria), *Leishmania, Entamoeba, Trypanosoma,* and *Cryptosporidium.* The vertebrate sequences can help in assembling human genes by clarifying the positions of introns. It can be quite difficult to identify intron boundaries when you only have a DNA sequence of a single species, but it is much easier with a human–mouse or human–fish pair. The GSSdb is a very valuable resource.

The last section is the HTGS section.[10] The HTGS section has sequence that is not finished. Once an HTGS sequence is finished it gets annotated and moved to the

---

[17] www.ncbi.nlm.nih.gov/dbGSS/dbGSS_summary.html

TABLE V
NUMBER OF HIGH THROUGHPUT GENOMIC SEQUENCE ENTRIES[a]

Organism	Finished sequence (nr)	Organism	Working draft (HTGS section)
Human	11,894	*Giardia*	53,667
C. elegans	2,363	Human	24,442
Drosophila	1,881	*Drosophila*	5,165
Arabidopsis	769	Mouse	1,909
Mouse	231	*Rice*	391
Rice	154	*C. elegans*	201
Other	142	Other	496
Total	17,435		86,309

[a] In GenBank (5/20/01).

nr database so HTG sequences can be found in both nr and HTGS. Table V shows the top six species for HTG sequences either in finished form or working draft sequences. The top four in the finished category are from model organisms that are completely sequenced or human that is mostly sequenced. On the working draft side there is a surprise in *Giardia* with 53,000 entries. *Giardia* is the parasite that causes hiker's diarrhea and these reads are short reads of about 500–800 bp. The human and mouse entries are much larger, being up to 200,000 bp and including many fragments. Do not be fooled by large numbers of entries, it is base pairs that really count. *Giardia* is anaerobic and does not appear to contain any P450 genes.

Other Databases and Systematic Search Strategy

GenBank is not the only place to go for sequence data. There are many sequencing facilities around the world that have searchable databases for specific organisms. A prime example is the *Dictyostelium discoideum* genome project. This project is being carried out by multiple laboratories that have set up one common BLAST server for sequence data.[18] Most of this data is not in GenBank. The strategy for finding P450s in this data is a general strategy that can be applied to other species, with either genomic or cDNA sequences or preferably both. The initial step is to do BLAST searches with a wide variety of sequences that are appropriate. For plants, searches with one member of each plant family would be desired, but that would be over 50 searches, so this might be shortened by selecting one member from each plant clan. For vertebrates, one member from each mammalian family or 18 sequences would be recommended. *Dictyostelium* was searched with the 18 mammalian sequences to find all possible hits. For searches done at GenBank, the species should be selected and the HTGS, EST, nr GSS, and STS sections should be searched.

[18] www.sdsc.edu/mpr/dicty

After the first search, the output needs to be examined and the sequence fragments that are true P450s need to be collected in a file. I make two files for this data: one is just a copy of the blast output, which eventually gets sorted into sequence bins and the other file is a FASTA format file of just the sequence fragments by themselves with an identifier line at the top of each sequence that begins with >. Each sequence is given an arbitrary number, 1, 2, 3, etc., to identify that sequence bin. A third file is made that has the accession numbers listed in alphabetical order, followed by the arbitrary sequence number. Once the FASTA file is made from the first search, it is used at the Proweb.org website as a database file for blast searching.[19] Here, all the fragments are searched against this file one by one to identify overlapping fragments. These are combined. Their sequence numbers are also combined, so if sequences 8 and 12 are really the same protein, that sequence bin becomes 8, 12 and the sequences are joined in the FASTA file to make a longer sequence. This process builds up contigs. In a species like human with many named P450 genes, a FASTA file can be used that has all the known human sequences present so any blast hits can be identified quickly.

The second BLAST search output is compared with the accession number list from the first search. Any new accession numbers that are real P450 sequences get added to the master accession number list. The sequences get added to the FASTA file and checked for any overlaps as before. If the sequence is new, it gets a new bin number. If the sequence is a match, it gets combined in an existing bin. The first three to five searches usually identify new sequence bins, but after that the new sequences become quite rare. It is unusual to find new sequence hits after seven or eight rounds of this process. After all the test sequences have been searched and the contigs sorted into the smallest number of bins, it is time to try to finish out the individual contigs. This process is different for EST and genomic data.

ESTs can be extended as far as the ends of the sequence will permit. The blast search output probably stopped before reaching the end of coding sequence in the EST so it is necessary to go get the EST and translate it in the three frames of the correct strand. Two useful sites for this are the JustBio.com[20] translator or the EMBOSS translator[21] at EBI. The translations need to be examined for possible extensions of the sequence out to start methionines or stop codons. Frameshifts should be identified by comparison to the most similar reference sequence, which may be from the species being searched. The result may not be completely correct, but errors will correct themselves later when more accurate sequence is available. Extensions should be checked against the FASTA file to see if any contigs can be joined. The names of sequences may be helpful in constructing contigs. Often, clone names indicate whether they are sequenced from the 5' or 3' end of a clone. Clones with the same prefix come from the same clone and these

[19] www.proweb.org/proweb/Tools/WU-blast.html
[20] www.justbio.com
[21] www.ebi.ac.uk/emboss/transeq

are probably from opposite ends of the same gene. The examples JC1b30e03.s1 and JC1b30e03.r1 are standard and reverse reads from the same *Dictyostelium* clone and from the same gene. The set of alphabetized accession numbers should be examined for these clone pairs and the sequence bins should probably be combined. The only problem that may occur in doing this is due to gene clusters in genomic DNA or chimeras in EST data. These will probably resolve themselves later.

Genomic data should be fetched and the closest reference sequence should be used in a TBLASTN search against just that one DNA sequence using the Proweb.org server[19] rather than the whole database. This may reveal weak matches not reported earlier for the middle, less conserved part of the gene. This will also reveal gene clusters by giving multiple alignments to the same region of the reference sequence. These should have been detected before. The middle of a P450 is the hardest to find. These regions can be found in genomic DNA by locating the conserved C-helix region or N-terminal proline-rich region and the I-helix region that are easier to recognize. Once the middle region is bracketed by these sequence blocks, it is possible to translate the region in between in three frames and look for possible exons. It is very helpful to have a complete gene sequence with known intron–exon boundaries from the same species to help in this process. The exons are mostly conserved in size and phase in a given family, so looking for the GT–AG pairs is easier. At this point, EST data are very useful to spot these boundaries, even if the EST is not a 100% match. If a complete gene can be assembled, the FASTA file should be searched again with the new sequence to try to join contigs. This process is an iterative one.

Dictyostelium P450s have been sorted into 55 contig bins with 705 accession numbers.[22] There are a minimum of 43 P450s in *Dictyostelium* based on overlapping C-terminal sequence regions. Twenty six P450s are complete sequences. An alignment of these sequences is posted.[23] The sequences fall into 15 families CYP51, 508, 513–525. Currently, there are 12 fragments not named because they are too short. There are eight pseudogenes. CYP51 is the only family shared with plant and animal species; however, CYP524 shows 34% identity to CYP710A1 and A2. It is possible that the common ancestor with plants had two P450s, a CYP51 and CYP710/CYP524 ancestor.

The Human Genome

This strategy has been applied to the human HTGS section of GenBank. It has not been carried out on nr, EST, or GSS sections. With over 3.5 million ESTs for human, it is expected that there will be thousands of accession numbers for

[22] drnelson.utmem.edu/LowerEuks.html
[23] drnelson.utmem.edu/dicty54aln.html

ESTs of P450s. A similarly high number will exist in nr because of duplicate entries and multiple accessions for exons. The result from the HTGS section done in September 2000 was 180 sequence entries on 130 accession numbers. Some accession numbers have more than one entry because they contain more than one gene. For example, AC020705 has CYP1A1 and 1A2. The results of this search have been sorted by CYP name,[24] chromosome,[25] and accession number.[26] There are 57 human P450s that are not pseudogenes.[27] The number of pseudogenes is not determined yet and it is going to be difficult to decide what constitutes a pseudogene. Some lone exons are found outside complete genes. Are they pseudogenes or part of the intact gene? There may be a need for some new terms to describe the detritus surrounding human genes.

The count of human P450s has increased by seven in the past 18 months. The new genes are 27C1 (April 25, 2000), 2U1 (April 25, 2000), 4V2 (April 25, 2000), 26C1 (July 19, 2000), 2W1 (Sept. 26, 2000), 20 (March 9, 2001), and 4A22 (May 23, 2001). The dates given are the dates I became aware of the sequences, not the first appearance in the database. Since the last systematic hunt for new human P450 genes (September 2000), only two new genes have been found. CYP4A22 is 95% identical to CYP4A11 (see alignment[28]) and there is some question whether it is an allele of 4A11. All mRNA sequences match 4A11, but three independent genomic sequences match 4A22 (AL390073, AF208532,[29] AL135960). The genomic sequence that matches the 4A11 mRNA is not available yet. CYP20 was identified in GenBank as a possible new P450 from human pheochromocytoma in September 2000 (AF183412). It did not turn up in the searches because it is a poor match to existing families (only 23% to CYP3A4 over 437 amino acids). The heme signature is short by one amino acid, but it does have the PERF motif and the EXXR motif. It does not have the typical AGX(D,E)T sequence at the I-helix oxygen-binding region, so the substrate for this P450 may have oxygen included as in the allene oxide synthase. I actually found this sequence when I did a text search for the term P450 in new GenBank entries. More data including sequences and accession numbers are available at the P450 home page.

One significant outcome of the human genome project has been the completion of known P450 fragments. At present, all of the 57 known human P450s are complete in their coding regions. This is a very recent occurrence, with the last missing pieces being added to the database on May 2, 2001 (AL358613.11 finished CYP26C1). Only one fragment (exon 5 of CYP27C1) is known solely based

[24] drnelson.utmem.edu/hum.htgs.byname.pdf
[25] drnelson.utmem.edu/hum.htgs.bychr.pdf
[26] drnelson.utmem.edu/hum.htgs.pdf
[27] drnelson.utmem.edu/hum.html
[28] drnelson.utmem.edu/4a22and4a11.html
[29] H. Kawashima, T. Naganuma, E. Kusunose, T. Kono, R. Yasumoto, K. Sugimura, and T. Kishimoto, *Arch. Biochem. Biophys.* **378**, 333 (2000).

on Celera Genomics public data. The rest are found in public project data. This does not mean that all genes are complete in a single contig. Many are still split in numerous unjoined DNA sequences. The long wait to complete the CYP26C1 and CYP27C1 genes was exacerbated by the fact that neither of these genes had any human or mouse ESTs in the db EST. The fragmentary structure and lack of ESTs might suggest that they are pseudogenes, but that possibility has been eliminated by finding highly similar EST sequences from bovine for CYP26C1 (95%) and from *Xenopus* (AW637606, BE669236 74%), bovine (BE723057 89%), and chicken (BE139968 71%) for 27C1. The absence of mouse or human ESTs from among more than five million mouse and human ESTs in the database suggests that these genes are expressed in a limited manner. They might be developmentally significant.

Mouse P450s

Even though the mouse genome project is not as far along as the human project, there are more P450 sequences known from the mouse. The most recent compilation shows 85 mouse P450 genes[8] and only three of these are pseudogenes. The main compilation was done April 19, 2001 with new genes being added up to May 17, 2002. There is still one human gene without a mouse ortholog discovered yet, but this is expected to exist (CYP27C1) so the mouse total should be 83 genes plus the three pseudogenes. That is 26 more than seen in human. One of the curious facts about the mouse is the presence of intact genes for Cyp2g1 and Cyp2t4. These are only found as pseudogenes in humans. The mouse also has nine intact Cyp4f genes and one pseudogene (Cyp4f13–18, 4f28, 4f29p, 4f30, 4f31), whereas humans have six intact 4fs and eight pseudogenes (CYP4F2, 3, 8, 9P, 10P, 11, 12, 22, 23P–28P). This may be due to the incomplete nature of the mouse genome sequence. More pseudogenes will probably be found as the mouse project matures. Subfamilies in the mouse that have more than a single member are usually larger than the same subfamily in humans. The 2C subfamily in humans has only four genes but mouse has eight plus two pseudogenes. The expansion of P450 subfamilies in the mouse is probably due to a more varied diet and a greater dependence on olfaction.

Fish, *Xenopus*, and Sea Urchin

Evolution of the P450 families in vertebrates can only be determined by sampling outside the mammals in an evolutionary spectrum. This is being done for several fish species, including zebrafish, *Tetraodon nigroviridis* (freshwater pufferfish), and *Takifugu rubripes* (Japanese pufferfish). The Fugu project was advertised as generating 3× coverage and about 95% of the genome represented by

April 2001.[30] The Fugu BLAST server[31] had 222 million bases of sequence in 412,012 sequences on May 31, 2001. Only 189,000 GSS sequences are in Gen-Bank, so this is another example of a non-GenBank sequence database. As a test case, a BLAST with 27C1 found the first three exons of the Fugu 27C1 sequence. A systematic search of this data will provide a valuable set of P450s from a non-mammalian vertebrate source. In the meantime, two searches have been done on smaller data sets that are of similar interest. *Xenopus* EST data have been searched exhaustively for all P450 hits as described earlier for *Dictyostelium*. Searches of 91,558 *Xenopus* ESTs resulted in finding 120 with the P450 sequence. These were sorted into 44 contigs.[32] There are at least 27 sequences represented from 11 of the 18 mammalian families. Seven protein sequences are complete (1A6, 1A7, 2Q1, 4T4, 17, 19, 26A). Three others are nearly complete (4T3, 27A, 51). Nineteen contigs belong to the CYP2 family, including a clear CYP2R1 ortholog. Based on sequence alignments, there are a minimum of 11 different CYP2 P450s. One contig is in the CYP3 family. Seven contigs are in the CYP4 family. There are at least six members in the *Xenopus* CYP4 family Two fragments are in the CYP8B subfamily and two more are in the CYP46 family. The 26 family has three different sequences corresponding to orthologs of 26A, 26B, and 26C of mammals. There is a clear 27C1 ortholog. Sequences are absent so far from CYP5, CYP7, CYP11, CYP20, CYP21, CYP24, and CYP39. This is a good sampling from only 92,000 ESTs.

The purple sea urchin *Strongylocentrotus purpuratus* has 76,012 GSS fragments available at GenBank. These have been searched for P450s, and 51 sequences[33] were found to contain P450 genomic DNA. The sequence similarity drops compared to fish and *Xenopus* sequences and it is harder to assign sequences to specific families and subfamilies. There are sequences that show good identity to CYP2D26, CYP2U1, and CYP4V3 sequences. Most other fragments are less similar. The problem with GSS fragments is their genomic nature. The introns in sea urchin are large enough that the GSS fragments only contain one exon or part of one exon each and these cannot be joined by overlap. Many more sequences will be needed along with ESTs or close homologues in other species to assemble these into whole genes.

Arabidopsis and Rice

The plant kingdom has many more P450 genes per species than animals. We know that from the complete set of 273 P450 genes[3–5] from *Arabidopsis*. The

[30] D. Cyranoski and P. Smaglik, *Nature* **408,** 6 (2000).
[31] bahama.jgi-psf.org/prod/bin/blast_fugu.cgi
[32] drnelson.utmem.edu/Xenopus.seqs.html
[33] drnelson.utmem.edu/seaurchin.html

next plant to be completely sequenced will be rice, probably in 2002. It is desirable to compare these monocot and eudicot genomes and see what the difference is between their P450s. This project has been started, but it is awaiting access to Monsanto 5× sequence data before it can be pushed forward. In September of 2000, BLAST searches were performed on the nr, EST, HTGS, and GSS sections of GenBank using one member from each P450 clan in plants (9 different sequences). After these 36 searches were done, 352 accession numbers of P450 sequences had been identified. These contained 389 P450 sequence fragments, as some accession numbers held clusters of 9 or even 10 P450 genes. All sequences from the BLAST output were compared against each other. The result was 209 contigs, of which 40 were full-length P450 sequences, and 169 were partial P450s. All 209 sequences were compared to a database of all *Arabidopsis* P450s plus five additional P450s from families not present in *Arabidopsis* (CYP80, CYP92, CYP99, CYP719, CYP723). This identified the fragments to specific families. Once this had been done, there were 20/50 plant P450 families that had no P450s present in the rice set. To be sure that no small fragments in the EST or GSS section of GenBank had been overlooked, especially fragments from less conserved regions, such as the extreme C-terminal, BLAST searches were done with one member from each of these 20 families against the EST or GSS section of GenBank limited to rice. This identified six additional entries in three different families. Three of these were identical sequences containing a small fragment upstream of the I-helix in CYP73. One was from a different region of CYP73. One was a single small exon including the heme-binding site of a CYP708 P450. One was from a CYP92 sequence. This increased the accession number count to 358 and the contig count to 213. Comparing families between rice and *Arabidopsis* genomes, 32 of 45 families (71%) were present in both species. CYP92 was found in rice but not in *Arabidopsis*. Because the rice genome is far from complete in public databases, it appears that most plant P450 families existed before the monocot–dicot divergence. As more data come in, the number of families missing between the two species will probably drop to a very small number. These may be specific to eudicots or even of more limited range. As mentioned earlier, five families are not seen in *Arabidopsis,* and this specialization may occur in each major lineage of plants. The detailed results of these searches are posted.[34]

Conclusions

Data from which P450 protein sequences can be derived are growing at a pace that is faster than it can be effectively searched. The process of systematically finding and assembling all P450 sequences for a given species is slow and tedious. I have posed this problem to computer scientists at The Tullahoma Space Institute

[34] drnelson.utmem.edu/rice.report.html

and Oak Ridge National Laboratories in Tennessee. They expressed interest in automating the steps of the procedure, but assumed that it would take on the form of an expert system, as some judgment is required during the process. Perhaps some of the more repetitive steps can be automated, saving time for people to make the harder decisions. This will be required to move forward in genome annotation because there are not enough people to dedicate themselves to this process for each protein family.

The P450 family has reached complete status in several eukaryotic species. Results from *Arabidopsis, Dictyostelium,* and three animals suggest that the evolutionary history of P450s in the crown group of eukaryotes is largely independent. The main lines starting from a billion or more years ago have been diverging from very few (two or three) P450 genes shared in the crown group common ancestor. The finding that diatoms have a CYP97 in common with plants points to CYP51 and CYP97 being in the last ancestor of plants and Stramenopiles. CYP524 in *Dictyostelium* may have a common ancestor with CYP710 in plants. Aside from these similarities across kingdoms, most P450s in different kingdoms seem to be derived independently. More complete data sets will clarify this and identify the precursor P450s present in the earliest eukaryotes. Deeper sampling of animals such as the sea urchin, *Ciona intestinalis* (the most primitive creature known with a notochord), proposed by the Department of Energy,[35] a radial animal like a jellyfish, and a sponge would make the evolutionary history of P450s (and our whole genomes) in animals apparent. I look forward to the accumulation of data and to the accumulation of better tools to analyze data.

[35] www.er.doe.gov/production/ober/berac/genome-to-life-rpt.html

[2] Sequence Alignments, Variabilities, and Vagaries

By Sandra E. Graham and Julian A. Peterson

One of the most critical steps in establishing phylogenetic relationships or for structural and homology modeling of a family of related proteins is the sequence alignment of the protein family members. Early in the study of the P450 gene superfamily, it was realized that there was remarkable sequence diversity from one family to another within the superfamily, yet there appears to be a highly conserved structural protein fold. With the advent of genome sequencing projects, many candidate sequences are being identified based on their sequence similarity to other P450s. These alignments of the candidate P450 sequences with other P450s are important not only in assigning the sequence to a gene family, but in constructing a three-dimensional model and in assessing the potential reactions that it might catalyze.

Early in the study of P450s, the identification of new members of the gene family relied on the presence of the characteristic absorbance band of the carbon monoxide complex at 450 nm. As the amino acid sequences of more P450s were determined, it was realized that there was an invariant cysteinyl residue in the carboxy-terminal 20% of the sequence, which was believed to be the coordinating ligand to the heme iron. An important milestone in this regard was the elucidation of the first atomic structure of a P450,[1] P450cam (CYP101), a soluble, bacterial P450, which confirmed the cysteinyl-containing region as the heme-binding region and gave us our first look at the P450 structural fold. P450cam, however, only has 414 amino acids, whereas most eukaryotic P450s, which are membrane associated, have approximately 500 residues. Even with the deletion of the 20–30 amino acids comprising the amino-terminal membrane anchor, there are still about 50 more amino acids in eukaryotic than prokaryotic P450s. The placement of these additional amino acids in a sequence alignment was difficult because of the low level of sequence similarity between P450cam and other P450 families. For example, a pairwise alignment of P450cam with CYP1A1 using the program Search at the PIR (Protein Identification Resource) found only "266 amino acids of overlap" with 25% identity, i.e., 266 amino acids of P450cam out of 414 could be aligned at the carboxyl-terminal half of the molecules, yet in that 266 amino acid span, there is only 25% identity. However, the amino-terminal portion of the sequences is hypervariable and could not be aligned at all.

These hurdles are being overcome. With the structural determination of the heme domain of P450BM3 (CYP102A1), i.e., P450BMP, and P450terp (CYP108),[2] it was shown that there is a conserved P450 structural fold and that even the hypervariable amino-terminal portion of these proteins is structurally conserved. Additionally, early sequence alignment programs, which relied on the initial computer algorithms, required a final adjustment of the alignment by hand. This was extremely time-consuming and subjective. Fortunately, from the P450BMP and P450terp structures, as well as several other new P450 bacterial structures,[3,4] and one eukaryotic structure (CYP2C5[5]), residues have been identified in the hypervariable amino-terminal region that can be used as landmarks when evaluating alignments.

[1] T. L. Poulos, B. C. Finzel, I. C. Gunsalus, G. C. Wagner, and J. Kraut, *J. Biol. Chem.* **260,** 16122 (1985).

[2] C. A. Hasemann, K. G. Ravichandran, J. A. Peterson, and J. Deisenhofer, *J. Mol. Biol.* **236,** 1169 (1994).

[3] J. R. Cupp-Vickery and T. L. Poulos, *Nature Struct. Biol.* **2,** 144 (1995).

[4] S. Y. Park, H. Shimizu, S. Adachi, A. Nakagawa, I. Tanaka, K. Nakahara, H. Shoun, E. Obayashi, H. Nakamura, T. Iizuka, and Y. Shiro, *Nature Struct. Biol.* **4,** 827 (1997).

[5] P. A. Williams, J. Cosme, V. Sridhar, E. F. Johnson, and D. E. McRee, *Mol. Cell* **5,** 121 (2000).

Over the past several years, the genomic sequencing efforts have begun to produce massive quantities of sequence information from diverse organisms, e.g., from human to plant to bacteria. While some of the newly identified sequences belong to preexisting P450 gene families, many belong to new gene families. Additionally, the use of diverse P450 protein sequences as probes in the database has allowed us to identify many new gene families in very distantly related organisms. In an effort to automate the process of identification and alignment of candidate P450 molecules as much as possible, a variety of new, more sophisticated, commercially and publicly available shareware has been examined for their utility in generating automated sequence alignments and in the subsequent analysis. This chapter discusses those programs that we have found most useful.

Identification of P450s in the Protein Database

To identify and then align all of the P450s present in the publicly maintained, nonredundant (nr) sequence database, sequences were retrieved from NCBI using the program PSI-BLAST available on their web site (www4.ncbi.nlm.nih.gov/BLAST/). PSI-BLAST was set with the maximum number of sequences returned at 500. The E-value for each sequence returned from the search is a measure of the similarity of the sequence to the target sequence. A sequence is considered to be similar if the E-value is less than 0.005. When using CYP1A1 as the target sequence, all 500 sequences retrieved were considered to be similar; however, not all known P450 sequences were returned by this search. In a similar fashion, when using the sequence of P450cam (CYP101) as the probe, approximately 200 sequences were returned that were considered to be similar. Thus, to adequately cover the "sequence space" encompassed by the P450 gene superfamily, we used 43 P450 sequences selected randomly from 43 different P450 gene families in our existing P450 database (Table I). This process returned a very large number of sequences more than once due to the redundant nature of the search process, in addition to the sequences that were retrieved only once. Standard database management software (e.g., Lotus123, MS Excel, MS Access) was used to remove the duplicate sequences. This left more than 3000 abstracted P450 sequences from the NCBI protein database.

Those sequences retrieved were examined to ascertain whether they were in fact P450s. Criteria used to establish whether a candidate sequence belonged to the P450 gene superfamily were: (1) Is there a cysteinyl residue in the carboxy-terminal 20% of the molecule containing the recognizable heme-binding region, and (2) is the conserved EXXR sequence present in the K helix? Generally, molecules that were retrieved only once did not meet this criteria; however, if they did, they were used with the PSI-BLAST program in another round of sequence retrieval from the NCBI database.

The retrieved sequences were grouped into gene families on the basis of the E-value returned by PSI-BLAST. In general, an E-value smaller than $2e^{-75}$ was

TABLE I
P450 FAMILIES USED TO PROBE SEQUENCE DATABASE

1	27	75	106
2	51	76	107
3	52	77	108
4	53	78	109
5	55	79	110
6	56	80	111
9	58	83	112
10	60	90	113
11	61	101	114
17	71	102	116
19	72	103	117
21	73	104	119
24	74	105	

used to establish family relationships automatically. This translates to a sequence identity between the probe and the returned sequence of about 40%. Sequences that were close to this cutoff value were examined to determine whether the P450 nomenclature committee had previously assigned the sequence to a particular P450 gene family. If the sequence had been assigned by the nomenclature committee, the sequence was moved to the assigned family. If the sequence had not yet been assigned by the committee and the E-value was smaller than the cutoff, the sequence was assigned to the current gene family. This process resulted in the identification of 47 P450 gene families containing five or more members with a total of 974 unique P450s as shown in Table II. Also, there are 5 P450 sequences retrieved from various *Streptomyces* with polyketide synthase activity, which cluster into one family designated in Table II as G1. Finally, although more than five members of the CYP107 family have been identified previously, when these sequences were aligned and clustered into families as described later, there seemed to be much more sequence diversity than would be tolerated by the 40% cutoff limit for identity, causing some of the members to be clustered into other bacterial families. Thus, this family was not included in subsequent alignments and analysis.

Multiple Sequence Alignments

Individual gene families were aligned with the VectorNTI software (InforMax, Inc., Bethesda, MD) using the Blossum62NT2 weighting matrix. The alignments were examined for faulty insertions or deletions that might be present in the sequence of individual family members because of sequencing errors or the inappropriate assignment of exon–intron boundaries from the genomic sequences. These insertions or deletion were designated as faulty either by their location in the secondary structure or because the reference citation indicated an automatic protein

TABLE II
COMPOSITION OF P450 FAMILY DATABASE

Family	Members	Family	Members
1	47	72	18
2	136	73	23
3	32	74	15
4	80	75	16
6	39	76	9
7	8	77	7
8	8	78	9
9	7	79	13
11	32	82	11
12	10	84	8
13	12	86	13
14	5	88	6
17	20	89	15
19	29	91	24
21	9	93	32
24	5	94	9
26	6	96	11
27	9	98	5
28	10	102	5
33	9	105	14
34	7	313	7
35	8	706	7
51	33	710	8
52	30	G1[a]	5
71	73		

[a] G1 is the group of bacterial sequences that did not appear
to have been assigned to a recognized gene family.

sequence generation by a genome project. These insertions or deletions usually were present in only a single member of the gene family, and thus this member was removed from our database. In all cases, the insertion or deletion would have been in a structural element rather than a loop or coil and would have severely disrupted the P450 fold.

Finally, if two sequences in the database were greater than 97% identical, their citations were examined. Frequently, sequences were in the database as a consequence of either corrections to preexisting sequences or point mutations of an identified P450 that resulted in changed or inactive protein. In those cases, these sequences were removed from the alignment database. It should be noted that there are approximately 200 P450 sequences that have not been included in the alignment because there were fewer than five members in a family. These proteins will provide a gold mine for individuals interested in P450 structure and function, as they may have unusual properties or functions, e.g., CYP119, a P450 from an acidothermophilic archaea.[6]

Rather than aligning all of the sequences in the database, we have chosen to align five sequences selected randomly from each gene family, thus giving us a more manageable and, more importantly, an evenly weighted multiple family sequence alignment. (*Note:* There were 136 sequences found in the CYP2 family but only 5 sequences found for the CYP24 family.) The alignment did not include P450cam because it is the only representative of CYP101, even though it is probably the best characterized; however, we did include CYP2C5,[5] CYP51A1,[7] and the heme domain of CYP102A1 (P450BMP[8]) for which protein structures have been determined. This alignment can be found on the Internet at www2.UTSouthwestern.edu/JAPlab, while the phylogenetic tree of sequence relatedness is shown in Fig. 1. Representative alignments extracted from the large alignment file can be seen in Fig. 2 containing the structurally determined CYP2C5, CYP51A1, and CYP102A1, along with CYP1A1 (a class I P450 requiring an FAD/FMN-containing NAPDH reductase), CYP11A1 (a class II P450 requiring an iron-sulfur protein and a reductase), and CYP8A1 (a class III P450 not requiring a redox partner). As discussed in the past, the addition of structural information to the multiple sequence alignment can be extremely informative in the conserved carboxyl-terminal region of the molecules, as well as in the hypervariable amino-terminal region. Generally, sequence alignments are adjusted manually to take into account structurally conserved residues; however, the alignments to be discussed and presented in Fig. 2 and on the Web have not been adjusted for instructive purposes. It should be remembered that the algorithms used in the alignment process usually become unreliable at residues adjacent to gaps, but these alignments can be improved by inspection and manual adjustments.

Inspection of the sequence alignment from the amino- to carboxy-terminal end in Fig. 2 and on the web shows the membrane-spanning segment at the N terminus in most eukaryotic P450s. The membrane anchor ends with a proline/glycine-rich hinge region (e.g., PPGP[9]), presumably producing a highly kinked secondary structure. This is followed by a pattern of hydrophobicity indicative of an amphipathic helix, which usually ends with XG, where X is usually an amino acid with a large hydrophobic side chain—most frequently F or Y. In structurally determined P450s, this is helix A. The next two secondary structural elements are β1-1 and

[6] M. A. McLean, S. A. Maves, K. E. Weiss, S. Krepich, and S. G. Sligar, *Biochem. Biophys. Res. Commun.* **252,** 166 (1998).
[7] L. M. Podust, T. L. Poulos, and M. R. Waterman, *Proc. Natl. Acad. Sci. U.S.A.* **98,** 3068 (2001).
[8] K. G. Ravichandran, S. S. Boddupalli, C. A. Hasemann, J. A. Peterson, and J. Deisenhofer, *Science* **261,** 731 (1993).
[9] K. Kusano, N. Kagawa, M. Sakaguchi, T. Omura, and M. R. Waterman, *J. Biochem.* (*Tokyo*) **129,** 271 (2001).

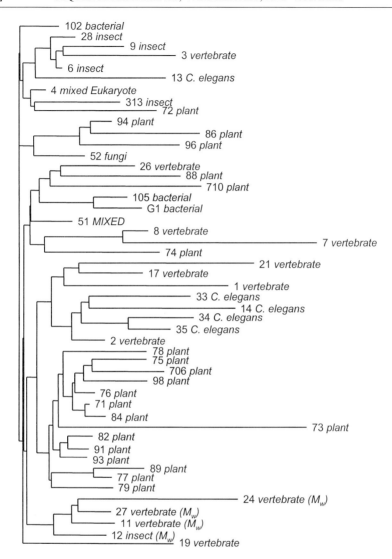

FIG. 1. A phylogenetic tree of the P450 gene superfamily. The tree was constructed using the VectorNTI program and tree construction subroutines. The phylogenetic tree calculation is based on a sequence distance method and does not imply evolutionary relationships. The origin of the majority of members is listed for each branch of the tree. The notation M_w indicates that these families are found in mitochondria.

```
                                                                                                    A
CYP102   ...........................TIKEMPQP................KTFGELKNLPLLNT....DKPVQALMKIADELGEIFKFEAPGRVTRYLSS
CYP051   ...................MSAVALPRVSG......................GHDEHGHLEEFRTDP....IG.LMQRVRDECG.DVGTFQLAGKQVVLLS
CYP002C5 .......MDPVVVLVLGLCCLLLLSIWKQNSGRGK.............LPPGPTPFPIIGNILQIDAKD....ISKSLTKFSECYG.PVFTVYLGMKPTVVLH
CYP001A1 MLFPISMSATEFLLASVIFCLVFWVIRASRPQVPKGLKN..........PPGPWGWPLIGHMLTLGKN....PHLALSRMSQQYG.DVLQIRIGSTPVVVLS
CYP011A1 .MLAKGLPPRSVLVKGYQTFLSAPREGLGRLRVPTGEGAGISTRSRPFNEIPSPGDNGWLNLYHFWRETGTHKVHLHHVQNFQKYG.PIYREKLGNVESVYVI
CYP008A1 .........MAWAALLGLLAALLLLLLLS.RRRTRRPGEP........P.LDLGSIPWLGYALDFGKD.....AASFLTRMKEKHG.DIFTILVGGRYVTVLL

         B                         B'                C                           D
CYP102   QRLIKEACDESRFDKNLSQA....LKFVRDFAGDGLFTSWTHEKNWKKAHNILLPSFSQQAMKGY.....H..AMMVDIAVQLVQKWERLNAD...EHIEV
CYP051   GSHANEFFFRAGDDDLDQAKAYP...FMTPIFGEGVVFDASPER....RKEMLHNAALRGEQ..........MKGHAATIEDQVRRMIADWGEAG.EIDL
CYP002C5 GYEAVKEALVDLGEEFAGTG....SVPILEKVSKGLGIAFSNAKTWKEMRRFSLMTLRNFGMG.......KRSIEDRIQEEARCLVEELRKTN...SPCDP
CYP001A1 GLDTIRQALVRQGDDFKGRP..DLYTFTLISNGQSMSFSPDSGPVWAARRRLAQNGLKSFSIASDPASSTSCYLEEHVSKEAEVLISTLQELMAG.GHFNP
CYP011A1 DPEDVALLFKSEGP.NPERFLIPPWVAYHQYYQRPIGVLLKKSAAWKKDRVALNQEVMAPEATKNFLPLLDAVSRDFVSVLHRRIKKAGSGN....YSGDI
CYP008A1 DPESYDAVVWEPRTRLDFHAY.......AIFLMERIFDVQLPHYSP....SDEKARMKLTLL........HRELQALTEAMYTNLHAVLLGDAT..EA..G

         E                      F                               G
CYP102   PEDMTRLTLDTIGLCGFNYRFNSFYRDQPHPFITSMVRALDEAMNKL....QRANPDDPAYDENK.RQFQEDIKVMNDLVDKIIADRKASGEQ........
CYP051   LDFFAELTIYTSSACLIGKKFR...........DQLDGRFAKLYHELERGTDPLAYVDPYLPIESFRRRDEARNGLVALVADIMNGRIANPPTDKS.....
CYP002C5 TFILGCAPCNVICSVIFHNRFDYKD..EEFLKLMESLNENVRILSSPWLQVYNNFPALLDYFPGIHKTLLKNADYIKNFIMEKVKEHEKLLDVN......N
CYP001A1 YRYVVVSVTNVICAICFGRRYDHN.HQELLSLVNLNNNFGEVVGSG...NPADFIPILRYLPNPSLNAFKDLNEKFYSFMQKMVKEHYKTFEKG......H
CYP011A1 SDDLFRFAFESITNVIFGERQGMLEEVVN.PEAQRFIDAIYQMFHTSVPMLNLPPDLFRLFRTKTWKDHVAAWDVIFSKADIYTQNFYWELRQKG....SV
CYP008A1 SGWHEMGLLDFSYSFLLRAG...........YLTLYGIEALPRTHE...........SQAQDRVHSADVFHTFRQLDRLLPKLARGSLS.VGDK..DHMCS

         H                             I                        J
CYP102   SDDLLTHMLNGKDPE................TGEPLDDENIRYQIITFLIAGHETTSGLLSFALYFLVKNPHVLQKAEEEAARVLVD........PV
CYP051   DRDMLDVLIAVKAETG................TPRFSADEITGMFISMMFAGHHTSSGTASWTLIELMRHRDAYAAVIDELDELYGDG.......RS
CYP002C5 PRDFIDCFLIKMEQEN...........NLEFTLESLVIAVSDLFGAGTETTSTTLRYSLLLLLKHPEVAARVQEEIERVIGRHR.......S
CYP001A1 IRDITDSLIEHCQEKQ..........LDENANVQLSDEKIINIVLDLFGAGFDTVTTAISWSLMYLVMNPRVQRKIQEELDTVIGRS.......RR
CYP011A1 HHDYRGMLYRLLGDS.................KMSFEDIKANVTEMLAGGVDTTSMTLQWHLYEMARNLKVQDMLRAEVLAARHQAQG.......
CYP008A1 VKSRLWKLLSPARLARRAHRSKWLESYLLHLEEMGVSEEMQARALVLQLWATQGNMGPAAFWLLLFLLKNPEALAAVRGELESILWQAEQPVSQTTT

         J'        K
CYP102   PSYKQVKQLKYVG.MVLNEALRLWPTAPAFSLYAKEDTVLGGE....YPLEKGDELMVLIPQLHRDKTI.WGDDVEEFRPERFEN...........PSAIPQ
CYP051   VSFHALRQIPQLEN.VLKETLRLHPPLIILMRVAKGEFEVQG......HRIHEGDLVAASPAISNRIPED..FPDPHDFVPARYEQPRQ.......EDLLNR
CYP002C5 PCNQDRSRMPYTD.AVIHEIQRFIDLLPTN..LPHAVTRDVRFRN..YFIPKGTDIITSLTSVLHDEKAFPN..PKVFDPGHFLDESG.......NFKKS
CYP001A1 PRLSDRSHLPYME.AFILETFRHSSFVPFT..IPHSTTRDTSLKG..FYIPKGRCVFVNQWQINHDQKLWVN..PSEFLPERFLTPDG........AIDKVLS
CYP011A1 .DMATMLQLVPLLKASIKETLRLHPISVT..LQRYLVNDLVLRDY..MIPAKTLVQVAIYALGREP..TFFFDPENFDPTRWLSKDKN.........ITY
CYP008A1 LPQKVLDSTPVLD.SVLSESLRLT.AAPFITREVVVDLAMPMADGREFNLRRGDRLLLFPFLSPQRDPE.IYTDPEVFKYNRFLNPDGSEKKDFYKDGKRLK

         L
CYP102   HAFKPFGNGQRACIGQQFALHEATLVLGMMLKHFDFEDHTNY.ELDIKET..LTLK..PEGFVVKAKSKKIPLGGI
CYP051   WTWIPFGAGRHRCVGAAFAIMQIKAIFSVLLREYEFEMAQPP..ESYRNDHSKMVVQLAQPACVRYRRRTGV....
CYP002C5 DYFMPFSAGKRMCVGEGLARMELFLFLTSILQNFKLQSLVEPKDLDITAVVNGFVSVPPSYQLCFIPI.......
CYP001A1 EKVIIFGMGKRKCIGETIARWEVFLFLAILLQRVEFSVPLG.VKVDMTPIYGLTMKHACCEHFQMQLRS.......
CYP011A1 FRNLGFGWGVRQCLGRRIAELEMTIFLINMLENFRVEIQHLS....DVGTTFNLILMPEKPISFTFWPFNQEATQQ
CYP008A1 NYNMPWGAGHNHCLGRSYAVNSIKQFVFLVLVHLDLELINAD.VEIPEFDLSRYGFGLMQPEHDVPVRYRIRP...
```

FIG. 2. A representative alignment of sequences. Sequences for the proteins shown were extracted from the multiple sequence alignment that is available on the web. α Helices are indicated in bold letters, whereas β strands are underlined. Names of the helical elements are indicated above the alignment.

β1-2. These strands sit just at the putative "mouth" of the substrate access channel and the β turn between these strands (usually containing a glycine) is on the B'-helix face of P450s, i.e., the edge between distal and proximal faces containing B' and G helices. R47 that has been shown to be important in fatty acid recognition in P450BM3 is found in the turn between these two strands. These β strands are followed by the B helix.

The B' helix is the most variable region in P450s. In the initial report of the structure of P450cam at 2.6 Å resolution,[1] the presence of this helix was not recognized because it lacked apparent order, but in subsequent higher resolution structures, it was determined to be a helix. It has also been identified in other high-resolution structures (e.g., P450terp,[2] P450BM3,[8] P450eryF[3]), but its position in three-dimensional space is quite variable. Finally, in the structure of CYP2C5,[5] it is a coil rather than a helix. Likewise, alignment of sequences in this region as seen in Fig. 2 is problematic and should be viewed with great care when used for the construction of homology models. For example, in P450BM3, the unadjusted alignment as seen in Fig. 2 has put a gap in the B′ helix.

The B′–C loop (i.e., the loop between B′ and C helices) and the C helix itself is much easier to align. In the middle of this helix is a highly conserved tryptophan residue and a basic residue (WXXXR). In P450BM3, these residues are WXXXH in which the tryptophan nitrogen and the histidine serve to charge neutralize one of the buried propionate side chains of the heme. In the alignment shown in Fig. 2, there is a break in the C helix of CYP51A1, most likely due to the lack of the WXXXR sequence. This conserved WXXXR sequence is absent in families 7, 8, 26, 51, 74, 88, 105, 710, and G1. In fact, in our alignments, these families cluster together as shown in the evolutionary trace in Fig. 1. Interestingly, families 7, 8, 74, and 710 do not contain the conserved threonine, and 7, 8, 51, and 74 do not contain the conserved acidic amino acid in the I helix, as will be mentioned later.

Helices D and E are both amphipathic and are approximately the same length. In P450BMP in Fig. 2, the D helix has a gap that is probably due in part to its bend; otherwise these helices are unremarkable. Following is the E–F loop, which, along with the G–H loop (including the H helix), has been characterized as a hinge[10] that participates in opening and closing the F–G helix region over the substrate-binding site. The E–F and G–H loops are extremely variable in length, probably due to the extent to which the F–G helix region is required to open like "jaws" of a mouth to accept or release substrate. The F helix and, to a lesser extent, the G helix are also variable in length, as would be expected from their involvement in substrate binding, i.e., depending on the size and shape of the substrate. The G helix frequently ends in two or more positively charged residues. In some P450 families, e.g., CYP7, CYP8 (Fig. 2), and CYP74, the loop after the H helix is approximately 20 amino acids in length. It should be noted that the substrates for CYP74 and CYP8 are an alkyl hydroperoxide and an endoperoxide, respectively. It is not known why a loop on the surface of a P450 in a location so far from the substrate-binding region might be diagnostic of these families. Might it be related to the fact that these proteins do not need a redox partner and that this loop on the proximal face might in fact cover the redox partner interaction domain? Only experiments directed to this question will provide the answer.

The remainder of the P450 sequence is relatively easy to align, as the C-terminal half of P450s is more highly conserved. The approximately 30 amino acid long I helix, which is predominately hydrophobic, contains a highly conserved acidic residue followed by a threonine. The acidic amino acid is not conserved in the closely clustered 7, 8, 51, and 74 (see Fig. 1), whereas the threonine is not conserved in families 7, 8, 74, and 710. (Remember that families 7, 8, 51, and 74 also are missing the conserved WXXXR in the C helix.) Of these families, it is

[10] D. C. Haines, D. R. Tomchick, M. Machius, and J. A. Peterson, *Biochemistry* **40,** 13456 (2001).

known that CYP8 and CYP74 do not require an external source of electrons, rather they catalyze the rearrangement of endo- or hydroperoxides. One might postulate that the CYP710 family from rice and *Arabidopsis,* which was gleaned from the genomic sequencing projects, may have a similar mechanism of action. Additionally, the CYP7 family may have an additional function other than 7α-hydroxylase activity. Another family that also does not have the conserved threonine is CYP79, whereas family 88 contains a serine rather than the threonine, and family 33 contains a serine instead of the acidic residue. Finally, in the CYP51A1 structure from *Mycobacterium tuberculosis,* the I helix has been found to be broken into an I and I′ helix and may be the case in other 51 related or 51 clustered P450s.

Eukaryotic-like sequences have an insertion after the J helix of a J′ helix that is missing in most prokaryotic sequences. The presence of the K helix and its alignment is relatively easy to define with the absolutely conserved EXXR in the carboxy-terminal portion of this helix. This sequence is buried in the interior of the molecule and participates in a hydrogen bond network, which we have called the ERR triad where R378 of the meander region of P450BM3 is the second "R" in the triad. This second cationic residue is in a consensus sequence "PERF." In some P450s, the R is replaced by a histidine residue.

As can be seen, the L helix is defined easily with the conserved cysteine residue of the heme-binding loop just N-terminal of the helix. The consensus heme-binding sequence is FG/SXGXRXCXGXXXA/G. It should be noted that the R in this sequence is present with a probability of 0.91 and serves to charge neutralize one of the propionate side chains of the heme ring. In those cases where R is replaced by some other residue, it is usually either an H or a K that probably serves a similar function. The H at this position in CYP51 also charge neutralizes the heme propionate. In all mitochondrial P450s, the L helix begins with two arginines, whereas in microsomal P450s it might begin with a glutamate, a glutamine, or a lysine. In mitochondrial P450s, the two arginines have been proposed to be involved in electrostatic interactions with the iron–sulfur protein, which is the redox partner for those class II P450s.[10,11]

While most of the landmarks described earlier have been noted previously, most multiple alignments failed to recognize these residues because of the hypervariability around them. Fortunately, the current alignment method assigned all of them without manual intervention!

Structural Implications from Analysis of Alignments

The conservation of particular amino acids can be expressed in a variety of ways. One that we have found particularly helpful in analyzing the P450 family

---

[11] S. E. Graham-Lorence and J. A. Peterson, *FASEB J.* **10**, 206 (1996).

of proteins was developed by Lockless and Ranganathan.[12]

$$\text{Conservation} = \sqrt{\sum_{aa}\left[\ln\left(\frac{P_{aa\,i}}{P_{aa\,(database)}}\right)\right]^2} \tag{1}$$

where $P_{aa\,i}$ is the observed probability of amino acid (aa) being at position i in the alignment and $P_{aa\,(database)}$ is the probability of the aa being found in the entire protein sequence database. Basically, Eq. (1) is used to evaluate the probability that the composition of amino acids at a particular residue number in an alignment will deviate from that found in the protein database as a whole. That is, in the protein database, there is a certain frequency at which each amino acid occurs. Deviation from this random frequency yields a conservation index. This conservation index does not distinguish whether a particular amino acid is similar to another in the alignment but simply whether the composition is random. The values for the index can vary from zero (random) to 19.55. The value 19.55 is achieved when the composition is invariant at a particular position. This analysis can be done both in families or across families.

A graph showing a plot of the conservation index versus the residue number across all families from the generated multiple sequence alignment is shown in Fig. 3. To calculate this value, each P450 family was aligned with P450BMP (the heme domain of CYP102A1) using VectorNTI as described earlier and the amino acid composition was input into an Excel database. Because of size limitations in the previous version of the VectorNTI software, not all of the sequences could be aligned simultaneously. This limitation has been removed. In general, the "pairwise" family alignment of any given family with CYP102 was the same as that shown in Fig. 2 and on the Web.

In these alignments, there were residues in a family that did not have comparable sequences with P450BMP (i.e., there were more residues in family sequences than in the P450BMP sequence). Usually these residues were found to be located in regions corresponding to surface loops of P450BMP. We believe these insertions to be due to the random sequence variability across a P450 family and, in fact, any protein family. These loop regions might be considered as tethers between conserved structural elements, with the important feature being the length rather than a specific sequence. Therefore, those residues were excluded from the compilation for the graph in Fig. 3 and from further analysis.

The probability calculation in Eq. (1) takes into account the total number of amino acid residues present at each position in the alignment with P450BMP. There were no instances where the number of amino acids present at a position was less

---

[12] S. W. Lockless and R. Ranganathan, *Science* **286**, 295 (1999).

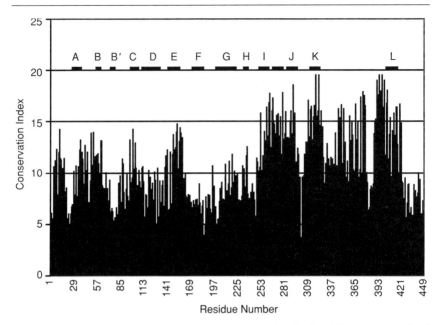

FIG. 3. A plot of the conservation index versus residue number for the P450 gene superfamily of proteins. The helical secondary structural elements of P450BMP are indicated at the top by black bars.

than 50% of the total number of sequences. There was no manual adjustment of the sequence alignment. With this analysis, we are able to confirm mathematically the conservation at the C-terminal half of P450s with a maximum in the K helix and the heme-binding region. In the N-terminal half, there is some conservation in β1-1 and β1-2, the C helix, and the E helix, but what is more important is the lack of conservation, i.e., the randomness seen in B', F, and G helices.

This analysis can be further appreciated by the transposition to the three-dimensional space of the conservation index. In Fig. 4, each residue in P450BMP was assigned the corresponding calculated conservation index number from the multiple sequence alignment as described earlier. Figure 4 shows only the peptide backbone atoms for greater clarity. Residues with values that are highly variable are colored white; those that are highly conserved are colored black; and those with intermediate values are colored in shades of gray. Figure 4 is a slice through this model parallel to the I helix and just in front of the heme ring, which is shown as a black stick model. The darkest residues, i.e., those that are most conserved, are those forming the heme pocket (those residues binding the heme group—not necessarily those in the substrate pocket) and many of those in the putative redox partner-binding pocket on the distal face, especially the K helix,

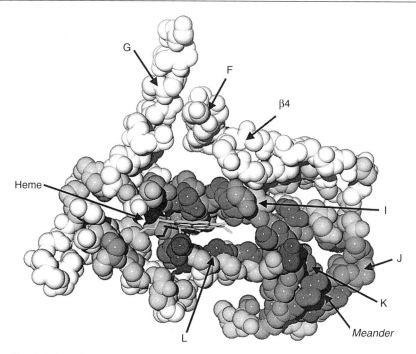

FIG. 4. A three-dimensional representation of the conservation index across families. The conserva-tion index values plotted in Fig. 3 were used to determine the color of the P450BMP template (Protein Data Bank identification code, 1JPZ). Only the backbone atoms in this space-filling representation were used with the most highly conserved residues shaded in black and the least conserved residues shown in white. The structure is sliced through the molecule, exposing the heme group and the interior of the molecule. Structural features of the molecule are indicated. The model was constructed with the program Swiss PDB Viewer (www.expasy.ch/spdv/mainpage.html).

meander, heme-binding region, and L helix. This is not surprising, as these re-gions are those that are involved in the catalytic mechanism: oxygen activation and electron transfer.

Summary

It seems as if the algorithms and weighting matrices for multiple sequence alignments of the highly divergent members of the P450 gene superfamily have advanced to the point that unknown proteins can be aligned to structurally known members with reasonable accuracy. As stated earlier, the alignment tends to break down at gaps in the sequence alignments, but these regions can be improved manually. This type of alignment and analysis is especially useful for extracting and analyzing the various genome databases.

Variations of the conservation analysis can be used to identify charged and uncharged residues that may be important in domain/domain interactions with redox partners or effector molecules (e.g., cytochrome b_5). From these alignments and with comparative analysis within families and across P450 families, one can readily obtain an estimation of those residues that might be involved in substrate binding, in redox partner interaction, and in the catalytic mechanism.

Acknowledgment

Supported in part by grants from the NIH (GM43479 and GM50858) and the American Heart Association (0150611N).

[3] Human CYP Allele Database: Submission Criteria Procedures and Objectives

By MAGNUS INGELMAN-SUNDBERG and MIKAEL OSCARSON

Introduction

The interindividual variability in xenobiotic metabolism and drug response is extensive. The drug levels in plasma can vary more than 1000-fold between two individuals with the same weight and receiving the same drug dosage. The causes for this variation are of genetic, physiological, pathophysiological, and environmental origin. In the past decade, genetic factors for this variability have received much emphasis. One could envision that these would account for about 20–40% of the interindividual differences in drug metabolism and response,[1] but for certain drugs or classes of drugs, genetic factors will be of utmost importance for the outcome of the drug therapy. Cytochromes P450, responsible for 70–80% of all phase I-dependent metabolism of clinically used drugs, play a central role.[2] Furthermore, the interindividual differences in susceptibility for carcinogens and environmental toxins are extensive, and cytochrome P450s play a key role in their bioactivation.

Adverse drug reactions (ADRs) are a greater problem for drug treatment and drug development than previously thought. It has been estimated that ADRs cause more than 100,000 deaths annually in the United States, 7% of all hospital admissions in the United Kingdom, and 13% of all admissions to internal medicine wards in Sweden (see Marshall[3] for references). The costs for ADRs, including on average

---

[1] M. Ingelman-Sundberg, *J. Intern. Med.*, **250**, 186 (2001).

[2] R. J. Bertz and G. R. Granneman, *Clin. Pharmacokinet.* **32**, 210 (1997).

[3] A. Marshall, *Nature Biotechnol.* **15**, 1249 (1997).

0076-6879/02 $35.00

2 days of prolonged hospitalization and reduced productivity, have been estimated in the United States to be about 100 billion dollars annually.[3] Thus, it is advisable to take the genetic constitution into account for drug therapy in many cases. The genetics of drug-metabolizing enzymes, in particular cytochromes P450, are critical to design a better drug therapy. In addition, the drug industry now takes these aspects into greater consideration when developing new drug candidates. For research and applications considering the interindividual variability in xenobiotic metabolism by polymorphic P450s, it is necessary to have a common nomenclature for genetic variants and a system that allows researchers to be updated rapidly within the field.

A Web Page for Human Cytochrome P450 Alleles: www.imm.ki.se/CYPalleles

In recent years, much research in the cytochrome P450 area has been focused on the identification and characterization of polymorphic human P450 genes. The rapid development in the field required the establishment of a web page with continuously updated information and recommended nomenclature regarding the various allelic variants of P450s. The aims were to encourage scientists worldwide to speak the same language and to avoid home-made allelic designations that can confuse the nomenclature system and the scientific literature. In addition, a rapid publication of new alleles would prevent unnecessary doubling of research efforts aimed at characterizing alleles already described. For this purpose, an initiative was taken to create a web page at a server at the Karolinska Institute with Mikael Oscarson as the Webmaster with an editorial board of the Allele Nomenclature Committee consisting of Magnus Ingelman-Sundberg, Ann Daly, and Dan Nebert and an international advisory board with Jürgen Brockmöller, Michel Eichelbaum, Seymour Garte, Joyce A. Goldstein, Frank J. Gonzalez, Fred F. Kadlubar, Tetsuya Kamataki, Urs A. Meyer, David R. Nelson, Michael R. Waterman, Anna Wedell, and Ulrich M. Zanger as members. The basis for the nomenclature system chosen was based on the general guidelines provided,[4] the system for allelic designations for *CYP2D6* described in Daly *et al.*,[5] and recommendations for numbering of nucleotides and style for assigning mutations as proposed by Antonarakis and the Nomenclature Working Group.[6]

The web page covers the nomenclature for polymorphic alleles of *CYP1A1, CYP1A2, CYP1B1, CYP2B6, CYP2C8, CYP2C9, CYP2C19, CYP2D6, CYP2E1,*

---

[4] T. B. Shows, P. J. McAlpine, C. Boucheix, F. S. Collins, P. M. Conneally, J. Frezal, H. Gershowitz, P. N. Goodfellow, and J. G. Hall, *Cytogenet. Cell Gene.* **46,** 11 (1987).

[5] A. K. Daly, J. Brockmoller, F. Broly, M. Eichelbaum, W. E. Evans, F. J. Gonzalez, J. D. Huang, J. R. Idle, M. Ingelman-Sundberg, T. Ishizaki, E. Jacqz-Aigrain, U. A. Meyer, D. W. Nebert, V. M. Steen, C. R. Wolf, and U. M. Zanger, *Pharmacogenetics* **6,** 193 (1996).

[6] S. E. Antonarakis, and the Nomenclature Working Group, *Hum. Mutat.* **11,** 1 (1998).

CYP2J2, CYP3A4, CYP3A5, CYP3A7, CYP5A1, CYP8A1, and *CYP21,* although in the near future nomenclature systems for all polymorphic CYPs are to be incorporated. An example of a representative web page is given in Fig. 1 showing nucleotide changes, their functional consequences, and easy accessible links to the PubMed citation(s) describing the allele. The CYPallele page is now well established and visited with over 100 hits per working day. About one-third of the visitors come from industry. A letter has been published in *Pharmacogenetics* describing some general decisions taken with respect to the nomenclature system and features of the page.[7] In addition, a letter has been published in *Clinical Epidemiology and Biomarkers* explaining the nomenclature and the page features for scientists primarily in the field of molecular epidemiology.[8] Similar web pages have also been established for other genes encoding drug-metabolizing enzymes, such as the UDPglucuronosyltransferases (http://www.unisa.edu.au/pharm_medsci/gluc_trans/), aldehyde dehydrogenases (http://www.uchsc.edu/sp/sp/alcdbase/aldhcov. html), and arylamine *N*-acetyltransferases 1 and 2 (http://www.louisville.edu/medschool/pharmacology/NAT.html/) allele web pages.

Human Cytochromes P450

Completion of the draft sequence of the human genome revealed the presence of about 90 different cytochrome P450 genes as determined by David Nelson (drnelson.utmem.edu/CytochromeP450.html): 55 of these are functional, whereas 35 are pseudogenes. The majority of genes among the xenobiotic-metabolizing P450s in gene families 1–3 are polymorphic (see Table I), whereas polymorphic variants are more rare in families 4–51, partly because of a more important endogenous role for these enzymes not allowing such a genetic drift as seen with the genes encoding xenobiotic-metabolizing P450s. Among P450s in families 1–3, only *CYP1A1* and *CYP2E1* are relatively well conserved, and in essence, no functionally important polymorphisms are present in these genes. Concerning some genes such as *CYP2J2, CYP2R1, CYP2S1, CYP2U1,* and *CYP2W1,* which are relatively recently identified, no polymorphisms have yet been described, but are likely to appear in the literature in the future. Only a few important endogenous substrates have been described for any of the polymorphic P450s in families 1–3, and their primary function is most likely the metabolism of dietary components explaining their extensive variability. Most allelic variants are distributed with pronounced interethnic differences.

The web page takes all allelic variants into account and describes both coding region SNPs (cSNPs) and SNPs in introns and regulatory regions of the *CYP* genes. cSNPs resulting in missense mutations can yield enzyme products with abolished,

---

[7] M. Ingelman-Sundberg, A. K. Daly, M. Oscarson, and D. W. Nebert, *Pharmacogenetics* **10,** 91 (2000).

[8] M. Ingelman-Sundberg, M. Oscarson, A. K. Daly, S. Garte, and D. W. Nebert, *Cancer Epidemiol. Biomarkers Prev.* **10,** 1307 (2001).

CYP2A6 allele nomenclature

Allele	Protein	Nucleotide changes	Trivial name	Effect	Enzyme activity		References
					In vivo	In vitro	
CYP2A6*1A	CYP2A6.1	None			Normal	Normal	Yamano et al, 1990
CYP2A6*1B	CYP2A6.1	gene conversion in the 3' flanking region					Oscarson et al, 1999b; Ariyoshi et al, 2000
CYP2A6*1X2	CYP2A6.1			CYP2A6 gene duplication			Rao et al, 2000
CYP2A6*2	CYP2A6.2	479T>A	v1	L160H	None	None	Yamano et al, 1990; Hadidi et al, 1997; Oscarson et al, 1998
CYP2A6*3	CYP2A6.3	CYP2A6/CYP2A7 hybrid	v2	.	?	?	Fernandez-Salguero et al, 1995
CYP2A6*4A		CYP2A6 deleted	CYP2A6del, E-type	CYP2A6 deleted	None		Oscarson et al, 1999a; Nunoya et al, 1999a; Nunoya et al, 1999b; Ariyoshi et al, 2000
CYP2A6*4B		CYP2A6 deleted	D-type	CYP2A6 deleted	None		Nunoya et al, 1998
CYP2A6*4C							Identical to the CYP2A6*4A allele (see above)
CYP2A6*4D		CYP2A6 deleted		CYP2A6 deleted	None		Oscarson et al, 1999b
CYP2A6*5	CYP2A6.5	1436G>T	.	G479V	None	None	Oscarson et al, 1999b
CYP2A6*6	CYP2A6.6	383G>A		R128Q		Decreased	Kitagawa et al, 2001
CYP2A6*7	CYP2A6.7	1412T>C; gene conversion in the 3' flanking region		I471T	(Decreased)	Decreased	Ariyoshi et al, 2001; Xu et al, 2002
CYP2A6*8	CYP2A6.8	1454G>T; gene conversion in the 3' flanking region		R485L	Normal		Ariyoshi et al, 2001; Xu et al, 2002
CYP2A6*9	CYP2A6.1	-48T>G		TATA box		Decreased	Pitarque et al, 2001
CYP2A6*10	CYP2A6.10	1412T>C; 1454G>T; gene conversion in the 3' flanking region		I471T; R485L	Decreased		Yoshida et al, submitted; Xu et al, 2002
CYP2A6*11	CYP2A6.11	670T>C		S224P	Decreased	Decreased	Daigo et al, in press

FIG. 1. Presentation of a representative Human CYPallele web page.

TABLE I
HUMAN CYTOCHROME P450 GENES[a]

CYP gene	Chromosome	cSNPs creating missense mutations	CYP gene	Chromosome	cSNPs creating missense mutations
1A1	15	YES	4F11	19	nd
1A2	15	YES	4F12	19	nd
1B1	2	YES	4F22	19	nd
2A6	19	YES	4F28	21	nd
2A7	19	nd	4V2	4	nd
2A13	19	nd	4X1	1	nd
2B6	19	YES	5A1	7	YES
2B8	10	YES	7A1	8	nd
2C9	10	YES	7B1	8	nd
2C18	10	YES	8A1	20	YES
2C19	10	YES	8B1	3	nd
2D6	22	YES	11A1	15	nd
2E1	10	YES	11B1	8	nd
2F1	19	nd	11B2	8	nd
2J2	8	YES	17	10	YES
2R1	11	nd	19	15	YES
2S1	19	nd	21A2	6	YES
2U1	4	nd	24	20	nd
2W1	7	nd	26A1	10	nd
3A4	7	YES	26B1	2	nd
3A5	7	YES	27A1	2	YES
3A7	7	nd	27B1	2	nd
3A43	7	nd	27C1	2	nd
4A11	1	nd	39A1	6	nd
4A20	1	nd	46	1	nd
4B1	1	nd	51	7	nd
4F2	19	nd			
4F3	19	nd			
4F8	19	nd			

[a] Shaded genes are currently included on the human CYPallele web page. nd, Not described.

reduced, altered, or increased enzyme activity. One of the reasons for abolished enzyme activity is that the whole gene has been deleted, but this is also seen as a consequence of mutations causing altered splicing, stop codons, abolished transcriptional start sites, and deleterious amino acid changes. In addition, *CYP* genes contain many conservative substitutions and silent mutations that do not influence function in addition to the SNPs located in the introns and in the regulatory 5′ and 3′-flanking regions of the genes. The functional importance of many SNPs in regulatory regions is usually dubious and it is difficult to directly prove their influence in association studies, in *in vivo* phenotyping studies, or in functional *in vitro* studies.

Inclusion Criteria for CYP Alleles

The designation of an allele would ideally require determination of all SNPs in the gene, *i.e.*, the complete haplotype. Because of lack of technology to determine haplotypes at a reasonable cost and efficiency, we are faced with the problem that most CYP alleles are characterized incompletely and the complete distribution of all SNPs in the genes is indeed unknown. The nomenclature system for human CYP alleles currently emphasizes alleles carrying polymorphisms that cause missense mutations in genes that are receiving a unique allelic number. This simplifies the matter somewhat, as the highest emphasis is put on the functional variation. In due time, however, when technology has advanced, it might be necessary to refine the current system and take the complete haplotype into account.

The current system for naming P450 alleles originates, as mentioned, from that developed for CYP2D6 alleles.[5] Discussions were held among the members of the editorial and advisory committees for a year before inclusion criteria (Table II) were agreed on and the web page became effective in June 1999. Some of the arguments in the debate have been published in the literature.[9,10]

Initially, we considered that it would be a clear disadvantage to rename existing alleles and therefore the web page contains the published allelic designations at the time of start of the page in order to avoid confusion, but the naming of all alleles since then follows the same general rules as detailed in Table II. In order to receive a unique allelic number, the allele should contain nucleotide changes that have been shown to affect transcription, splicing, translation, posttranscriptional, or posttranslational modifications or result in at least one amino acid change. Additional nucleotide changes and combinations of nucleotide changes in the gene will be given letters (e.g., CYP2D6*6B). When several missense polymorphisms are present on the same allele, the allelic number given is based on the polymorphism that causes the most severe functional consequence, e.g., a splicing defect, and other mutations in such an allele receive a letter that adheres to the allelic number of the allele with the polymorphism resulting in the most serious alteration. The first allele sequenced (reference allele) is designated *1 (or *1A, *1B, etc.). The *1 allele will thus not be the major allele in every ethnic group.

The number of SNPs described in P450 genes and their neighborhood increases as a result of the action of the SNP consortia, and for their correct designations, one has to consider how much DNA around each transcribed region should be considered to belong to the gene. For a gene in which the 5′ regulatory region has been characterized incompletely, we consider that at least 3 kb upstream from the transcription start constitutes the gene. If regulatory sequences are known to exist further 5′-ward, the 5′ end of such an element would be defined as the start of

[9] D. W. Nebert, M. Ingelman-Sundberg, and A. K. Daly, *Drug Metab. Rev.* **31**, 467 (1999).
[10] D. W. Nebert, *Pharmacogenetics* **10**, 279 (2000).

TABLE II
Criteria for Inclusion of Alleles and SNPs on the Home Page of the Human
Cytochrome P450 (CYP) Allele Nomenclature Committee

1. On the web page, only human CYP alleles are considered
2. The gene and allele are separated by an asterisk followed by Arabic numerals and uppercase Roman letters with less than four characters to name the allele (e.g., *CYP1A1*3, CYP1B1*22, CYP2D6*10B*)
3. A gene is considered as the sequence from 5 kb upstream from the transcription start site to 500 bp downstream of the last exon. However, if a regulatory element has been characterized at a more distant part of the gene, this area also belongs to the gene
4. To be assigned as a unique allele, it should contain nucleotide changes that have been shown to affect transcription, splicing, translation, posttranscriptional, or posttranslational modifications or result in at least one amino acid change
5. Additional nucleotide changes and combinations of nucleotide changes in the gene will be given letters (e.g., *21A, *21B*). Thus, in cases where silent mutations occur or mutations are present in regulatory parts or introns with unclear function, the allelic name should adhere to the closest functionally characterized allele by subgroup assignments, such as *CYP2D6*4A*. Allelic variants can be defined as combinations of up to three letters (e.g., *CYP2D6*2ABC*), thereby allowing room for $22 \times 22 \times 22 = 10,648$ different variants for each allelic number. The letters I, O, X, and Y are excluded because of indexing problems
6. For extra gene copies (n) placed in tandem, the entire allelic arrangement should be referred to, e.g., *CYP2D6*2Xn*
7. Numbering of nucleotides in the allele should be as described in Antonarakis and the Nomenclature Working Group (1998). Base A in the initiation codon ATG is denoted $+1$ and the base before A is numbered -1
8. For reasons of indexing, the names for proteins should have a period between the name of the gene product and number (e.g., CYP2D6.2A)
9. The wild-type allele is defined as the sequence of the first alleles sequenced and should be designated as *1 (or *1A and *1B in case of slightly variant sequences)
10. SNPs that are not assigned easily to a specific allele will be listed at the bottom of the corresponding nomenclature page with relevant literature references
11. Submission of new alleles should be done with information sufficient to fulfill the criteria to be assigned a unique allele as under criterion 4 or a letter as described under criterion 5. For incorporation into the web page as a unique allele, all exons and exon–intron borders should have been sequenced. If a new allele has been detected on the cDNA level, verification of the mutation(s) on the genomic level is required. For acceptance of a new SNP given a separate letter, evidence for its presence on the genomic level is required
12. No temporary allelic numbers or letters are provided, and information about any new allele submitted will be published continuously on the web page. In case an author does not want to release the information on the web page before publication, the Webmaster can usually provide the author with an allelic designation but not release the information on the web page until the manuscript has been accepted or published

the gene. On most cases, 150 bp downstream of the last exon is considered to be sufficient to be included as the gene.

Procedures for Submission of New Alleles

Submission of new alleles should be done to the Webmaster with information sufficient to fulfill the criteria to be assigned a unique allele. For incorporation into the web page as a unique allele with a new specific number, preferentially all exons and exon–intron borders should have been sequenced. If a new allele has been detected on the cDNA level, verification of the polymorphism(s) on the genomic level is required. For acceptance of a new SNP given a separate letter, evidence for its presence on the genomic level is required. No temporary allelic numbers or letters are provided, and information about any new allele submitted is published continuously on the web page. In case an author does not want to release the information on the web page before publication, the Webmaster can usually provide him or her with an allelic designation but not release the information on the web page until the manuscript has been accepted or published.

Controversies and Problems in the Nomenclature System

The current system for naming CYP alleles favors that missense and nonsense, mutations as well as insertions and deletions, which cause alteration in function, receive higher attention because the nomenclature system with numbers is based on this property, whereas alleles with SNPs of unclear function receive a letter adhering to the *1 allele (e.g., CYP1A1*1F). For scientists involved in studies of evolution and anthropology, a system where all alleles have the same ranking in naming might be clearer. However, for clinicians and scientists involved in the functional properties of the alleles, the current hierarchy in naming offers advantages because much fewer allelic numbers have to be dealt with for each CYP and it is easier to remember alleles that have functionally important differences.

As mentioned, a major disadvantage and problem with the naming system today is incomplete knowledge about the complete haplotypes. We think that this problem will be diminished in magnitude in the years to come.

Conclusions

The Human CYPallele web page offers a rapid online publication of new CYP alleles describing their major properties. This is of benefit for all scientists being able to speak the same language and of benefit for industry having an easy accessible source of information of importance in drug development and for design of clinical trials with respect to genetic polymorphic factors. Furthermore, the existence of pages with newly updated information prevents unnecessary doubling of research efforts aimed at characterizing alleles already described.

Acknowledgment

The constructive and sometimes lively discussions held within the editorial and advisory boards of the Human CYPallele page are gratefully acknowledged.

[4] Fine-Scale Mapping of *CYP* Gene Clusters: An Example from Human *CYP4* Family

By SUSAN M. G. HOFFMAN and DIANE S. KEENEY

Introduction

Now that various genome projects have produced immense quantities of sequence from both the mouse and the human, with other species in progress, we have an unprecedented opportunity to analyze cytochrome P450 gene families. The available sequences, however, are neither complete nor perfectly assembled at this time, nor are they likely to be so in the immediate future. Because P450 genes in vertebrates frequently occur in tandem arrays that can include many pseudogenes,[1] they are particularly likely to be misassembled. It is incumbent on the community of P450 specialists to examine and annotate sequence data within such gene clusters, because available genome assembly software and nonspecialist workers are unlikely to detect problems at a detailed level.

Determining the number of loci in a cluster of P450 genes is a vital step toward understanding and manipulating those genes. Since P450 genes within a cluster are usually members of the same or related subfamilies, they typically have highly similar sequences that can confound both polymerase chain reaction- and hybridization-based analyses. A clear picture of the number of genes and pseudogenes in a cluster allows transcripts to be assigned confidently to loci, clarifies orthologous relationships, permits accurate genotyping, and sometimes elucidates the mechanistic processes that create significant polymorphisms.[1]

We have used the *CYP4A11* gene in humans as an example of how to analyze genomic sequences for the presence of gene clusters. The series of steps described in this article explains how to search for additional related loci near *CYP4A11*, identify genes and pseudogenes, and evaluate the genomic sequence assembly

---

[1] S. M. G. Hoffman, D. R. Nelson, and D. S. Keeney, *Pharmacogenetics* **11,** 687 (2001).

currently available in the public database. This general method can be adapted to the fine-scale mapping of any P450 gene cluster in any organism for which there is a substantial collection of genomic sequence available through GenBank. It relies throughout on data and software that are freely available to all researchers.

Procedure

Identifying and Comparing Human CYP4A Sequences in Databases

An Entrez query of GenBank nucleotide data[2] with the key word "CYP4A11" (as of July 2001) yields a list of several different cDNA sequences derived from human kidney and liver (e.g., S67580), a single 15-kb segment of a BAC genomic clone (AF208532), and several genomic sequences from human chromosome 1 assembled from smaller fragments of a single (AL135960) or multiple (NT_004525) BAC clones. Expanding the search by using "CYP4A* AND human" yields two additional cDNAs. Care must be taken to read the annotation notes associated with each sequence from such a search, since one database entry is often simply an alternative version of another entry (e.g., NM_000778 is derived from L04751 and S67580).

These sequences can be compared using pairwise BLAST[3] or placed in a multiple alignment by any appropriate software. Such an analysis reveals the first complication illustrated by this example. All of the *CYP4A11* cDNA sequences are completely or nearly identical, and all of the genomic sequences are almost identical, but the cDNA sequences are somewhat different from the genomic sequences. BLAST comparisons of these two groups, cDNA versus genomic sequences, indicate that they are about 98% identical overall at the nucleotide level, but only about 95% identical in terms of predicted amino acid sequence. This is consistent with the cDNA and genomic sequences being from two different loci rather than being two alternative alleles of *CYP4A11*[4] —both possibilities must be considered.

Next, compare results of these searches with the extremely useful "Cytochrome P450 Homepage" maintained by Nelson,[5] which extracts and organizes information on P450 genes from multiple sources. On the "Human P450 sequence collection" page,[6] Nelson suggests that the two sequence groups are too different to represent alleles; he designates the cDNA sequences as representing *CYP4A11* and the genomic sequences as representing a new locus, *CYP4A22*. He also lists another

[2] www.ncbi.nlm.nih.gov/entrez/query.fcgi?db=Nucleotide
[3] www.ncbi.nlm.nih.gov/blast/bl2seq/bl2.html
[4] P. Fernandez-Salguero, S. M. G. Hoffman, S. Cholerton, H. W. Mohrenweiser, H. Raunio, O. Pelkonen, J. Huang, W. E. Evans, J. R. Idle, and F. J. Gonzalez, *Am. J. Hum. Genet.* **57**, 651 (1995).
[5] drnelson.utmem.edu/cytochromeP450.html
[6] drnelson.utmem.edu/human.P450.seqs.html

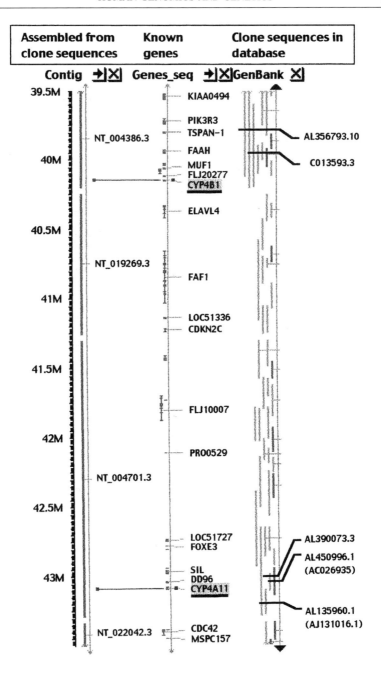

4A subfamily member in humans, *CYP4A20*, which is known only from the draft genomic sequence from chromosome 1. This locus is rather distantly related to *CYP4A11* and may eventually be placed in another subfamily. The question of how many loci are in the *CYP4A* subfamily, and whether they are physically clustered, is clearly an unresolved issue.

Examining Relevant Genomic Sequences and Evaluating Sequence
Assemblies in Databases

The next step is to check putative locations for *CYP4A* loci on human chromosome 1. The current state of the public genome project assembly can be observed via the Map Viewer browser.[7] Setting the chromosome window to 1, using the key word CYP4A11, clicking on the Gene_seq link, and adjusting the settings as given (Fig. 1) yields a map in the p33 band of chromosome 1 with *CYP4A11* (now *CYP4A22*) highlighted. Typing "CYP4B1" into the "Find in this View" box of the browser highlights the *CYP4B1* locus three megabases nearer to the p telomere than *CYP4A11*. The two loci are in different assembled contigs (NT_004701 versus NT_004386), which are separated by the small contig NT_019269. Note that this Map View display can change significantly over time, as the large contigs are reassembled with each update of sequence data.

A different organization of the same region can be examined using the HGP Gateway browser,[8] by searching with "CYP4A11" as the key word, zooming out 45×, and adjusting the settings at the bottom of the window as given (Fig. 2). This arrangement places a clone (AL356793) containing the *CYP4B1* locus immediately adjacent to a clone (AL135960) containing *CYP4A11*. The discrepancy between the Map View and Gateway arrangements is discussed later.

The "Human P450s sorted by chromosome" table of the CYP Homepage[9] lists additional loci from the *CYP4* family as being on chromosome 1. The *CYP4X1*

---

[7] www.ncbi.nlm.nih.gov/cgi-bin/Entrez/map_search
[8] genome.ucsc.edu/index.html
[9] drnelson.utmem.edu/hum.htgs.bychr.pdf

FIG. 1. Images from Map View browser. Human chromosome 1 is oriented with the p terminus (pter) at the top rather than at the left, with the distance from the terminus shown on the ruler in megabases. The "GenBank" part of the map shows the individual BAC clones being sequenced, whereas the "contig" part of the map shows large blocks of draft sequence assembled from clone sequences. On the web site, clones are shown in the browser as blue if the sequence is finished and as orange if it is draft. Most clone numbers have been eliminated for clarity; clones in parentheses are redundant with preceding clones. AC026935 is an older version of AL450996. The display is set for Contig (with ruler), Clone, Gene_seq, and GenBank, using the "Display Settings" window. Note that *CYP4B1* and *4A11* (*4A22*) are separated by 3 Mbp in this organization and that the contig containing *CYP4A11*, DD96, and SIL has the opposite orientation (SIL closest to pter) from the organization in the Gateway (Fig. 2) view (*CYP4A11* closest to pter).

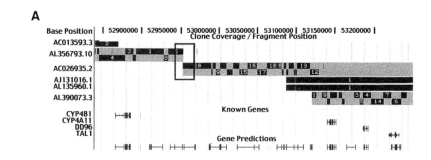

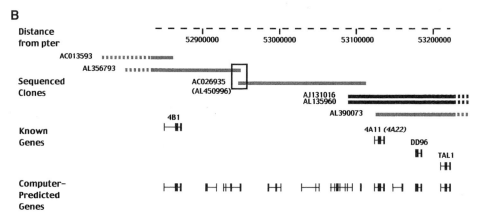

FIG. 2. (A) Image as it appears on HGP (Human) Gateway browser for the October 2000 data freeze. "CYP4A11" was used as the search key word, followed by clicking on one of the listed mRNAs. The display was modified by zooming out 10×, 3×, and then 1.5× and then shifting once to the left with the single arrow (<) button. The initial display was then simplified by setting the following options to "hide"—Chr band, STS Markers, Gap, Full mRNA, Exofish, Overlap SNPs, Random SNPs, Affy Genes, Ensembl Genes, RepeatMasker, Spliced ESTs—and setting coverage to "full," and then hitting the "refresh" button. Note that because both software and data assemblies at this site are updated periodically, these settings may need to be modified. The text font is enlarged for clarity; the scale across the top is based on distance from the p terminus in base pairs. Sequenced clones are shown as a patchwork of gray (spaces) and black (contigs) if sequencing is unfinished; numbers along those clones represent individual sequence contigs in putative order. Finished clones are shown as solid black lines. Genes are represented by vertical hash marks for each exon along a horizontal bar. Computer prediction of genes is shown for one of three possible algorithms (F-Gnesh). AC026935 is an older version of AL450996. The region of false overlap between clones AL356793 and AL450996/AC026935 is highlighted by a box overlaid on the figure. (B) Simplified version of the Gateway figure. Compare to Fig. 3.

locus is listed as being on the same BAC clone as *CYP4A20,* and a pseudogene (*CYP4A20P*) is on the same BAC clone as *CYP4B1.* Whether they are ultimately placed in the same subfamily, all of these loci are more closely related to each other than to other *CYP4* family members, as shown by the amino acid identities on the "Human P450 sequence collection" page.[6] Thus, there are a number of loci (*CYP4A11, 4A20, 4A22, 4A20P, 4X1, 4B1;* see Fig. 3) located on human chromosome 1 at about 1p33-34 that potentially are part of a cluster of related *CYP4* genes.

Next, the clones actually sequenced to produce the assemblies should be evaluated (these are usually BACs but may be cosmids). They are listed in "Human P450s sorted by chromosome"[9] and are also displayed by the Map Viewer[7] and Gateway browsers.[8] The accession numbers of all the clones from the general region should be noted, as well as the organizations given for them (done most easily by printing a hard copy of the displays, as in Figs. 1 and 2).

Each sequenced clone needs to be assessed separately—their accession numbers should be used as key words in an Entrez-Nucleotide[2] search of the NCBI Sequence Viewer to look up sequencing information in the database. The sequencing status of each clone is given in its "Definition" line—it may be finished, ordered draft or unordered draft. A clone labeled "complete" or "finished" should consist of a single contiguous block of sequence that is highly reliable. A clone labeled "draft sequence" may consist of "n *ordered* pieces," which means that sequenced contigs have been placed into a linear order based on unspecified additional evidence. In our experience, the ordering of contigs is generally good, and sequences within the contigs are also highly reliable, but there are gaps (usually small) between contigs that may eliminate desirable sequence. Finally, a clone labeled "draft sequence" and as having "n *unordered* pieces" consists of sequenced contigs placed into an *arbitrary* order. Each contig usually contains good quality sequence, but they may be in any order and may also be reversed. Such clones can provide useful information, but only after considerable processing by the user. Generally, both ordered and unordered draft clones are in the process of being refined by the sequencing centers, but researchers may have to wait months or years for a given clone to be finished.

In the case of our *CYP4A* example, the relevant clones are the BACs AL356793, AL450996, AC013593, AL390073, and AL135960. Of these, only the last is sequenced completely, while BLAST results indicate that AL390073 is completely redundant with AL135960. The remaining three sequences (AL356793, AL450996, and AC013593) consist of 4, 16, and 3 *unordered* contigs, respectively. The current organizations of these clones in the two different browsers are shown in Figs. 1 and 2. Because there is a significant discrepancy between the two organizations displayed by the two browsers, an evaluation of the assembly is particularly important in this example, but some review of the assemblies should be done in any case.

Both browser assemblies show considerable overlap between clones AC013593 and AL356793, and again between AL450996 and AL135960. These overlaps can be verified by doing a pairwise BLAST[3] between the linked clones. Such comparisons should always be done using a small section of one clone (100–2000 bp) as the query sequence, rather than comparing two complete BACs, a very slow process that gives confusing results. For an unordered clone, the base pair numbering seen in its Entrez-Nucleotide entry can be used to compare individual contigs to the subject sequence. For example, the 16th contig of AL450996 (bp 176563–178628) can be compared to AL135960, yielding a 98% match across 2048 bases that indicates a true overlap. Query sequences that align with multiple short target sequences usually prove to contain repetitive element DNA; in those cases, another section of the first clone should be chosen as the query sequence.

The Gateway browser assembly shows a small overlap between the AC013593/ AL356793 and AL450996/AL135960 clone pairs. Pairwise BLAST does not confirm this linkage; it appears to be a false match based on repetitive element sequence in clones AL356793 and AL450996. The Map Viewer, instead, places another contig (NT_019269) in between the assembled sequence containing AC013593/ AL356793 and that containing AL450996/AL135960. The specific evidence supporting this arrangement (probably FISH data) is not completely clear, but the general methods used are discussed on the help pages of the web site. This Map Viewer arrangement is somewhat suspect in that it spreads out the *CYP4A*-related loci across three megabases, whereas the Gateway organization puts all potentially related loci within less than half a megabase (Fig. 2). A check of the syntenic region in the mouse[10] shows *Cyp4b1* and several *Cyp4a* genes as being closely linked, which also supports the Gateway arrangement. A final decision on which is the correct arrangement, however, cannot be made until more data are available. For the purposes of this example, we will accept the general organization shown by the Gateway browser (Fig. 2), while discounting the false overlap between AL356793 and AL450996.

Localizing Genes Inside Sequence Assemblies

Once a general organization for a region has been accepted, the placement of known loci on the assembled contigs can be checked and a search made for additional related loci. Pairwise BLASTs of a known P450 cDNA sequence against genomic clones should give 98–100% identities for each exon if the match is gene-specific. Less specific matches can sometimes be found by using amino acid sequences instead of cDNAs as queries against genomic sequences; for such comparisons, the pairwise BLAST program should be set to "tblastn" (protein

[10] www.informatics. jax.org

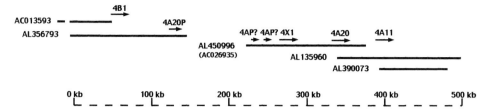

FIG. 3. Current interpretation of mapping data. The size of the gap shown in the middle is a very rough estimate based on STS and FISH data shown in the Gateway viewer. Distances are shown in kilobases relative to the end of clone AL356793, rather than to the p terminus. AC026935 is an older version of AL450996.

against translation of nucleotides). It is most efficient to start by BLASTing with the first and last exons separately to bracket complete loci. Multiple matches with the same exon, or matches of <95%, usually indicate that additional related loci have been detected. For example, exon 1 (bases 1–177) of the CYP4A11 cDNA L04751 aligns with one segment of BAC AL135960 at the site now identified with *CYP4A22*.[5] Exon 12 (bases 1398–1589 of L04751), however, aligns with two different segments of BAC AL135960: a 98% match to the *CYP4A22* region and a 75% match to the region (around base 160100 of clone AL135960) that overlaps clone AL450996. Segments that show such secondary matches should then be tested against other related cDNAs, e.g., the CYP4B1 sequence, to find the best match. In this case, the region detected in clone AL450996 has already been associated with the *CYP4A20* locus.[5]

Results of a complete series of pairwise BLASTs of *CYP4* cDNA exons against all relevant clones are summarized in Fig. 3. The previously defined loci *CYP4A20, 4A22, 4A20P, 4X1,* and *4B1* are all localized to the BAC clones, while the genomic sequence of the true *CYP4A11* locus (as currently defined by Nelson) is still missing. Two small regions that match exon 12 of the *CYP4A22* locus at 75% are indicated by 4AP?—these presumably are partial pseudogenes, since no other exons are detected upstream and they do not closely match any known transcripts. Figure 3 makes it clear why the overall arrangement shown by the Gateway browser is preferable to that of the Map Viewer browser: the former holds all of the related loci in a coherent cluster, with the *CYP4A20P* pseudogene fairly near to its closest relative, the *CYP4A20* gene, whereas the latter splits them up. Some of the named loci in Fig. 3 correspond to the unnamed "Computer-Predicted" genes shown in Fig. 2.

Completing a Map

All of the maps shown in Figs. 1–3 are incomplete: there is at least one gap between the contig containing *CYP4B1* and *4A20P,* and the contig containing *CYP4A20, 4A22,* and *4X1.* It is not clear that any attempt is being made to fill this gap, as the institution sequencing this part of chromosome 1 (Sanger Centre) still displays the questionable overlap between clones AL356793 and AL450996 on its web site.[11] In other words, the sequencers may be unaware that there is a gap. This situation illustrates the importance of having specialists review genomic sequences in the database—they are more likely to detect such anomalies. Anyone discovering a sequence or assembly problem can draw it to the attention of the responsible sequencing institution, which is named in the Sequence Viewer[2] entry of each clone.

Researchers can also continue to monitor a chromosomal region and incorporate improved sequences into their provisional maps. A small region that does not contain any repetitive element sequence can be selected from the end of a clone flanking a gap and used to search the database periodically for new, potentially overlapping sequences. Searches using cDNA sequences from unlocalized genes can also be done periodically. In this case, the true *CYP4A11* locus is still missing. Since there is a large contig extending out from AL135960 (right-hand side of Fig. 3) that contains several non-P450 genes, it is unlikely that the missing *4A11* locus will be found there. Instead, the internal gap shown in Fig. 3 is the most likely place for an additional related locus, and more sequences in this region will be needed to finalize the *CYP4* cluster map.

Acknowledgments

The authors thank Kyle Donley for critically reading the manuscript. SMGH was supported in part by NIH Grant 1R15GM55951-01A1. DSK was supported in part by the Office of Research and Development, Medical Research Service, Department of Veterans Affairs, and NIH Grants AR47357 and AR41943.

---

[11] www.sanger.ac.uk/HGP/Chr1

[5] Detection of Single Nucleotide Polymorphisms in *CYP2B6* Gene

By ULRICH M. ZANGER, JOACHIM FISCHER, KATHRIN KLEIN, and THOMAS LANG

Introduction

CYP2B6 is the human ortholog to the phenobarbital-inducible rat CYP2B1/2 and mouse CYP2B9/10.[1] In contrast to these well-investigated murine forms, the human isozyme has been less well studied due to its initially underestimated expression but also because of the lack of suitable probes.[2] Studies with new antibodies of higher sensitivity and specificity have shown that CYP2B6 is expressed in the liver of most or all individuals, but at extremely different levels, ranging from nearly undetectable up to about 80 pmol/mg of microsomal protein, which corresponds to about one-fifth of the total P450 content of the liver.[3–9] Xenobiotic substrates of CYP2B6 are chemically diverse and include procarcinogens such as aflatoxin B_1[10] and dibenzo [*a,h*]anthracene,[11] drugs of widespread abuse such as nicotine[12] and the methylenedioxyalkylamphetamines MDMA ("ecstasy") and MDE ("Eve"),[13] as well as several clinically used drugs. The most important of these are the prodrugs cyclophosphamide and ifosfamide, which depend on C-4 hydroxylation for further chemical decomposition to the alkylating phosphoramide

[1] S. Yamano, P. T. Nhamburo, T. Aoyama, U. A. Meyer, T. Inaba, W. Kalow, H. V. Gelboin, O. W. McBride, and F. J. Gonzalez, *Biochemistry* **28**, 7340 (1989).

[2] S. Ekins and S. A. Wrington, *Drug Metab. Rev.* **31**, 719 (1999).

[3] E. L. Code, C. L. Crespi, B. W. Penman, F. J. Gonzalez, T. K. Chang, and D. J. Waxman, *Drug Metab. Dispos.* **25**, 985 (1997).

[4] T. J. Yang, K. W. Krausz, M. Shou, S. K. Yang, J. T. Buters, F. J. Gonzalez, and H. V. Gelboin, *Biochem. Pharmacol.* **55**, 1633 (1998).

[5] S. Ekins, M. Vandenbranden, B. J. Ring, J. S. Gillespie, T. J. Yang, H. V. Gelboin, and S. A. Wrighton, *J. Pharmacol. Exp. Ther.* **286**, 1253 (1998).

[6] L. Gervot, B. Rochat, J. C. Gautier, F. Bohnenstengel, H. Kroemer, V. de Berardinis, H. Martin, P. Beaune, and I. de Waziers, *Pharmacogenetics* **3**, 295 (1999).

[7] D. M. Stresser and D. Kupfer, *Drug Metab. Dispos.* **27**, 517 (1999).

[8] K. Venkatakrishnan, L. L. von Moltke, M. H. Court, J. S. Harmatz, C. L. Crespi, and D. J. Greenblatt, *Drug Metab. Dispos.* **28**, 1493 (2000).

[9] T. Lang, K. Klein, J. Fischer, A. K. Nüssler, P. Neuhaus, U. Hofmann, M. Eichelbaum, M. Schwab, and U. M. Zanger, *Pharmacogenetics* **11**, 399 (2001).

[10] T. Aoyama, S. Yamano, P. S. Guzelian, H. V. Gelboin, and F. J. Gonzalez, *Proc. Natl. Acad. U.S.A.* **87**, 4790 (1990).

[11] M. Shou, K. R. Korzekwa, K. W. Krausz, J. T. Buters, J. Grogan, I. Goldfarb, J. Hardwick, F. J. Gonzalez, and H. V. Gelboin, *Mol. Carcinog.* **4**, 241 (1996).

[12] H. Yamazaki, K. Inoue, M. Hashimoto, and T. Shimada, *Arch. Toxicol.* **73**, 65 (1999).

[13] K. Kreth, K. A. Kovar, M. Schwab, and U. M. Zanger, *Biochem. Pharmacol.* **59**, 1563 (2000).

mustards[14]; the antidepressant bupropion, which is also in use as an oral anti-smoking agent[15]; the platelet aggregation inhibitor clopidogrel[16]; the potent cyclic nucleotide phosphodiesterase type IV inhibitor RP 73401[17]; and the commonly used anesthetic propofol,[18,19] as well as various benzodiazepines.[20] Because S-mephenytoin N-demethylation[21] and bupropion hydroxylation[22] have been proposed as marker activities for CYP2B6, it should be easier to define additional metabolic roles for this isozyme in the future.

Like animal CYP2B isozymes, the human form was also found to be inducible by barbiturates and some other agents, including the CYP3A4 inducers rifampicin and dexamethasone.[2] Investigation of phenobarbital response elements of animal and human genes revealed that they are composed of binding sites for the (constitutively activated receptor (CAR) and other nuclear receptors.[23] CYP2B6 is also inducible by cyclophosphamide via an unknown mechanism.[6] Extrahepatic expression was shown to occur at lower levels in kidney, intestine, lung, uterine endometrium, bronchoalveolar macrophages, and peripheral blood lymphocytes.[2,6]

Genetic Polymorphism of CYP2B6 Gene

The CYP2B6 gene spans a region of approximately 28 kb on the long arm of chromosome 19 within a 350-kb gene cluster, together with a number of other genes and pseudogenes of the CYP2A, 2B, and 2F subfamilies, and codes for nine exons and a protein with 491 amino acids.[24] The human CYP2B subfamily consists of the functional CYP2B6 gene, the nonfunctional CYP2B7 gene, and a CYP2B7-like pseudogene. A systematic mutational analysis in the CYP2B6 gene revealed five single nucleotide polymorphisms (SNP) that result in amino acid substitutions and that occur with frequencies between 0.5 and over 30% in the Caucasian population.[9] Several additional silent and intronic mutations were also identified.

[14] T. K. Chang, G. F. Weber, C. L. Crespi, and D. J. Waxman, Cancer Res. 23, 5629 (1993).
[15] L. M. Hesse, K. Venkatakrishnan, M. H. Court, L. L. von Moltke, S. X. Duan, R. I. Shader, and D. J. Greenblatt, Drug Metab. Dispos. 28, 1176 (2000).
[16] A. J. Coukell and A. Markham, Drugs 54, 745 (1997).
[17] J. C. Stevens, R. B. White, S. H. Hsu, and M. Martinet, J. Pharmacol. Exp. Ther. 3, 1389 (1997).
[18] Y. Oda, N. Hamaoka, T. Hiroi, S. Imaoka, I. Hase, K. Tanaka, Y. Funae, T. Ishizaki, and A. Asada, Br. J. Clin. Pharmacol. 51, 281 (2001).
[19] M. H. Court, S. X. Duan, L. M. Hesse, K. Venkatakrishnan, and D. J. Greenblatt, Anaesthesiology 94, 110 (2001).
[20] S. Ono, T. Hatanaka, S. Miyazawa, M. Tsutsui, T. Aoyama, F. J. Gonzalez, and T. Satoh, Xenobiotica 26, 1155 (1996).
[21] H. Heyn, R. B. White, and J. C. Stevens, Drug Metab. Dispos. 24, 948 (1996).
[22] S. R. Faucette, R. L. Hawke, E. L. Lecluyse, S. S. Shord, B. Yan, R. M. Laethem, and C. M. Lindley, Drug Metab. Dispos. 28, 1222 (2000).
[23] T. Sueyoshi and M. Negishi, Annu. Rev. Pharmacol. Toxicol. 41, 123 (2001).
[24] S. M. Hoffman, P. Fernandez-Salguero, F. J. Gonzalez, and H. W. Mohrenweiser, J. Mol. Evol. 6, 894 (1995).

Haplotype analysis allowed to establish at least six mutant alleles coding for protein products with one, two, or three amino acid changes compared to the wild-type protein (CYPallele Nomenclature home page: www.imm.ki.se/CYPalleles). The potential functional relevance of these polymorphisms is still largely unknown, but investigations in human liver samples demonstrated that they are associated with significant differences in expression and enzymatic function. One variant detected in the Japanese population was shown to be functionally different from the wild type by recombinant expression analysis.[25]

Detection of *CYP2B6* Single Nucleotide Polymorphisms by PCR-RFLP

For each of the five nonsynonymous mutations, a specific diagnostic test was developed (Fig. 1). Each test consists of three steps: (1) specific polymerase chain reaction (PCR) amplification of the respective exon using a *CYP2B6*-specific pair of primers located in exon-flanking sequences; (2) digestion with an appropriate restriction enzyme to yield a fragment pattern diagnostic for the presence or absence of each mutation; and (3) agarose gel electrophoresis and interpretation of the result.

Like for other *CYP* loci, the existence of closely related *CYP2B* pseudogenes represents a serious danger because co-amplification may occur. The amplification primer sequences presented in this article have been optimized carefully in order to yield amplicons derived specifically from *CYP2B6*. However, when setting up these assays, conditions may not be absolutely comparable from one to another laboratory or from one to another PCR cycler. It may therefore be advisable to control the origin of the amplified product by sequencing.[9]

Materials

General Reagents and Equipment

High-quality leukocyte DNA isolated by standard methods from blood samples is used to establish all assays. *Taq* DNA polymerase is from Promega (Madison, WI). dNTPs are from Eppendorf (Hamburg, Germany). All amplifications are performed on a PTC200 thermal cycler (MJ Research Inc., Watertown, MA). For purification of PCR products, the Qiaquick PCR purification kit is used (Qiagen, Hilden, Germany). Restriction enzymes are from Roche (Basel, Switzerland) (*Hae*II, *Bgl*II, and *Sty*I) and from New England Biolabs (Beverly, MA) (*Bsr*I) and are used with the buffers supplied by the manufacturer. Agarose gel electrophoresis is performed according to standard procedures using ultrapure agarose

[25] N. Ariyoshi, M. Miyazaki, K. Toide, Y. Sawamura, and T. Kamataki, *Biochem. Biophys. Res. Commun.* **281,** 1256 (2001).

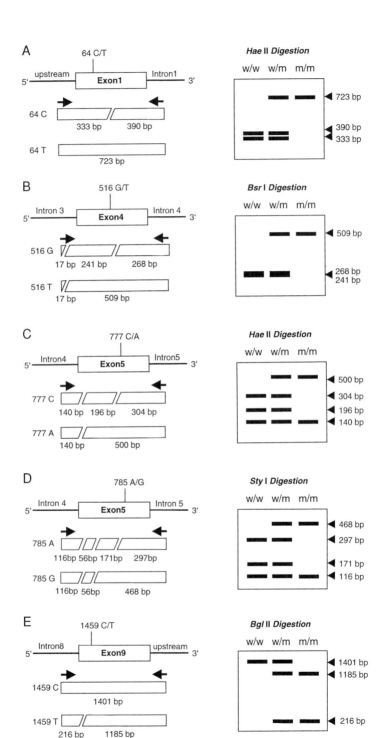

and a 1-kb DNA ladder (GIBCO-BRL, Eggenstein, Germany). DNA in agarose gels is visualized by staining with ethidium bromide and UV illumination.

Genomic Sequences and Oligonucleotide Primers

Genomic and cDNA sequences of the *CYP2B6* and *CYP2B7* genes are available from public databases. GenBank accession numbers are X06399, X06400, X13494, and J02864 for mRNA-derived sequences and AC023172 for the complete human *CYP2B6* sequence determined in the course of the Human Genome Project. The nearly complete human *CYP2B7* sequence (lacking exon 9) can be found with accession number AC008537. The reported CYP2B7-like pseudogene has not yet been determined. Oligonucleotide primer sequences are derived from these genomic sequences and are obtained from MWG-Biotech (Ebersberg, Germany). Base numbering for oligonucleotides is derived from the complete genomic sequence (GenBank AC023172) with the A of the initiation codon being base number +1. Single nucleotide polymorphism (SNP) positions are based on the cDNA sequence as described previously.[9]

Procedures

General Assay Procedure

All five PCR amplifications can be performed under similar conditions: approximately 100 ng of genomic DNA is added to a master mix consisting of 20 pmol of each primer in a final volume of 50 μl of 50 mM KCl, 10 mM Tris–HCl, pH 8.3, 1.5 mM MgCl$_2$, 0.5% Tween 20, 0.01% bovine serum albumin, and 200 μM dNTPs. After the addition of 0.5 U of *Taq* polymerase, PCR cycling is performed with 5 min of denaturation at 95°, 30 cycles of denaturation at 95° for 30 sec, annealing at 50 to 60° for 30 sec, and extension at 72° for 30 sec to 2 min followed by a final step of 10 min at 72°. Aliquots of each PCR product are then subjected to agarose gel electrophoresis to control for proper amplification. PCR products are then purified, and an aliquot of each sample is digested with the appropriate restriction enzyme and analyzed by agarose gel electrophoresis. Deviations from these standard conditions are indicated in the following individual protocols.

*1. Hae*II *Polymorphism in Exon 1 (64C → T: 22R → C)*

Forward primer (CYP2B6-1F, position −272 to −254 bp)
5′-ACATTCACTTGCTCACCT-3′

FIG. 1. Strategies used to detect nonsynonymous mutations at the human *CYP2B6* gene locus. Predicted sizes of restriction enzyme digestion products of PCR fragments are shown for samples derived from homozygous wild types (w/w), heterozygotes (w/m), and homozygous mutants (m/m).

Reverse primer (CYP2B6-1R, position 451 to 434 bp)
5′-GTAAATACCACTTGACCA-3′
Annealing temperature: 50°
Extension time: 1 min
Amplification product: 723 bp

Restriction digestion with *Hae*II results in two fragments of 333 and 390 bp from wild-type DNA and in an uncleaved 723-bp fragment from mutant DNA (Fig. 1A).

2. *Bsr*I Polymorphism in Exon 4 (516G → T: 172Q → H)

Forward primer (CYP2B6-4F, position 15,370 to 15,388 bp)
5′-GGTCTGCCCATCTATAAAC-3′
Reverse primer (CYP2B6-4R, position 15,894 to 15,874 bp)
5′-CTGATTCTTCACATGTCTGCG-3′
Annealing temperature: 56°
Extension time: 40 sec
Amplification product: 526 bp

Restriction digestion with *Bsr*I for 1 hr at 60° results in three fragments of 241, 268, and 17 bp from wild-type DNA and in two fragments of 509 and 17 bp from mutant DNA (Fig. 1B).

3. *Hae*II Polymorphism in Exon 5 (777C → A: 259S → R)

Forward primer (CYP2B6-5F, position 17,707 to 17,727 bp)
5′-GACAGAAGGATGAGGGAGGAA-3′
Reverse primer (CYP2B6-5R, position 18,346 to 18,324 bp)
5′-CTCCCTCTGTCTTTCATTCTGT-3′
Annealing temperature: 60°
Extension time: 1 min
Amplification product: 640 bp

Restriction digestion of the wild-type amplicon with *Hae*II results in three fragments of 140, 196, and 304 bp, whereas a 500- and a 140-bp fragment are observed in the case of mutant DNA (Fig. 1C).

4. *Sty*I Polymorphism in Exon 5 (785A → G: 262K → R). The PCR fragment generated for assay 3 is also used in this assay. Digestion of wild-type DNA with *Sty*I results in fragments of 56, 116, 171, and 297 bp. Presence of the 785A → G mutation leads to only three fragments of 56, 116, and 468 bp (Fig. 1D). Analysis should be performed using a 2.5% MetaPhor agarose gel (FMC, Rockland, ME) to resolve fragments 171 and 116 bp.

5. *Bg*lII *Polymorphism in Exon 9 (1459C → T: 487R → C)*

Forward primer (CYP2B6-9F, position 24,314 to 24,336 bp)
5′-TGAGAATCAGTGGAAGCCATAGA-3′
Reverse primer (CYP2B6-9R, position 25,714 to 25,690 bp)
5′-TAATTTTCGATAATCTCACTCCTGC-3′
Annealing temperature: 60°
Extension time: 1 min   30 sec
Amplification product: 1401 bp

Restriction digestion of the wild-type amplicon with *Bgl*II results in the un-cleaved product of 1401 bp, whereas mutant DNA is cleaved into fragments of 216 and 1185 bp (Fig. 1E).

Detection of 1459C → T (487R → C) by Denaturing HPLC

Denaturing HPLC (DHPLC) is a rather new technology for mutation detection that is based on the differential elution of homo- and heteroduplex DNA by reversed-phase chromatography under partially denaturing conditions.[26] The presence of SNPs or other small changes in DNA sequence (deletions or insertions of a few bases) can be detected with high sensitivity in amplified DNA fragments of up to 1 kb length in a highly automated fashion. Although the main application of DHPLC is screening for the presence of unknown mutations, which have then to be confirmed by sequencing, the high specificity and reproducibility of the method also permit reliable detection of known mutations.[27]

DHPLC Reagents and Equipment

We use the WAVE DNA fragment analysis system (Transgenomic, Inc., San Jose, CA) equipped with a DNA Sep column as the stationary phase and a gradient mixed of 0.1 *M* triethylammonium acetate (TEAA), pH 7.0, with 0.025% acetoni-trile (buffer A) and with 25% (v/v) acetonitrile (buffer B) as the mobile phase. WAVEMAKER software is an integral part of the WAVE system and is used to determine melting temperatures and gradient profiles.

Method

Forward primer (DHPLC-ex9-F: position 25,348 to 25,369 bp)
5′-GATTTGTCTTGGTGAAGGCATC-3′
Reverse primer (DHPLC-ex9-R: position 25,601 to 25,584 bp)
5′-GGGGAGTCAGAGCCATTG-3′

---

[26] W. Xiao and P. J. Oefner, *Hum. Mutat.* **17,** 439 (2001).
[27] E. Schaeffeler, T. Lang, U. M. Zanger, M. Eichelbaum, and M. Schwab, *Clin. Chem.* **47,** 548 (2001).

PCR results in amplification of a 254-bp fragment surrounding position 1459 in exon 9 of the *CYP2B6* gene. Reactions are carried out in 96-well microtiter plates (Peqlab Biotechnology, Erlangen, Germany) in a total volume of 50 μl containing ~50–100 ng of genomic DNA, 200 μM of each dNTP, 20 μM of each primer, and 1 U of *Taq* polymerase in 1× *Taq* polymerase reaction buffer (Qiagen). Following 5 min of denaturation at 95°, 30 cycles of 30 sec at 92°, 30 sec at 60°, and 30 sec at 72° and a final step of 7 min at 72° are performed in a PTC 200 thermal cycler (MJ Research Inc.).

DHPLC Analysis. Five microliters of unpurified PCR product is injected into the column preheated at 63° and eluted at a flow rate of 0.9 ml per minute. The optimized elution profile starts with an equilibration step of 48% of buffer B for 0.5 min followed by a linear gradient from 53 to 62% buffer B within 5 min. A washout step of 100% buffer B for 0.5 min removes remaining DNA from the column. The cycle ends with 0.2 min at 48% buffer B. To detect homozygotes, a second DHPLC run is performed with 20 μl of a mixture containing 10 μl of each PCR product and 10 μl of PCR product from any known wild-type sample. The mixture is denatured for 5 min at 95° and is cooled down to 65° over 30 min in a PCR machine to allow formation of heteroduplexes. These samples are analyzed by DHPLC as described earlier.

Interpretation of Results

A typical column temperature optimization is shown in Fig. 2A, and elution profiles of a homozygous 1459 CC wild-type sample and a 1459 CT heterozygous sample are shown in Fig. 2B. The homoduplex elutes as a single, almost symmetric peak from the column, whereas the heteroduplex is eluted faster with a more complex profile. Heteroduplex profiles are usually highly reproducible and characteristic for a particular mutation.[27] Amplicons that elute with a different

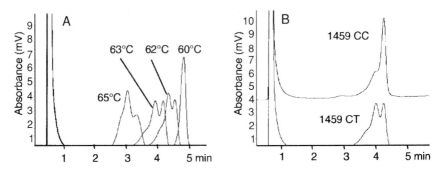

FIG. 2. DHPLC analysis of the 1459 C/T polymorphism in exon 9 of *CYP2B6*. (A) Optimization of column temperature for heteroduplex analysis using a heterozygous 254-bp amplicon and (B) analysis of amplicons obtained from a wild-type (1459 CC) and a heterozygous (1459 CT) DNA sample.

profile from the column most likely contain either a different or an additional
mutation and should be sequenced for confirmation.

Acknowledgments

The authors acknowledge the excellent technical assistance of Andrea Zwicker and Igor Lieber-
mann. This work was supported by Grant 01 GG 9846 from the German Federal Ministry of Education
and Science and by the Robert Bosch Foundation, Stuttgart, Germany.

[6] Genotyping Human Cytochrome: P450 1B1 Variants

By Carrie Hayes Sutter, Zheng Qian, Yeon-Pyo Hong,
Jenna S. Mammen, Paul T. Strickland, and Thomas R. Sutter

Introduction

Human cytochrome P450 1B1 (CYP1B1) was first isolated by differential
hybridization as a 2,3,7,8-tetrachlorodibenzo-*p*-dioxin (TCDD)-responsive cDNA
clone from a human keratinocyte cell line treated with TCDD.[1] Analysis of the
complete cDNA sequence of this mRNA identified a new gene subfamily of cytochrome P450, CYP1B1, based on 40% sequence homology to other polycyclic
aromatic hydrocarbon (PAH)-inducible isoforms, CYP1A1 and CYP1A2.[2]
  Initial characterization of the human CYP1B1 gene[3] described the DNA sequence of a 12-kb genomic clone corresponding to the entire 5.1-kb CYP1B1
cDNA and containing 3.0 kb of upstream DNA. Comparison of these sequences
revealed the location of three exons (371, 1044, and 3707 bp) and two introns
(390 and 3032 bp), with the CYP1B1 open reading frame spanning exons 2 and 3.
High-resolution chromosome mapping confirmed the previous somatic cell hybrid
analysis[2] and placed the CYP1B1 gene at 2p21–22 of human chromosome 2.[3]
Comparison of the human CYP1B1 genomic and cDNA sequences, obtained independently from two cell lines derived from different individuals, revealed three
sequence differences, including an amino acid change at valine-432. Concurrent
human genetic studies to identify one of two loci for primary congenital glaucoma

[1] T. R. Sutter, K. Guzman, K. M. Dold, and W. F. Greenlee, *Science* **254**, 415 (1991).
[2] T. R. Sutter, Y. M. Tang, C. L. Hayes, Y.-Y. P. Wo, E. W. Jabs, X. Li, H. Yin, C. W. Cody, and W. F.
  Greenlee, *J. Biol. Chem.* **269**, 13092 (1994).
[3] Y. M. Tang, Y.-Y. P. Wo, J. Stewart, A. L. Hawkins, C. A. Griffin, T. R. Sutter, and W. F. Greenlee,
  J. Biol. Chem. **271**, 28324 (1996).

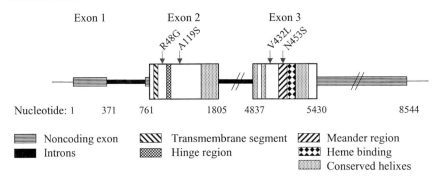

FIG. 1. Schematic of the human CYP1B1 gene. The four coding region single nucleotide polymorphisms are identified by arrows and indicate the corresponding change in the amino acid sequence. The nucleotide numbered 1 is the reported transcription start site[3] and corresponds to nucleotide 3045 of the GenBank/EBI Data Bank sequence with the accession number U56438. Other characteristics, including five conserved protein regions, are indicated.

(PCG) led to the mapping of the PCG locus GCL3A to human chromosome 2 at the 2p21 region.[4] Subsequent investigations have led to independent reports that identify distinct CYP1B1 gene mutations that segregate with the GCL3A phenotype in PCG families.[4-7]

Based on extensive DNA sequence analysis of the translated regions of the CYP1B1 gene in 22 PCG families and 100 randomly selected normal individuals,[6] the Val432-Leu CYP1B1 polymorphism was confirmed. Three additional polymorphisms predicting variant amino acid sequences were identified: Arg48-Gly, Ala119-Ser, and Asn453-Ser. The frequency of each wild-type allele, calculated for the reported sample of 100 normal individuals,[6] is as follows: Arg-48, 0.71; Ala-119, 0.71; Val-432, 0.28; and Asn-453, 0.76. The identification of several frequently occurring polymorphisms of the human CYP1B1 gene demonstrates the strength of direct genomic sequence analysis as a method to identify single nucleotide polymorphisms.

This article describes the methods used in our laboratory to genotype the four common single nucleotide polymorphisms found in the coding region of CYP1B1 (Fig. 1).

[4] I. Stoilov, A. N. Akarsu, and M. Sarfarazi, *Hum. Mol. Genet.* **6**, 641 (1997).

[5] M. Sarfarazi, *Hum. Mol. Genet.* **6**, 1667 (1997).

[6] I. Stoilov, A. N. Akarsu, I. Alozie, A. Child, M. Barsoum-Homsy, M. E. Turacli, M. Or, R. A. Lewis, N. Ozdemir, S. Brice, S. G. Aktan, L. Chevrette, M. Coca-Prados, and M. Sarfarazi, *Am. J. Hum. Genet.* **62**, 573 (1998).

[7] B. A. Bejjani, R. A. Lewis, K. F. Tomey, K. L. Anderson, D. K. Duekler, M. Jabak, W. F. Astle, B. Otterud, M. Leppert, and J. R. Lupski, *Am. J. Hum. Genet.* **62**, 325 (1998).

Genotyping CYP1B1

Isolation of Genomic DNA

Human genomic DNA can be isolated from blood using the Qiagen Genomic-tip System (Qiagen, Valencia, CA). This purification system uses an anion-exchange resin, which isolates high molecular weight DNA conveniently and quickly. Leukocytes are the only cells in the blood that contain nuclei and can be separated from other components of blood by centrifugation.

Collect at least 1 ml of blood in a heparinized Vacutainer tube (Becton Dickinson, Franklin Lakes, NJ). Keep samples at 4° and fractionate by centrifugation within 6 hr of collection. Fractionate whole blood by centrifugation for 10 min at 3300g at room temperature using a swinging bucket rotor. The blood is separated into three layers. The middle layer, known as the buffy coat, contains leukocytes; the upper layer is plasma; and the lower level contains erythrocytes. Isolate the buffy coat and keep frozen until DNA is purified. At time of purification, adjust the leukocyte concentration to the recommended concentration for the appropriate Qiagen Genomic tip. Following the protocol of the Qiagen Genomic-tip system, lyse the cells, bind DNA to the resin, wash the resin, and elute and precipitate the DNA. Finally, resuspend the DNA in TE (10 mM Tris, 1 mM EDTA, pH 8) and store at 4° until genotype analysis. Yields from this procedure should be approximately 15 μg DNA/ml of whole blood. Check the quality of the DNA by agarose gel electrophoresis; it should be 25–50 kb in size.

Amplification of CYP1B1 Fragments from Genomic DNA

For polymerase chain reaction (PCR)-based restriction fragment length polymorphism (RFLP) assays[8–11] of exon 2 and 3 of the CYP1B1 gene, use three sets of PCR primers (Table I). Make or purchase (Integrated DNA Technologies, Inc., Coralville, IA) the following primers.

Primer set 1: forward 5′-tctctgcacccctgagtgtc-3′, reverse complement 5′-tagtg gccggtacgttctcc-3′
Primer set 2: forward 5′-aattggatcaggtcgtgg-3′, reverse complement 5′-atttcag cttgcctcttgc-3′
Primer set 3: forward 5′-cacctctgtcttgggctacc-3′, reverse complement 5′-atttca gcttgcctcttgcttc-3′.

[8] K. B. Mullis, F. Faloona, S. J. Scharf, R. K. Saiki, G. T. Horn, and H. A. Erlich, *Cold Spring Harb. Symp. Quant. Biol.* **51,** 263 (1986).
[9] R. K. Saiki, D. H. Gelfand, S. Stoffel, S. J. Scharf, R. Higuchi, R. G. T. Horn, K. B. Mullis, and H. A. Erlich, *Science* **238,** 4839 (1988).
[10] H. Kiko, E. Niggemann, and W. Ruger, *Mol. Gen. Genet.* **173,** 303 (1979).
[11] M. H. Skolnick and R. White, *Cytogenet. Cell. Genet.* **32,** 58 (1982).

TABLE I
LOCATIONS OF KNOWN CODING POLYMORPHISMS[a] IN HUMAN CYP1B1

Location	Codon	Primer	Enzyme and recognition sequence	Allele Wild type	New	Base pairs (total, fragments) and amplified regions
Exon 2	48	F[b]: 5′-tctctgcacccctgagtgtc-3′	CspI	CGG[c]	GGG	1015, 829/186
		RC: 5′-tagtggccggtacgttctcc-3′	cg▾g(a/t)ccg			nt 718–1732
Exon 2	119	F: 5′-tctctgcacccctgagtgtc-3′	NgoMIV	GCC[c]	TCC	1015, 616/399
		RC: 5′-tagtggccggtacgttctcc-3′	g▾ccggc			nt 718–1732
Exon 3	432	F: 5′-aattggatcaggtcgtgg-3′	Eco57I	GTG	CTG[c]	579, 341/238
		RC: 5′-atttcagcttgcctcttgc-3′	ctgaag(n)$_{16}$▾			nt 4870–5448
Exon 3	453	F: 5′-cacctctgtcttgggctacc-3′	MwoI	AAC	AGC[c]	437, 297/140
		RC: 5′-atttcagcttgcctcttgcttc-3′	gcnnnnn▾nngc			nt 5012–5448

[a] There is also a silent polymorphism at codon 449 (GAT → GAC).
[b] F, Forward primer; RC, reverse complement primer.
[c] Codon present in the nucleotide sequence that is recognized by the indicated restriction endonuclease.

Use the *Taq* PCR Core Kit (Qiagen) to amplify genomic DNA. Place a 50-μl reaction containing 100 ng of genomic DNA, 200 μM of each of the four deoxynucleotide triphosphates, 1× reaction buffer with 1.5 mM MgCl$_2$, 1× Q-solution, 20 pmol of each primer, and 2.5 units *Taq* polymerase into a GeneAmp PCR System 9600 thermocycler (PerkinElmer, Foster City, CA). Alternatively, we have found that the Q solution can be replaced with 2 mM MgCl$_2$ in the reaction mixture. For genotype analysis, we recommend a PCR unit with a heated cover so that the use of an oil overlay is avoided. After an initial denaturing step for 3 min at 94°, repeat the following cycle 30 times: 94° for 1 min, 60° for 30 sec, and 72° for 1 min, followed by a final 10 min elongation cycle at 72° and storage at 4°. To check for laboratory template contamination, include a negative control reaction (without DNA) in each set of PCR reactions. Analyze 9 μl of each PCR product, including the control, by gel electrophoresis (1.5% agarose gel) in order to confirm the expected size of the product. Remove amplification primers and primer dimers from the PCR reaction using the Wizard PCR Preps DNA purification system (Promega, Madison, WI). This step allows visualization of low molecular weight fragments without the possible interference by primer dimers.

Although the conditions described earlier have repeatedly amplified the target CYP1B1 sequence in our laboratory, successful PCR conditions will vary and are dependent on the length and nucleotide composition of the template target sequence, as well as on the length and nucleotide composition of the primer sequences. Conditions can be optimized by varying the concentrations of MgCl$_2$, primers, and template, as well as by varying annealing temperatures. In addition, adding DNA denaturants such as formamide, DMSO, and Q solution (Qiagen) can improve amplification by modifying the melting temperature of DNA.[12]

[12] Alkami Biosystems, Inc., www.alkami.com (1999).

Digestion of CYP1B1 PCR Fragments with Restriction Endonucleases

Analysis of the reported CYP1B1 gene sequence revealed that each of the four polymorphisms in the coding region of CYP1B1 could be evaluated using four different restriction endonucleases: *Csp*I (Promega), *Ngo*MIV (New England Biolabs, Beverly, MA), *Eco*57I (MBI Fermentas, Hanover, MD), and *Mwo*I (New England Biolabs). To visualize DNA fragments, combine the restriction endonuclease reaction solutions with 1× loading buffer (10% Ficoll and 0.25% xylene cyanol) and separate the DNA by electrophoresis through a 2% NuSieve 3 : 1 agarose gel (FMC Bioproducts, Rockland, ME) in 1× TAE buffer (40 m*M* Tris–acetate, 1 m*M* EDTA, pH 8.5) at 5 V per cm length of the gel. Stain the gels with the fluorescent SYBR Gold nucleic acid gel stain (Molecular Probes, Eugene, OR), visualize the stained DNA by UV transillumination, and document by photography using a SYBR Green/Gold gel stain photographic filter (Molecular Probes). SYBR Gold gel stain is more sensitive than ethidium bromide staining, allowing for visualization of small DNA fragments that are often difficult to detect. Each sample is genotyped based on the relative size of the observed fragment(s), compared to the base pair standard (Fig. 2).

Digest 20 μl of each PCR product with the indicated restriction endonucleases (Table I). Digestions of samples with 5 units of *Csp*I reveal the presence of a polymorphism in exon 2 (codon 48). This restriction endonuclease site is present in a homozygous wild-type sample, and two bands corresponding approximately to 829 and 186 bp are seen on the gel (Fig. 2A, lane 1). A homozygous polymorphic sample does not have the restriction site and thus is not digested, and a

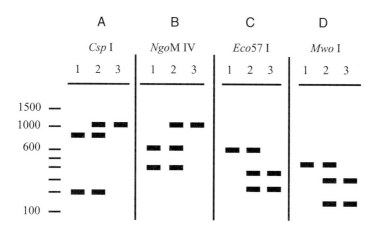

FIG. 2. Depiction of restriction fragments from PCR products in agarose gels. Fragments were generated by the digestion of PCR products with the indicated restriction endonuclease (see text for detailed explanation). Lane 1, homozygous wild-type individual; lane 2, heterozygous individual; and lane 3, homozygous polymorphic individual. DNA size standards (bp) are on the left-hand side.

single band appears at approximately 1015 kb (Fig. 2A, lane 3). A heterozygote sample, which has both a wild-type and a polymorphic allele, contains all three bands, corresponding to 1015, 829, and 186 bp (Fig. 2A, lane 2). Digestions of samples with 5 units of *Ngo*MIV reveal the presence of a polymorphism in exon 2 (codon 119). This restriction endonuclease site is present in a homozygous wild-type sample, and two bands corresponding approximately to 616 and 399 bp are seen on the gel (Fig. 2B, lane 1). A homozygous polymorphic sample does not have the restriction site and thus is not digested, and a single band appears at approximately 1015 kb (Fig. 2B, lane 3). A heterozygote sample, which has both a wild-type and a polymorphic allele, contains all three bands, corresponding to 1015, 616, and 399 bp (Fig. 2B, lane 2). Digestions of samples with 2 units of *Eco*57I reveal the presence of a polymorphism in exon 3 (codon 432). This restriction endonuclease site is absent in a homozygous wild-type sample, and a single band corresponding to 579 bp is seen on the gel (Fig. 2C, lane 1). A homozygous polymorphic sample has the restriction site, and two bands appear at approximately 341 and 238 bp (Fig. 2C, lane 3). A heterozygote sample, which has both a wild-type and a polymorphic allele, contains all three bands, corresponding to 579, 341, and 238 bp (Fig. 2C, lane 2). Digestions of samples with 5 units of *Mwo*I reveal the presence of a polymorphism in exon 3 (codon 453). This restriction endonuclease site is absent in a homozygous wild-type sample, and a single band corresponding to 437 bp is seen on the gel (Fig. 2D, lane 1). A homozygous polymorphic sample has the restriction site, and two bands appear at approximately 297 and 140 bp (Fig. 2D, lane 3). A heterozygote sample, which has both a wild-type and a polymorphic allele, contains all three bands, corresponding to 437, 297, and 140 bp (Fig. 2D, lane 2).

As a positive control for the activity of the restriction endonucleases, digest amplified genomic DNA that has previously been determined by DNA sequence analysis to be homozygous for the restriction site under investigation.

Conclusions

Genotyping by RFLP analysis will soon be replaced by more rapid, high-throughput assays based on electrochemical or fluorescent detection methods, including oligonucleotide microarrays. In the interim, the procedures described here should facilitate CYP1B1 genotyping of banked DNA samples.

Acknowledgments

This research was supported by grants from the National Institutes of Health (ES 08148, ES 06071, and ES 03819) and the W. Harry Feinstone Center for Genomic Research.

[7] Genotyping Human *CYP2A6* Variants

By SHARI D. GOODZ and RACHEL F. TYNDALE

Substrates of CYP2A6

Cytochrome P450 2A6 (CYP2A6) is the major enzyme in humans involved in the inactivation of nicotine to cotinine and the further metabolism of cotinine to 3'-hydroxycotinine.[1,2] CYP2A6 is not known to metabolize many commonly used drugs, except nicotine, but it can activate many toxicants and procarcinogens.[3] Large species differences exist in the substrate specificity of the CYP2A family, limiting the use of rodent models as predictors of human *CYP2A6*-mediated metabolism.[4]

Phenotyping of CYP2A6 Activity

Phenotyping studies suggest significant interindividual as well as interethnic variability in CYP2A6 activity.[3,5] Most studies have used the 7-hydroxylation of coumarin as the *in vivo* phenotypic measure.[6–8] Nicotine and SM-12502, a newly identified platelet-activating factor antagonist, have also been used as *in vivo* probes for assessing CYP2A6 activity.[7,9,10]

Genetic Variants of *CYP2A6*

The *CYP2A6* gene contains nine exons and is mapped to 19q12–19q13.2.[11] *CYP2A6* is located within a 350-kbp gene cluster containing the highly homologous *CYP2A7* gene, *CYP2A13* gene, and *CYP2A7* pseudogenes [including *2A7P(C)* or

[1] M. Nakajima, T. Yamamoto, K. Nunoya, T. Yokoi, K. Nagashima, K. Inoue, Y. Funae, N. Shimada, T. Kamataki, and Y. Kuroiwa, *J. Pharmacol. Exp. Ther.* **277,** 1010 (1996).

[2] M. Nakajima, T. Yamamoto, K. Nunoya, T. Yokoi, K. Nagashima, K. Inoue, Y. Funae, N. Shimada, T. Kamataki, and Y. Kuroiwa, *Drug Metab. Dispos.* **24,** 1212 (1996).

[3] O. Pelkonen, A. Rautio, H. Raunio, and M. Pasanen, *Toxicology* **144,** 139 (2000).

[4] P. Fernandez-Salguero and F. J. Gonzalez, *Pharmacogenetics* **5,** S123 (1995).

[5] M. Oscarson, *Drug Metab. Dispos.* **29,** 91 (2001).

[6] K. Kitagawa, N. Kunugita, M. Kitagawa, and T. Kawamoto, *J. Biol. Chem.* **276,** 17830 (2001).

[7] N. Ariyoshi, Y. Sawamura, and T. Kamataki, *Biochem. Biophys. Res. Commun.* **281,** 810 (2001).

[8] A. Rautio, H. Kraul, A. Kojo, E. Salmela, and O. Pelkonen, *Pharmacogenetics* **2,** 227 (1992).

[9] Y. Rao, E. Hoffmann, M. Zia, L. Bodin, M. Zeman, E. M. Sellers, and R. F. Tyndale, *Mol. Pharmacol.* **58,** 747 (2000).

[10] K. Nunoya, Y. Yokoi, K. Kimura, T. Kodama, M. Funayama, K. Inoue, K. Nagashima, Y. Funae, N. Shimada, C. Green, and T. Kamataki, *J. Pharmacol. Exp. Ther.* **277,** 768 (1996).

[11] J. S. Miles, W. Bickmore, J. D. Brook, A. W. McLaren, R. Meehan, and C. R. Wolf, *Nucleic Acids Res.* **17,** 2907 (1989).

TABLE I
ESTABLISHED *CYP2A6* VARIANTS[a]

Allele	Type of mutation	Enzyme activity	Refs.
CYP2A6*1A (wt, wild type)	None	Normal	b
CYP2A6*1B	Gene conversion containing 58 bp of *CYP2A7* in the 3′ flanking region; likely the allele named the conversion type in some papers	Normal	c–e
CYP2A6*1X2 (duplication, Y)	Gene duplication believed to be due to an unequal crossover event with *CYP2A7* in a 268-bp region that extends between intron 8 and the 3′ flanking region complementary to the reverse crossover found in the *CYP2A6*4D* deletion. Duplication variant exists in addition to wild-type *CYP2A6* and *CYP2A7* genes	Increased	f
CYP2A6*2 (v1)	L160H amino acid substitution due to a single base change (T479A) in exon 3.	Absent	b, g
CYP2A6*3 (v2)	Conversions between *CYP2A6* and *CYP2A7* in exons 3, 6, and 8. This variant has been detected in genomic clones; however, it is found in extremely low frequencies in human genotyping studies	Absent	g, h
CYP2A6*4A (CYP2A6del, E variant)	*CYP2A6* gene deletion identified by RFLP as a *Sac*I (E-type) of *Sph*I (C′-type) variant. It is thought to result from an unequal crossover event in the "deletion junction region," an 82-bp region of identical sequence in the 3′ flanking regions or *CYP2A6* and *CYP2A7,* resulting in a *CYP2A7/CYP2A6* hybrid	Absent	i–k
CYP2A6*4B (D variant)	*CYP2A6* gene deletion first detected as a *Sac*I (D-type) or *Sph*I (A′-type) RFLP variant; may be identical to the *CYP2A6*4D* variant	Absent	l
CYP2A6*4C	*CYP2A6* gene deletion reported to be identical to the *CYP2A6*4A* variant	Absent	e, k
CYP2A6*4D	*CYP2A6* gene deletion due to an unequal crossover event in intron 8 or exon 9 resulting in a *CYP2A7/CYP2A6* hybrid	Absent	d
CYP2A6*5	G479V amino acid substitution due to single base change (G1436T) in exon 9	Absent	d
CYP2A6*6	R128Q amino acid substitution due to a single base change (G383A) in exon 3. Activity tested *in vitro*	Decreased	m
CYP2A6*7	I471T amino acid substitution due to a single base change (T1412C) in exon 9. Activity tested *in vitro*	Decreased for coumarin, absent for nicotine	n

TABLE I (*continued*)

Allele	Type of mutation	Enzyme activity	Refs.
*CYP2A6*8*	R485L amino acid substitution due to a single base change (G1454T) in exon 9	Unknown	*n*
*CYP2A6*9*	Single nucleotide substitution (T-48G) in the TATA box of the *CYP2A6* promoter. Activity tested *in vitro*.	50% decreased	*o*

[a] Modified version of *CYP2A6* allele nomenclature @www.imm.ki.se/CYPalleles/cyp2a6.htm by Dr. M. Oscarson.

[b] S. Yamano, J. Tatsuno, and F. J. Gonzalez, *Biochemistry* **29**, 1322 (1990).

[c] S. M. Miyamoto, Y. Umetsu, H. Dosaka-Akita, Y. Sawamura, Y. Yokota, H. Kunitoh, N. Nemoto, K. Sato, N. Ariyoshi, and T. Kamataki, *Biochem. Biophys. Res. Commun.* **261**, 658 (1999).

[d] M. Oscarson, R. A. McLellan, H. Gullsten, J. A. Agundez, J. Benitez, A. Rautio, H. Raunio, O. Pelkonen, and M. Ingelman-Sundberg, *FEBS Lett.* **460**, 321 (1999).

[e] N. Ariyoshi, Y. Takahashi, M. Miyamoto, Y. Umetsu, S. Daigo, T. Tateishi, S. Kobayashi, Y. Mizorogi, M. A. Loriot, I. Stucker, P. Beaune, M. Kinoshita, and T. Kamataki, *Pharmacogenetics* **10**, 687 (2000).

[f] Y. Rao, E. Hoffmann, M. Zia, L. Bodin, M. Zeman, E. M. Sellers, and R. F. Tyndale, *Mol. Pharmacol.* **58**, 747 (2000).

[g] P. Fernandez-Salguero, S. M. Hoffman, S. Cholerton, H. Mohrenweiser, H. Raunio, A. Rautio, O. Pelkonen, J. D. Huang, W. E. Evans, J. R. Idle, and F. J. Gonzalez, *Am. J. Hum. Genet.* **57**, 651 (1995).

[h] G. F. Chen, Y. M. Tang, B. Green, D. X. Lin, F. P. Guengerich, A. K. Daly, N. E. Caporaso, and F. F. Kadlubar, *Pharmacogenetics* **9**, 327 (1999).

[i] M. Oscarson, R. A. McLellan, H. Gullsten, Q. Y. Yue, M. A. Lang, M. L. Bernal, B. Sinues, A. Hirvonen, H. Raunio, O. Pelkonen, and M. Ingelman-Sundberg, *FEBS Lett.* **448**, 105 (1999).

[j] K. I. Nunoya, T. Yokoi, K. Kimura, T. Kainuma, K. Satoh, M. Kinoshita, and T. Kamataki, *J. Pharmacol. Exp. Ther.* 289, 437 (1999).

[k] K. Nunoya, T. Yokoi, Y. Takahashi, K. Kimura, M. Kinoshita, and T. Kamataki, *J. Biochem. (Tokyo)* **126**, 402 (1999).

[l] K. Nunoya, T. Yokoi, K. Kimura, K. Inoue, T. Kodama, M. Funayama, K. Nagashima, Y. Funae, C. Green, M. Kinoshita, and T. Kamataki, *Pharmacogenetics* **8**, 239 (1998).

[m] K. Kitagawa, N. Kunugita, M. Kitagawa, and T. Kawamoto, *J. Biol. Chem.* **276**, 17830 (2001).

[n] N. Ariyoshi, Y. Sawamura, and T. Kamataki, *Biochem. Biophys. Res. Commun.* **281**, 810 (2001).

[o] M. Pitarque, O. von Richter, B. Oke, H. Berkkan, M. Oscarson, and M. Ingelman-Sundberg, *Biochemistry* **284**, 455 (2001).

CYP2A7P1, 2A7P(T) or *CYP2A7P2*] as well as members of the *CYP2B, CYP2F,* and *CYP2S* gene subfamilies.[12–14]

Several genetic variants of *CYP2A6* have been identified (see Table I and Fig. 1), which are associated with altered enzymatic activity and have different frequencies

[12] J. Sheng, Z. Hua, J. Guo, M. Caggana, and X. Ding, *Drug* **29**, 4 (2001).

[13] T. Rylander, E. P. Neve, M. Ingelman-Sundberg, and M. Oscarson, *Biochem. Biophys. Res. Commun.* **281**, 529 (2001).

[14] S. M. Hoffman, P. Fernandez-Salguero, F. J. Gonzalez, and H. W. Mohrenweiser, *J. Mol. Evol.* **41**, 894 (1995).

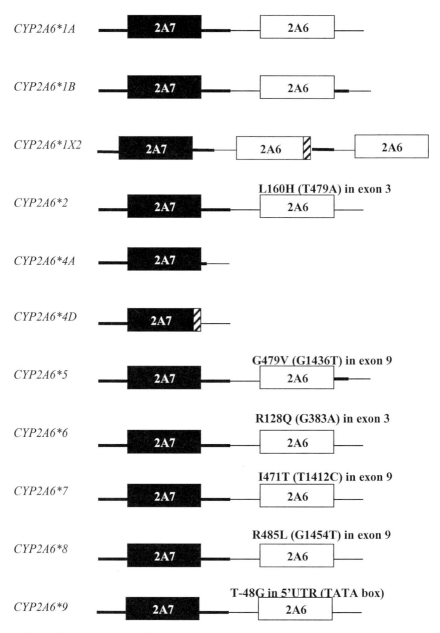

FIG. 1. Schematic representation of *CYP2A6* genetic variants. Boxes represent coding regions including introns (white, *CYP2A6*; black, *CYP2A7*) and lines represent 5′ and 3′ flanking regions (thin, *CYP2A6*; thick, *CYP2A7*). Hatched boxes represent crossover junctions between *CYP2A6* and *CYP2A7* in which the exact crossover position cannot be determined due to identical or similar sequences between the two genes.

among ethnic groups. Studies correlating genotype with phenotype have shown that the occurrence of *CYP2A6*1X2, *2, *3, *4, *5*, and *7* alleles provides some explanation for the observed interindividual and interethnic variability found in CYP2A6 activity, but also indicate that additional alleles remain unidentified. In addition, other factors, including environmental inducers and/or inhibitors, may also contribute to the observed variability in CYP2A6 activity.[3]

CYP2A6, Smoking, and Lung Cancer

Because of the role of CYP2A6 in the metabolism of nicotine, the primary addictive compound in tobacco,[15] and its ability to activate several tobacco smoke procarcinogens,[3] the *CYP2A6* genetic variants may influence smoking behavior and susceptibility to tobacco-related cancers.[16,17] Previous studies have presented conflicting results on the roles of CYP2A6, which may be due to several factors, including the genotyping assay used, the presence of unknown allelic variants, and the use of mixed ethnic groups.[5,17]

Conclusion

Future studies using improved genotyping techniques can provide insight on the roles of CYP2A6 and into the reasons for interindividual as well as interethnic differences in rates of substrate metabolism, smoking behavior, and incidence of tobacco-related cancers.

Methods

PCR Amplification

General Genotyping Assay

Genomic DNA is extracted from whole blood according to the manufacturer's instructions (GenElute Mammalian Genomic DNA Kit, Sigma-Aldrich Co., Canada Ltd., Oakville, Ontario) and stored at $-20°$. The two-step polymerase chain reaction (PCR) assays for *CYP2A6*1B, *1X2, *2, *4*, and *5* variants, as well as the PCR restriction fragment length polymorphism (RFLP) assay for the *CYP2A6*3* allele, are outlined in the following section. These assays are reproducible within our own and other laboratories. The two-step assay for the *CYP2A6*4* allele will detect all currently identified *CYP2A6*4* variants listed in Table I. Assays for the recently discovered *CYP2A6*6, *7, *8*, and *9* alleles are currently being modified in our laboratory; however, the original assays have been published.[6,7,18]

[15] J. E. Henningfield, K. Miyasato, and D. R. Jasinski, *J. Pharmacol. Exp. Ther.* **234,** 1 (1985).

[16] T. Kamataki, K. Nunoya, Y. Sakai, H. Kushida, and K. Fujita, *Mutat. Res.* **428,** 125 (1999).

[17] R. F. Tyndale and E. M. Sellers, *Drug Metab. Dispos.* **29,** 548 (2001).

[18] M. Pitarque, O. von Richter, B. Oke, H. Berkkan, M. Oscarson, and M. Ingelman-Sundberg, *Biochemistry* **284,** 455 (2001).

Two-Step PCR Method
STEP I

i. For *CYP2A6*1B, *2, *3*, and **5* assays, the primers are gene specific and are used to amplify specific regions of the *CYP2A6* gene (see Tables II and III).

ii. For the *CYP2A6*1X2* assay, the forward primer will amplify both *CYP2A6* and *CYP2A7* and the reverse primer is specific for *CYP2A7*. This allows a region of the *CYP2A7* gene, as well as the *CYP2A6/CYP2A7* duplication variant (if present), to be amplified. For the *CYP2A6*4* assay, the forward primer will amplify both *CYP2A6* and *CYP2A7*, and the reverse primer is specific for *CYP2A6*. This allows a region of the *CYP2A6* gene, as well as the *CYP2A7/CYP2A6* deletion variant (if present), to be amplified (see Tables II and III).

PCR Reaction mixture. DNA samples are heated at 65° for 5 min followed by cooling to room temperature. One microliter of DNA ($0.05\ \mu g/\mu l$ in 1XTE, 10 mM Tris–HCl, 1 mM EDTA, pH 8.0) is added to the bottom of each PCR tube. The reaction mixture is prepared as a master mix with the forward and reverse primers, dNTPs (Amersham Pharmacia, Canada), *Taq* polymerase (Invitrogen Life Technologies), MgCl$_2$ (50 mM, Invitrogen Life Technologies), 1XPCR buffer (10X, Invitrogen Life Technologies), and autoclaved water (see Table III). Sufficient reaction mixture for samples and controls (three positive controls with different genotypes and a negative control without DNA) is prepared and mixed by vortexing. The master mix (24 μl) is then added to each PCR tube already containing DNA for a final volume of 25 μl.

PCR CONDITIONS. After initial denaturation at 95° for 1 min, a set number of cycles are performed, each consisting of DNA denaturation, annealing of primers, and extension followed by a final extension period (see Table IV) in the PTC-200 Peltier thermal cycler (MJ Research, Boston, Massachusetts).

STEP II

i. For *CYP2A6*1B, *2*, and **5* assays, two separate reactions (a and b) are performed using combinations of primers consisting of a common reverse primer and allele-specific forward primers selective for either (a) the **1* allele allowing amplification of the wild-type *CYP2A6* gene as a positive control or (b) the variant allele (see Tables II and III).

ii. For *CYP2A6*1X2* and **4* assays, two separate reactions (a and b) are performed using combinations of primers consisting of a common reverse primer and different gene-specific forward primers. For the *CYP2A6*1X2* assay, the *CYP2A7*-specific reverse primer is combined with (a) a *CYP2A7*-specific forward primer, allowing amplification of the *CYP2A7* gene as a positive control, or (b) a *CYP2A6*-specific primer, allowing a nested region of the *CYP2A6/CYP2A7* duplication variant (if present) to be amplified. For the *CYP2A6*4* assay, the *CYP2A6*-specific

TABLE II
PRIMER SEQUENCES AND LOCATION

Primer name	Primer sequence	Location	Specificity	Assay	Refs.
2Aex7F	5'-GGC CAA GAT GCC CTA CAT G-3'	Exon 7	Nonspecific	*1X2, *4	a, b
2A6ex1F	5'-GCT GAA CAC AGA GCA GAT GTA CA-3'	Exon 1	CYP2A6	*2, *3	c
2A6ex8F	5'-CAC TTC CTG AAT GAG-3'	Exon 8	CYP2A6	*1B, *1X2, *4, *5	b
2A7ex8F	5'-CAT TTC CTG GAT GAC-3'	Exon 8	CYP2A7	*1X2, *4	b
2A6*1BwtF	5'-ACT GGG GGC AGG ATG GC-3'	3' flanking	CYP2A6	*1B	d
2A6*1BmutF	5'-AAT GGG GGG AAG ATG CG-3'	3' flanking	CYP2A6	*1B	d
2A6*2wtF	5'-CTC ATC GAC GCC CT-3'	Exon 3	CYP2A6	*2	c
2A6*2mutF	5'-CTC ATC GAC GCC CA-3'	Exon 3	CYP2A6	*2	c
E3F	5'-GCG TGG TAT TCA GCA ACG GG-3'	Exon 3	CYP2A6	*3	e
2A6*5wtF	5'-CCC CAA ACA CGT GGG-3'	Exon 9	CYP2A6	*5	d
2A6*5mutF	5'-CCC CAA ACA CGT GGT-3'	Exon 9	CYP2A6	*5	d
2A6ex4R	5'-GGA GGT TGA CGT GAA CTG GAA GA-3'	Exon 4	CYP2A6	*2, *3	c
E3R1	5'-AAC GCG CGC GGG TTC CTC GT-3'	Intron 3	CYP2A6	*2, *3	f
2A6R1	5'-GCA CTT ATG TTT TGT GAG ACA TCA GAG ACA A-3'	3' flanking	CYP2A6	*1B	b
2A6R2	5'-AAA ATG GGC ATG AAC GCC C-3'	3' flanking	CYP2A6	*1B, *4, *5	b
2A6R3	5'-GGA ATA GGT GCT TTT TAA GAA TC-3'	3' flanking	CYP2A6	*4, *5	Unpublished
2A7R1	5'-GCA CTT ATG TTT TGT GAG ACA TCA GAT AGA G-3'	3' flanking	CYP2A7	*1X2	g
2A7R2	5'-AAA ATG GGC ATG AAC GCT T-3'	3' flanking	CYP2A7	*1X2	g

[a] Single base change from original primer sequence.

[b] M. Oscarson, R. A. McLellan, H. Gullsten, Q. Y. Yue, M. A. Lang, M. L. Bernal, B. Sinues, A. Hirvonen, H. Raunio, O. Pelkonen, and M. Ingelman-Sundberg, *FEBS Lett.* **448**, 105 (1999).

[c] M. Oscarson, H. Gullsten, A. Rautio, M. L. Bernal, B. Sinues, M. L. Dahl, J. H. Stengard, O. Pelkonen, H. Raunio, and M. Ingelman-Sundberg, *FEBS Lett.* **438**, 201 (1998).

[d] M. Oscarson, R. A. McLellan, H. Gullsten, J. A. Agundez, J. Benitez, A. Rautio, H. Raunio, O. Pelkonen, and M. Ingelman-Sundberg, *FEBS Lett.* **460**, 321 (1999).

[e] P. Fernandez-Salguero, S. M. Hoffman, S. Cholerton, H. Mohrenweiser, H. Raunio, A. Rautio, O. Pelkonen, J. D. Huang, W. E. Evans, J. R. Idle, and F. J. Gonzalez, *Am. J. Hum. Genet.* **57**, 651 (1995).

[f] H. Gullsten, J. A. Agundez, J. Benitez, E. Laara, E. J. M. Ladero, M. Diaz-Rubio, P. Fernandez-Salguero, F. J. Gonzalez, A. Rautio, O. Pelkonen, and H. Raunio, *Pharmacogenetics* **7**, 247 (1997).

[g] Y. Rao, E. Hoffmann, M. Zia, L. Bodin, M. Zeman, E. M. Sellers, and R. F. Tyndale, *Mol. Pharmacol.* **58**, 747 (2000).

TABLE III
PCR REACTION MIXTURES FOR EACH GENOTYPING ASSAY

Assay Reaction mixture	*1B Step 1	*1B Step 2	*1X2 Step 1	*1X2 Step 2	*2 Step 1	*2 Step 2	*3 Step 1	*3 Step 2	*4 Step 1	*4 Step 2	*5 Step 1	*5 Step 2
DNA^{a,b}	50 ng	0.8 μl	50 ng	1.0 μl	50 ng	0.8 μl	50 ng	1.0 μl	50 ng	0.8 μl	50 ng	0.8 μl
Primer (μM)	0.25	0.25	0.25	0.25	0.25	0.25	0.25	0.25	0.25	0.25	0.25	0.25
Primer combinations	3, 14	a. 5, 15 b. 6, 15	1, 17	a. 4, 18 b. 3, 18	2, 12	a. 7, 13 b. 8, 13	2, 12	9, 13	1, 16	a. 3, 15 b. 4, 15	3, 16	a. 10, 15 b. 11,15
dNTP (mM)	0.2	0.1	0.2	0.2	0.2	0.1	0.2	0.2	0.2	0.1	0.2	0.1
MgCl$_2$(mM)	1.0	1.0	1.2	1.8	1.2	1.0	1.2	1.5	1.5	1.0	1.2	1.0
Taq polymerase (units)	0.6	0.3	1	1	0.6	0.2	0.6	0.5	1.25	0.5	1.25	0.3
PCR buffer	1X	1X	1X	1X	1X	1X	1X	1X	1X	1X	1X	1X
Total volume (μl)	25	25	25	25	25	25	25	20	25	25	25	25

^{a} Step I: genomic DNA.
^{b} Step II: PCR product.

TABLE IV
PCR Conditions for Each Genotyping Assay

Assay PCR conditions[a]	*1B Step 1	*1B Step 2	*1X2 Step 1	*1X2 Step 2	*2 Step 1	*2 Step 2	*3 Step 1	*3 Step 2	*4 Step 1	*4 Step 2	*5 Step 1	*5 Step 2
Initial denaturation	95°, 1 min		95°, 1 min	95°, 1 min	95°, 1 min	95°, 1 min	95°, 1 min	94°, 3 min	95°, 1 min	95°, 1 min	95°, 1 min	95°, 1 min
Denaturation	95°, 15 sec	95°, 15 sec	95°, 15 sec	95°, 15 sec	95°, 15 sec	95°, 15 sec	95°, 15 sec	94°, 1 min	95°, 15 sec	95°, 15 sec	95°, 15 sec	95°, 15 sec
Annealing	56°, 20 sec	54°, 20 sec	60°, 20 sec	44°, 20 sec	60°, 20 sec	50°, 20 sec	60°, 20 sec	60°, 1 min	50°, 20 sec	52°, 20 sec	50°, 20 sec	54°, 20 sec
Extension	72°, 1.5 min	72°, 45 sec	72°, 3 min	72°, 4.5 min	72°, 3 min	72°, 45 sec	72°, 3 min	72°, 1 min	72°, 3 min	72°, 2 min	72°, 1.5 min	72°, 45 sec
Number of cycles	35	15	35	15	37	19	37	31	31	18	30	20
Final extension	72°, 7 min		72°, 7 min	72°, 10 min	72°, 7 min		72°, 7 min		72°, 7 min		72°, 7 min	

[a] These times are for the PTC-200 Peltier thermal cycler using block probes and may vary with different cyclers.

reverse primer is combined with (a) a *CYP2A6*-specific forward primer, allowing the amplification of the *CYP2A6* gene as a positive control, or (b) a *CYP2A7*-specific forward primer, allowing a nested region of the *CYP2A7/CYP2A6* deletion variant (if present) to be amplified (see Tables II and III).

iii. For the *CYP2A6*3* assay, forward and reverse primers are gene specific and are used to amplify a nested region of the *CYP2A6* gene (see Tables II and III).

PCR Reaction mixture. For all assays, excluding *CYP2A6*3*, the PCR product from step I is added to two PCR tubes. Two separate sets of reaction mixtures are prepared containing either the first [(a) wild type] or second [(b) variant] combination of primers, dNTPs, *Taq* polymerase, $MgCl_2$, PCR buffer, and autoclaved water (see Table III). The two reaction mixtures are vortexed lightly and centrifuged briefly. Aliquots of each reaction mixture are added to the separate sets of PCR tubes already containing the PCR I amplification product for a final volume of 25 μl. Therefore, each original DNA sample is amplified in two separate step II reactions to assay for the presence of either the wild-type and/or the allelic variant. The reaction mixture for the *CYP2A6*3* assay is prepared as described in step I using step II reagents.

PCR CONDITIONS. After initial denaturation, a certain number of cycles are performed, each consisting of DNA denaturation, annealing of primers, and extension (see Table IV).

Restriction Digestion

For the *CYP2A6*3* assay, a restriction enzyme digestion is performed in duplicate using 5 μl of the step II PCR product in 1X digestion buffer (10X, New England Biolabs, Beverly, MA) with 2.5 units of *Dde*I (10 U/μl New England Biolabs) or autoclaved water for the uncut control (final volume of 30 μl). The reaction proceeds at 37° for 4 hr.

Gel Electrophoresis

For all assays, excluding *CYP2A6*3*, an aliquot of 20 μl of each of the step II PCR products is mixed with 2 μl (0.25%) of bromphenol blue (*1B, *4, *1X2 assays) or xylene cyanol FF (*2, *5 assays) loading buffers (containing 30% glycerol). The DNA product resulting from amplification with the wild-type primer is loaded beside the product resulting from amplification with the variant primer and electrophoresed in 1.2% (*1B, *4, *5, Y assays) or 3% (*2 assay) agarose gels stained with ethidium bromide (10 mg/ml stock solution, 60 μg/100 ml agarose gel). For the *CYP2A6*3* assay, 20 μl of the digestion products (uncut and cut for each DNA sample) is mixed with 2 μl of xylene cyanol FF loading buffer and loaded side by side for electrophoresis in a 3% agarose gel stained with ethidium bromide. A 1-kb DNA ladder (1 μg/μl) is used to identify the band sizes. The

A

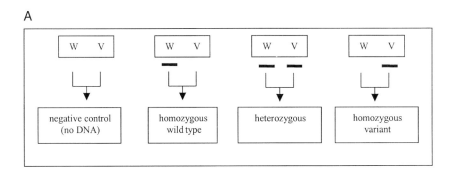

B

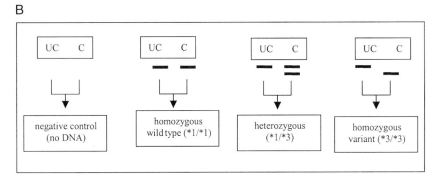

Fig. 2. Schematic representation of *CYP2A6* genotypes. (A) Two-step PCR assays (*CYP2A6* *1B*, *1X2*, *2*, *4*, and *5*). Following the second PCR amplification, the two products of each DNA sample, resulting from amplification with either an allele-specific wild type (w) or an allele-specific variant (v) primer, are loaded side by side. (B) PCR–RFLP assay (*CYP2A6* *3*). Following the restriction enzyme digestion, uncut (UC) and cut (C) products of each DNA sample are loaded side by side. For both assays, the genotype can be identified by the band pattern, which is a different depending on whether the individual is a homozygous wild-type, heterozygous, or homozygous variant.

samples are run at 100 V for 40–80 min in a horizontal gel electrophoresis apparatus in 1X TAE buffer (0.04 M Tris–acetate, 0.001 M EDTA). The DNA bands are viewed using a UV transilluminator and images are captured digitally.

Analysis of Results

For all the described assays, three different patterns may appear when viewing the gel representing either a homozygous wild type or a heterozygous or homozygous variant individual (see Fig. 2).

Section II

Structure and Mechanism

[8] Purification and Crystallization of N-Terminally Truncated Forms of Microsomal Cytochrome P450 2C5

By MICHAEL R. WESTER, C. DAVID STOUT, and ERIC F. JOHNSON

Eukaroytic P450s are generally membrane proteins. To date, only one structure for a mammalian membrane P450, a modified form of rabbit microsomal P450 2C5, has been published.[1] The successful crystallization of the modified enzyme, P450 2C5/3LVdH, was achieved through engineering the parental enzyme to improve its solubility and the appropriate use of detergents during purification and crystallization to improve the quality of the crystals for diffraction studies.

P450 2C5 was modified in three ways to facilitate purification and crystallization. The membrane-spanning leader sequence was deleted from the N terminus to reduce the strength of interactions between P450 2C5 and cellular membranes when the truncated enzyme was expressed in *Escherichia coli*. In addition, a four-histidine (His_4) tag was added to the C terminus to aid purification.[2] The resulting protein, P450 2C5dH, exhibits membrane interactions that are disrupted by increasing ionic strength. In addition, the protein is sufficiently soluble to purify the enzyme without the use of detergents. Substitution of five amino acid residues derived from a related enzyme, P450 2C3, for the corresponding residues in P450 2C5 further increased the solubility and monodispersity of the protein, P450 2C5/3LVdH, relative to that of P450 2C5dH.[3] The initial structure of the protein was determined from data collected to 3.0 Å resolution.[1] Subsequent studies have produced crystals that diffract to 2.1 Å. This article describes the preparation and crystallization of the enzyme and discusses procedures contributing to the improved resolution of the crystals that should be applicable to other modified membrane P450s.

Heterologous Expression

The objective of the expression protocol is to obtain suitable quantities of the enzyme for purification and crystallization without introducing microheterogeneity or promoting irreversible aggregation. Expression vectors for P450 2C5dH and related constructs are constructed using the pCWOri$^+$ vector. XL-1 Blue (Stratagene, La Jolla, CA) transformation-competent *E. coli* are routinely used as the host. For protein expression, overnight cultures are started in Superbroth (5 ml)

[1] P. A. Williams, J. Cosme, V. Sridhar, E. F. Johnson, and D. E. McRee, *Mol. Cell* **5**, 121 (2000).

[2] C. Von Wachenfeldt, T. H. Richardson, J. Cosme, and E. F. Johnson, *Arch. Biochem. Biophys.* **339**, 107 (1997).

[3] J. Cosme and E. F. Johnson, *J. Biol. Chem.* **275**, 2545 (2000).

containing 50 μg/ml ampicillin (Sigma, St. Louis, MO). To 500 ml of Terrific broth containing 50 μg/ml ampicillin, a 3.5-ml aliquot of the overnight culture is seeded, and the 1-liter Erlenmeyer flask containing the culture is incubated at 37° with shaking at 225 rpm. A total culture volume of 3 liters is generally sufficient to produce >150 nmol of purified protein. When the $OD_{600\ nm}$ of the culture reaches 0.35, the heme precursor aminolevinulic acid (Sigma) is added to a final concentration of 0.5 mM. Approximately 25 min later, when the light absorbance reaches 0.45, expression of the P450 is induced by the addition of 1 mM isopropyl-β-D-thiogalactopyranoside (Roche Molecular Biochemicals, Bosel, Switzerland). The rate of shaking is then reduced to 190 rpm, and the temperature is lowered to 30°. Expression is then allowed to proceed for 48 hr before the cells are harvested. Although the total yield of P450 will generally increase through 72 hr, the suitability of this material for purification may be diminished due to aggregation of the expressed P450.

At 48 hr, the cells are harvested by centrifugation at 3000g for 10 min at 4°. The cells are then resuspended in 10% of the original culture volume in 20 mM potassium phosphate buffer, pH 7.4, containing 20% glycerol, 1 mM phenylmethylsulfonyl flouride (PMSF), and 10 mM 2-mercaptoethanol. Spheroplasts are then produced by treating the suspension for 30 min with 0.2 mg/ml lysozyme with stirring at 4°. Following the lysozyme treatment, an equal volume of cold water is added slowly to the stirring mixture. After an additional incubation with stirring for 10 min, the resulting spheroplasts are pelleted by centrifugation at 5000g for 10 min at 4°. The spheroplasts are then lysed by freezing the pellets *in situ* using a liquid nitrogen bath. The frozen pellets are thawed and suspended in 20% of the original culture volume of 500 mM potassium phosphate buffer, pH 7.4, containing 20% glycerol, 0.5 mM PMSF, and 10 mM 2-mercaptoethanol. The concentration of the buffer is chosen to promote the dissociation of the enzyme from membranes and to prevent precipitation of the protein. A sonic dismembrator is employed to disrupt the spheroplasts using three 45-sec pulses with 60 sec of cooling on ice between pulses. The material is then centrifuged at 7000g for 10 min at 4°. The supernatant is then centrifuged at >105,000g for 90 min at 4° to remove membrane vesicles and insoluble material. Loss of a significant fraction of the P450 from the supernatant indicates potential problems with aggregation or residual membrane binding.

Purification in Absence of Detergents

Both P450 2C5dH and 2C5/3LVdH are sufficiently soluble that they can be purified in the absence of detergents. Conditions that lead to aggregation or precipitation should be avoided as they may interfere with the subsequent crystallization of the proteins. If specific detergents are found to improve the crystallization of the protein purified in the absence of detergent, their use in the purification of the protein can be examined in subsequent purification procedures.

The clarified cellular lysate containing the P450 can be applied directly to a metal ion-affinity resin, Ni-NTA Agarose (Qiagen, Chatsworth, CA). Approximately 1 ml of resin is used for each 90 nmol of P450. Prior to use, the resin is equilibrated with 500 mM potassium phosphate buffer, pH 7.4, containing 20% (v/v) glycerol, 0.5 mM PMSF, and 10 mM 2-mercaptoethanol. Following application of the lysate, the resin is washed with 50 ml of equilibration buffer. The resin is subsequently washed with 2 volume equivalents of 100 mM potassium phosphate, pH 7.4, containing 100 mM sodium chloride, 20% glycerol, 1 mM PMSF, and 10 mM 2-mercaptoethanol. Subsequent washes with 5 column volumes each of this buffer containing 50 mM glycine and then 10 mM histidine are used to reduce the nonspecific binding of undesired proteins to the resin. The bound P450 is then eluted using 10 mM potassium phosphate, pH 7.4, 500 mM NaCl, 20% glycerol, 10 mM 2-mercaptoethanol, 1 mM PMSF, and 40 mM histidine. This concentration of histidine displaces the Ni ion from the resin. The Ni-rich fractions are not included in the pooled P450-containing fractions. However, most of the P450 elutes slowly behind the initial Ni ion-rich fractions. The elution step and washes also exchange the high concentration phosphate buffer for a lower concentration suitable for the next chromatography step while increasing the concentration of NaCl. In these and subsequent steps, the salt concentrations should be sufficiently high to maintain the solubility of the protein. A poor yield is suggestive that the protein may not be sufficiently soluble in the absence of detergents for efficient binding and/or elution.

The pooled fractions from the metal ion-affinity column are then applied directly to a column containing hydroxylapatite (10 to 15 ml HA, Clarkson Chemical Co., Williamsport, PA) that has been equilibrated with elution buffer used for the metal ion-affinity column omitting the histidine. After loading, the hydroxylapatite is washed with 5 volume equivalents of the equilibration buffer. Elution is performed using a gradient of equilibration buffer to 500 mM potassium phosphate buffer containing the same components (10 column volume equivalents). Fractions containing the P450 protein as judged by their color are pooled. The protein is concentrated and the buffer is exchanged with 50 mM potassium phosphate, pH 7.4, 1 mM EDTA, 0.2 mM dithiothreitol (DTT), 500 mM NaCl, and 20% glycerol using a centrifugal ultrafiltration device (50 kD a cutoff, Centricon, Millipore, Bedford, MA). The concentration of the protein is adjusted to 600 μM for crystallization trials.

If the protein of interest behaves poorly during the purification procedure, detergents can be used to facilitate binding and elution of the protein from chromatography media. In the presence of detergents, salt concentrations can usually be reduced so that ion-exchange chromatography can be used for additional or alternative purification procedures. Ideally, the selection of the detergent should be based on its compatibility with protein stability, purification, and crystallization. In the absence of specific knowledge regarding crystallization, a detergent such as CHAPS can be used at a concentration that is sufficient to dissociate aggregates of the truncated protein. If most of the detergent can be removed prior to crystallization trials, it should be possible to screen for detergents that promote crystallization.

Purification in Presence of Detergents

The following procedures work well with P450 2C5/3LVdH leading to diffraction quality crystals. The initial steps follow the aforementioned protocol. However, the elution buffer for the metal ion-affinity column is adjusted to include the detergent and to lower the ionic strength to the minimum recommended for use with the affinity resin. For this procedure, 10 mM potassium phosphate buffer, pH 7.4, containing 100 mM sodium chloride, 20% glycerol, 10 mM 2-mercaptoethanol, 1 mM PMSF, 40 mM histidine, and the detergent of choice is used for the elution. In the presence of detergent, the enzyme should elute in a sharper band. The pooled eluate from the metal ion-affinity resin is diluted 10-fold with 5 mM potassium phosphate, pH 7.4, 1 mM EDTA, 0.2 mM DTT, 1 mM PMSF, and 20% glycerol containing the detergent prior to applying the sample to a column containing 20 ml CM-Sepharose, which has been equilibrated with 10 mM potassium phosphate, pH 7.4, containing 1 mM EDTA, 0.2 mM DTT, 1 mM PMSF, and 20% glycerol. The CM-Sepharose is then washed with 5 volume equivalents of the equilibration buffer omitting the detergent. P450s 2C5 and 2C5/3LVdH can be eluted with 50 mM potassium phosphate, pH 7.4, 1 mM EDTA, 0.2 mM DTT, 500 mM NaCl, and 20% glycerol. The purified enzyme is then concentrated using the centrifugal ultrafiltration as described earlier. The wash and elution buffers for the CM-Sepharose do not contain detergents, and this protocol is designed to remove most of the detergent from the protein prior to concentration of the protein for crystallization. Depletion of the detergent is beneficial because the ultrafiltration procedure is likely to concentrate detergent micelles that could, in turn, affect the subsequent crystallization of the protein. Although extensive washes of the column-bound P450 should reduce the amount of detergent present in the preparation, some detergent may be retained. P450 2C5/3LVdH has been purified using both of the methods described, and the resulting preparations have produced crystals that diffract equally well: 2.1 and 2.3 Å. The crystals diffract better when the detergent used in the purification procedure matches the one used for crystallization of the protein: 2.3 vs 3.4 Å for purification in the presence of CYMAL-5 vs CHAPS, respectively, P450 2C5dH can also be purified using both methods. In contrast to 2C5/3LVdH, 2C5dH exhibits a tendency to precipitate over time; however, this is reduced when 2C5dH is purified in the presence of a detergent.

Crystallization

The crystallization of P450 2C5/3LVdH was achieved by screening conditions that are generally employed for soluble proteins, and this methodology and approach are not described in detail here. Screening kits are available commercially, and kits from Hampton Research, Laguna Niguel, CA, and Emerald Biostructures, Bainbridge Island, WA, have been employed successfully by our laboratory to identify crystallization conditions.

Detergents were not present in the initial screens that identified conditions permitting the crystallization of P450 2C5/3LVdH. However, a screen of detergents used for the crystallization of membrane proteins (Hampton Research) was undertaken during the optimization of the crystallization conditions. P450 2C5/3LVdH crystallizes in the presence of several detergents. The crystals used to determine the structure of the enzyme were grown in the presence of CYMAL-5 (Anatrace, Inc., Maumee, OH). The protein crystallized more quickly in the presence of CYMAL-5, and the crystals exhibited greater size, better morphology, and enhanced resolution. Detergents identified in these screens have been incorporated into new iterations of the screening process.

Our laboratory uses a hanging drop method for setting up vapor diffusion crystallization screens because this method uses less protein than sitting drop methods, is easy to perform, and is relatively inexpensive. Initial screens using P450 protein stock concentrations of 600 μM have been effective, but should be optimized for individual proteins and crystallization conditions. The hanging drop routinely consists of 1.0 μl of protein, 0.25 μl of 24 mM CYMAL-5 detergent, and 1.25 μl of the solution from the reservoir. Volumes can be doubled in the case of sitting drops, for a drop size of 5 μl, and this may facilitate the growth of larger crystals for X-ray diffraction studies.

Cocrystallization of P450 2C5/3LVdH with a substrate can produce crystals that exhibit better X-ray diffraction than the substrate-free enzyme. This probably reflects the increased order exhibited by flexible portions of the structure that interact with the substrate. The limited solubility of most P450 substrates is problematic because the high concentrations of the enzyme that are employed during crystallization require equally high concentrations of the substrate. The effects of solvents used to prepare solutions of substrates are tested for their effects on the crystallization. Addition of up to 3% (v/v) ethanol or dimethyl sulfoxide (DMSO) had little effect on the crystallization of P450 2C5/3LVdH.

When a substrate is included in the setup, an equimolar ratio of substrate to P450 is used. Substrates have been introduced successfully to the crystallization setup in two ways. First, if the substrate is soluble in the protein buffer, a 2× stock is prepared in the protein buffer. The protein is also concentrated to 2× (1200 μM), and equal volumes of the protein and substrate solutions are combined to achieve a solution containing 600 μM of both P450 and substrate for use in the hanging drop setups. This may prove difficult with some substrates, as many P450 substrates may not be sufficiently soluble under these conditions. Second, it may be possible to prepare the substrate as a 4× solution (2400 μM) in the detergent solution used for the drop setup. Under the conditions used for the crystallization of CYP2C enzymes, small crystals can be observed within 48 to 72 hr following the initial drop setup. In other cases, crystals may form slowly.

TABLE I
EFFECTS OF CRYOPROTECTANT SOLUTIONS ON
DIFFRACTION OF CRYSTALS OF P450 2C5/3LVdH[a]

Cryoprotectant	Resolution
20% sucrose	2.1
20% sucrose	2.3
15% sucrose	2.6
20% glycerol	2.9
20% glycerol	3.0
15% PEG 400	3.4
15% PEG 400	No diffraction

[a] Crystals of a preparation of 2C5/3LVdH were grown in the presence of equimolar concentrations of the same substrate under similar conditions. Before freezing, crystals were treated with solutions containing the indicated cryprotectant as described in the text. PEG, polyethylene glycol.

Preparing Crystals for Diffraction Studies

Cryoprotectants can help preserve the quality of crystals when they are frozen and placed in the X-ray beam. This is particularly critical for crystals of P450 2C5 because they contain a high solvent content, roughly 70%, and are fragile. As the crystallization conditions and properties of specific proteins will vary, the optimal cryoprotectant should be determined empirically for each. In addition to the choice of cryoprotectant, concentration and equilibration time should be manipulated in order to achieve optimum results. Table I shows the diffraction resolution observed for several crystals prepared from the same preparation of 2C5/3LVdH crystallized under similar conditions. As shown in Table I, the identity of the cyroprotectant has a significant effect on the resolution limit of the diffraction pattern that is obtained.

The crystals were frozen using the following procedure. The crystal is removed from the hanging drop by either a pipette or a mounting loop and placed into a spot plate containing 200 μl of cryoprotectant solution for 1 min. The cryoprotectant solution is prepared by diluting the precipitant solution used in the well of the crystallization plate with a concentrated solution of the cryoprotectant, 85% (w/v) sucrose, to achieve the final concentration of the cryoprotectant. The crystal is then picked up in a nylon loop and is placed quickly into the nitrogen stream used to cool the crystal during the X-ray diffraction study. Alternatively, the crystal may be plunged directly into liquid nitrogen once it is extracted from the cryoprotectant solution with the nylon loop. Using either method, the cryoprotectant solution surrounding the crystal in the loop will freeze to an amorphous glass, minimizing both crystal damage and diffraction from ice crystals.

Summary

Engineering more soluble forms of P450 2C5 has contributed to the crystallization of the enzyme. When detergents are used in both crystallization and purification of the protein, the ability to control the content and identity of the detergent is dependent on the protein exhibiting a sufficient degree of solubility to permit its concentration in the absence of detergents. The production of concentrated solutions of the protein containing little or no detergent provides a means for screening crystallization conditions and the selection of detergents that facilitate crystallization. These detergents can then be used not only to improve the purification of the protein, but also to solublize substrates for the cocrystallization of enzyme–substrate complexes.

Acknowledgment

This work was supported by USPHS Grants GM31001 (efj) and GM59229 (cds).

[9] Molecular Replacement in P450 Crystal Structure Determinations

By HUIYING LI, JASON K. YANO, and THOMAS L. POULOS

Introduction

The cytochrome P450 superfamily has been growing rapidly, with over 1500 P450 sequences identified to date (drnelson.utmem.edu/CytochromeP450.html). This tremendous growth in recent years has been facilitated by the various genome projects. The number of P450 crystal structures solved has, of course, grown at a much slower pace, with 8 currently in the PDB (www.rcsb.org/pdb/index.html).

P450s occur throughout the biosphere from bacteria, archaebacteria, fungi, plants, insects, and mammals.[1,2] The functions of P450s cover a broad range, including antibiotic synthesis in bacteria[3,4]; hormone synthesis in plants[5]; and

[1] D. R. Nelson, L. Koymans, T. Kamataki, J. J. Stegeman, R. Feyereisen, D. J. Waxman, M. R. Waterman, O. Gotoh, M. J. Coon, R. W. Estabrook, I. C. Gunsalus, and D. W. Nebert, *Pharmacogenetics* **6**, 1 (1996).

[2] D. R. Nelson, *Arch. Biochem. Biophys.* **369**, 1 (1999).

[3] A. Shafiee and C. R. Hutchinson, *J. Bacter.* **170**, 1548 (1988).

[4] D. Stassi, S. Donadio, M. J. Staver, and L. Katz, *J. Bacter.* **175**, 182 (1993).

[5] K. Salchert, R. Bhalerao, Z. Koncz-Kalman, and C. Koncz, *Phil. Trans. R. Soc. Lond. B* **353**, 1517 (1998).

cholesterol biosynthesis,[6] vascular hemostasis through the arachidonic acid cascade,[7] xenobiotic detoxification,[8] and steroid synthesis[9] in mammals, to name a few. Physiological and biochemical studies on P450 enzymes are not only aimed at biomedical applications,[10] but also at environmental bioremediation.[11,12] Although P450s can catalyze an array of different reactions,[13] monooxygenation of various organic substances is the most commonly observed P450 enzymatic activity. The fact that a wide variety of chemicals can be metabolized by P450 enzymes has been reflected in the extraordinary variations in structural elements that form the substrate-binding site observed in the known P450 crystal structures.[14–22] Homology modeling based on the available P450 structures can surely provide valuable guidance to experimental work when crystal structures for the P450s of interest are not available.

However, variations in the P450 substrate-binding pocket make modeling difficult to achieve at the level of precision required for drug design or for optimization of metabolizing target pollutants. As with any new protein, growing crystals is the bottleneck for new P450 structure determinations, and P450s are especially difficult, as the majority of the most interesting P450s are membrane bound. The CYP2C5 crystal structure[22] became the first breakthrough to this barrier, which taught us some protein engineering techniques to make membrane proteins more soluble.

Despite the various difficulties in obtaining new P450 structures, the P450 structure database has been growing steadily and has provided essential models

---

[6] D. Rozman and M. R. Waterman, *Drug Metab. Disp.* **26,** 1199 (1998).

[7] K. Makita, J. R. Falck, and J. H. Capdevila, *FASEB J.* **10,** 1456 (1996).

[8] J. P. J. Whitlock and M. S. Denison, *in* "Cytochrome P450" (P. R. Ortiz de Montellano, ed.), p. 367. Plenum Press, New York, 1995.

[9] D. J. Waxman and T. K. H. Chang, *in* "Cytochrome P450" (P. R. Ortiz de Montellano, ed.), p. 391. Plenum Press, New York, 1995.

[10] D. J. Waxman, L. Chen, J. E. Hecht, and Y. Jounaidi, *Drug Metab. Rev.* **31,** 503 (1999).

[11] S. Harayama, *Curr. Opin. Biotech.* **8,** 268 (1997).

[12] D. G. Kellner, S. A. Maves, and S. G. Sligar, *Curr. Opin. Biotech.* **8,** 274 (1997).

[13] D. Mansuy, *Comp. Biochem. Physiol.* **121,** 5 (1998).

[14] T. L. Poulos, B. C. Finzel, and A. J. Howard, *J. Mol. Biol.* **195,** 687 (1987).

[15] K. G. Ravichandran, S. S. Boddupalli, C. A. Hasermann, J. A. Peterson, and J. Deisenhofer, *Science* **261,** 731 (1993).

[16] C. A. Hasemann, K. G. Ravichandran, J. A. Peterson, and J. Deisenhofer, *J. Mol. Biol.* **236,** 1169 (1994).

[17] J. R. Cupp-Vickery and T. L. Poulos, *Nature Struct. Biol.* **2,** 144 (1995).

[18] H. Li and T. L. Poulos, *Nature Struct. Biol.* **4,** 140 (1997).

[19] S. Y. Park, H. Shimizu, S. Adachi, A. Nakagawa, I. Tanaka, K. Nakahara, H. Shoun, E. Obayashi, H. Nakamura, T. Iizuka, and Y. Shiro, *Nature Struct. Biol.* **4,** 827 (1997).

[20] J. K. Yano, L. S. Koo, D. J. Schuller, H. Li, P. R. Ortiz de Montellano, and T. L. Poulos, *J. Biol. Chem.* **275,** 31086 (2000).

[21] L. M. Podust, T. L. Poulos, and M. R. Waterman, *Proc. Natl. Acad. Sci. U.S.A.* **98,** 3068 (2001).

[22] P. A. Williams, J. Cosme, V. Sridhar, E. F. Johnson, and D. E. McRee, *Mol. Cell* **5,** 121 (2000).

for homology modeling, as well as a better pool of search models to aid in solving new P450 structures by the molecular replacement (MR) technique. This article discusses the principle of the MR method and the application of this technique to P450 crystal structure determinations.

Principle of Molecular Replacement Method

The MR method has been widely used to determine a new protein structure whenever a homologous known structure is available as a search model. MR calculations are essentially the processes of orienting and positioning a model of the known structure in the unit cell of the target protein until a match is found with the unknown protein molecule.[23] The orientation and position of a molecule in the unit cell can be fully described by six parameters: three rotation angles and three translational coordinates. In principle, a match between model and target molecules can be determined by varying these six variables of a model until the structure factors, $|F_c|$, calculated with the search model agree with the observed structure factors, $|F_o|$, from the target structure. However, this type of six-dimensional search was not practical until more recently, as modern computing power has become faster and less expensive and the newly developed fast structure factor interpolation algorithm was incorporated into computer programs.[24]

To start an MR project requires only two pieces of information: the structure of the known and a set of X-ray intensity data from crystals of the new protein whose structure we wish to solve. The only useful information that can be derived from X-ray data alone without phases is the Patterson function, which is proportional to the square of the structure factor, $|F(hkl)|^2$, thus proportional to $I_{obs}(hkl)$, intensity data. Therefore, the Patterson function of the unknown structure can be computed directly from the intensity of the observed reflections, $I_{obs}(hkl)$, whereas the Patterson function of the search model can be calculated from the atomic coordinates.[25] The Patterson map is a vector map of distances between atomic centers and hence represents intramolecular vectors between atoms within a single protein molecule as well as intermolecular vectors between atoms in symmetry related molecules. Peaks in a Patterson map represent distances between atomic centers, and the peak height is roughly proportional to the atomic numbers of the atoms involved. Because proteins contain thousands of atoms of similar atomic weight (N, C, O), a Patterson vector map is hopelessly complex even though each structure will generate a unique Patterson map. However, two similar molecules will generate similar Patterson maps so the trick in the MR technique is to rotate and translate the Patterson of the known structure until it matches the Patterson of the unknown.

[23] E. Lattman, *Methods Enzymol.* **115,** 55 (1985).
[24] C. R. Kissinger, D. K. Gehlhaar, and D. B. Fogel, *Acta Cryst.* **D55,** 484 (1999).
[25] T. L. Blundell and L. N. Johnson, "Protein Crystallography." Academic Press, London, 1976.

Traditionally, the MR method involves two steps in which a Rossmann and Blow rotation function[26] search is followed by a Crowther and Blow translation function[27] search. Both rotation and translation functions are defined by the product of Patterson functions of both model and target structures. When the Patterson functions of model and target protein molecules are superimposed, both rotation and translation functions show their maximum values. Unfortunately, the amplitudes of rotation and translation functions are not very sensitive scores. In modern MR programs, correlation coefficients (CC) based on either Patterson functions[28] or structure factors[24,29] are used for selecting correct solutions against noise.

Available MR Programs

Many program packages are available for MR calculations. This chapter briefly describes a few that are widely used or newly developed.

Crystallography and NMR System (CNS)

The CNS[28] is a comprehensive program package covering most of the calculations used for both crystal structure and NMR solution structure determinations, including molecular replacement. The rotation function search uses the Patterson correlation coefficient (CC) to score the solutions.

In addition, a packing value is also reported for each solution, which is the percentage of the unit cell covered by the search model. The rotation search routine also allows a so-called "direct" rotation function[30] search where the search model itself, rather than the Patterson vectors, is rotated and the correlation coefficients of structure factors are calculated. There often are several rotation function solutions, which are difficult to differentiate based on correlation coefficients. As a result, it is usually necessary to apply several unique rotation solutions in a translational search. In CNS the translation function search can read in 10 or more rotation solutions at a time. Prior to the translation search, the program will carry out the Patterson correlation refinement on the selected rotation solutions. The correct translation solutions should have both a high CC value and a high packing value. The latter reflects optimized packing of the search model in the unit cell without overlap. When there is more than one molecule in the asymmetric unit, the found orientation and position of the first molecule can be fixed while searching for

[26] M. G. Rossmann and D. M. Blow, *Acta Cryst.* **15**, 24 (1962).
[27] R. A. Crowther and D. M. Blow, *Acta Cryst.* **23**, 544 (1967).
[28] A. T. Brunger, P. D. Adams, G. M. Clore, W. L. DeLano, P. Gros, R. W. Grosse-Kunstleve, J.-S. Jiang, J. Kuszewski, M. Nilges, N. S. Pannu, R. J. Read, L. M. Rice, T. Simonson, and G. L. Warren, *Acta Cryst.* **D54**, 905 (1998).
[29] J. Navaza, *Acta Cryst.* **A50**, 157 (1994).
[30] W. L. DeLano and A. T. Brunger, *Acta Cryst.* **D51**, 740 (1995).

translational solutions for additional molecules. All the final solutions can be fine-tuned by running a rigid body refinement. If the solutions of multiple molecules are not packed closely to each other, a shift routine can be used to apply the space group symmetry operations and/or translation along the unit cell to bring them together.

AMoRe

AMoRe,[29] one of the programs in the CCP4 suite,[31] is the most popular package used for MR calculation mainly because of its high speed and automation. A situation frequently encountered in MR is that the correct rotation function solution(s) is poorly contrasted from random noise. The rotation function search in AMoRe can output up to 99 solutions. The subsequent translation function search will accommodate all 99 rotation solutions and carry out the translational search on every one of the oriented models in a single run. The program uses the structure factor correlation coefficient (CC) as the main scoring criterion. After the translation search, a crystallographic R factor is also reported along with the CC value. When more than one molecule is expected in the asymmetric unit, the already known molecular orientation and position can be fixed while searching for additional molecules in translation function calculations. For multidomain proteins, AMoRe also utilizes multiple independent search models for different segments (domains) of the unknown structure. Again, once the solution of one domain is found, its orientation and position are fixed during the calculation to find the location for the remaining portion of the structure. An efficient rigid body refinement routine in AMoRe can be used to fine-tune the rotation angles and translational coordinates of the final solutions. When dealing with multiple solutions, rigid body refinement on the first solution found is recommended prior to the subsequent search for the other solutions. Recent versions of AMoRe have incorporated new algorithms into the translational search, such as phased translation (PT) and correlation coefficient (CC) in addition to the traditional Crowther and Blow (CB) method, which widens its capability in coping with various calculation demands.

EPMR

Most of the MR packages carry out rotational and translational searches sequentially in two separate steps. A systematic six-dimensional search is computationally demanding and costly. EPMR[24] implements an evolutionary algorithm that iteratively optimizes a population of trial solutions with respect to the correlation coefficient between observed and calculated structure factors. A rapid structure factor calculation algorithm other than the fast Fourier transform is incorporated into EPMR, which makes it an efficient six-dimensional search program. Determining MR solutions with both orientation and position simultaneously in EPMR

[31] Collaborative Computational Project, Number 4, *Acta. Cryst.* **D50,** 760 (1994).

improves the signal-to-noise ratio. Due to the nature of the evolutionary search, a reliable indicator of the correct solution found by EPMR is whether the solution is obtained repeatedly in multiple runs. EPMR can sequentially search for multiple molecules (or domains) in the asymmetric unit in a single input. EPMR should be used as an alternate approach to search for MR solutions if conventional rotation and translation function searches using other programs failed.

BEAST

The maximum likelihood algorithm has been widely used in the crystallographic phase and/or structure refinements in recent years. This superior statistical tool has been, for the first time, incorporated by Randy Read into an MR package called BEAST,[32] a modification of the author's earlier MR routine, BRUTE.[33] The program computes the log-likelihood gain (LLG) instead of correlation coefficient (CC) as the scoring value for search solutions. LLG has proven to be a much more sensitive score than CC. BEAST first carries out a coarse rotational search and then fine-tunes around the best found rotation solution with a much smaller rotational step size followed by a translational search. If there are multiple molecules or domains in the asymmetric unit, the rotation (and translation) solution of the first molecule or domain can be fixed while searching for the remaining solutions. Another useful feature of BEAST is utilizing multiple search models. When multiple homologous models are available, one usually generates a search model by superimposition and concatenation of the coordinates of multiple models prior to the MR calculation.[34,35] BEAST can read in multiple models and internally computes a statistically averaged search model, which then will be used in the subsequent MR calculations. This should prove particularly valuable as the number of P450 structures grows.

Some other MR programs are available in the crystallographic community. The older package MERLOT[36] has been outdated by faster and automated packages such as AMoRe. Other programs such as GLRF[37] are reported in the literature, but are not widely utilized. A newer version of MolRep[38] has been incorporated in the CCP4 suite, which includes some specific features, such as an electron density model, fitting two models, and multicopy search (www.ysbl.york.ac.uk/~alexei/molrep7.html). Which package to use for a P450 MR project may well be determined by the user's familiarity with a certain MR package or by the unique feature to a package that suits the nature of an MR calculation.

[32] R. Read, *Acta Cryst.* **D57,** 1373 (2001).
[33] M. Fujinaga and R. J. Read, *J. Appl. Cryst.* **20,** 517 (1987).
[34] D. J. Leahy, R. Axel, and W. A. Hendrickson, *Cell* **68,** 1145 (1992).
[35] U. Pieper, G. Kapadia, M. Mevarech, and O. Herzberg, *Structure* **6,** 75 (1998).
[36] P. M. D. Fitzgerald, "MERLOT, and Integrated Package of Computer Program for the Determination of Crystal Structures by Molecular Replacement." Rahway, New Jersey, 1988.
[37] L. Tong and M. G. Rossmann, *Methods Enzymol.* **276,** 594 (1997).
[38] A. Vagin and A. Teplyakov, *J. Appl. Cryst.* **30,** 1022 (1997).

P450 Sequence Divergence

The first step for a MR calculation is to choose a search model according to the sequence identity between structure-known and structure-unknown P450s. A careful, accurate sequence alignment is essential to the process and is beneficial to the subsequent MR calculations.

All P450s are categorized into a family and subfamily depending on whether their primary sequence identity is greater than 40 and 55%, respectively.[1] However, the sequence identity of eukaryotic P450s across the family or of prokaryotic P450s across species is generally very low (cpd.ibmh.msk.su or drnelson. utmem.edu/CytochromeP450.html). The sequence divergence of P450s often makes the automatic sequence alignment programs fail to give reliable results.[39] Therefore, visual inspection is necessary to correct mistakes. The C-terminal half of the P450 sequence shows the highest sequence homology, especially the signature sequence of the heme Cys ligand, FXXGX(H/R)XCXG. Computer programs can usually do a nice job aligning the C-terminal half (after helix I) of a P450 sequence. Making a good sequence alignment of the N-terminal half can be quite challenging. Table I lists some key residues, compiled from the known P450 structures and literature,[39,40] that can aid in identifying α helices, β strands, and turns in a P450. The sequence alignment prior to the C helix might still be ambiguous due to the extremely low sequence homology in the N-terminal segment among P450s.

Topology Conservation in Known P450 Crystal Structures

So far there are eight P450 crystal structures available in the PDB data bank. Among them, six are bacterial,[14–18, 20, 21] one is fungal,[19] and one is mammalian.[22] Despite the sequence diversity, the crystal structures of these P450s do share a very similar general fold consisting of almost the same set of secondary structure elements. Shown as an example in Fig. 1 is the schematic diagram of P450cam with all the secondary structure elements labeled. A P450 structure usually comprises about 12 α helices, labeled A to L,[14] and a few β sheets, numbered β-1 through β-5.[40]

The prosthetic heme group of a P450 is sandwiched by two major helices: the distal I helix and the proximal L helix. These two helices, along with a loop (β bulge) bearing the heme Cys ligand, are considered the signature core structure of a P450. Going from this core to the surface, P450 structures diverge. The most dramatic structural variations, which also coincide with the sequence divergence, are clustered in a few regions on the molecular surface. These include the N-terminal segment, the B$'$ helix and its flanking loops, F and G helices and the

[39] D. R. Nelson, in "Methods in Molecular Biology" (I. R. Phillips and E. A. Shephard, eds.), Vol. 107, p. 15. Humana Press, Totowa, NJ, 1998.
[40] C. A. Hasemann, R. G. Kurumbail, S. S. Boddupalli, J. A. Peterson, and J. Deisenhofer, *Structure* **3,** 41 (1995).

TABLE I
KEY RESIDUES THAT AID IN SEQUENCE ALIGNMENT OF CYTOCHROMES P450

Key residues[a]	Comments
G^{37}	Gly (or Pro) at the turn between helix A and β1-1
G^{46}	Gly at the turn between β1-1 and β1-2
$PYLS^{53}$	Corresponding to WXXT conserved in β1-2 in most bacterial P450
F^{67}	Phe, Leu, or Tyr in β1-5
HEK^{94}	Corresponding to DPP found in bacterial P450s prior to helix C
$WXXXH^{100}$	Trp (or His) and His (or Arg) in helix C that interact with heme propionate
E^{137}	More often a Gly right before β3-1
G^{157}	Gly at the C-terminal end of helix E
P^{172}	Often a Pro before the N-terminal end of helix F
R^{223}	Arg (or His) before the C-terminal end of helix G, making a salt bridge to an Asp at the N-terminal end of helix H
D^{232}	Asp at the N-terminal end of helix H salt bridged to an Arg (His) in helix G
$(A/G)GX(E/D)T^{268}$	Highly conserved active site residues forming a kink in the middle of helix I
P^{284}	Pro at the turn between helices I and J
$EXXR^{323}$	Highly conserved Glu and Arg in helix K forming an ERR-triad H bond network with an Arg (His) in the meander region
L^{333}	Often an Arg (or His) in β1-4 H bonded to heme propionate in bacterial P450
G^{343}	Gly at the turn between β2-1 and β2-2
G^{350}	Gly at the turn between β2-2 and β1-3
D^{363}	Highly conserved Asp at the C-terminal end of helix K' that is H bonded with its side chain to the backbone amide of the first residue in helix B
R^{378}	Arg or His in the meander region, part of the ERR triad
$FXXGX(R/H)XCXG^{402}$	The P450 signature sequence in the heme ligand Cys-binding loop, a conserved Arg (or His) two residues preceding Cys H bonds to heme propionate
E^{409}	Highly conserved Glu in helix L

[a] Residues in P450BM-3 with X denotes any residue.

loop in between, and the β-5 region. For instance, P450cam bears a long N-terminal segment, whereas CYP119 starts at the A helix without any N-terminal tail. The B' helix region is also known for its structural variations. It was nevertheless a surprise when the segment from the B helix to the C helix in CYP51 was found to adapt an unusual conformation, which opens another channel accessible to the heme active site.[21] The length and location of F and G helices vary substantially from one P450 to another. In P450BM-3, they participate in the open–closed motion of the substrate access channel on substrate binding.[18] In CYP119, the C-terminal end of the F helix can unwind to accommodate ligands with a different size and

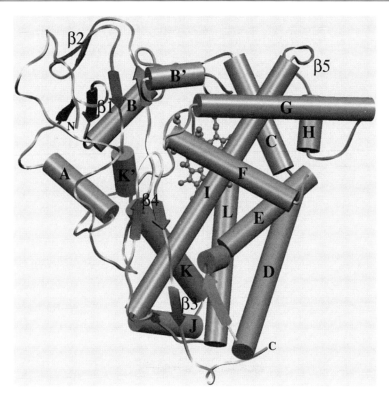

FIG. 1. A schematic drawing of the P450cam crystal structure. α helices, labeled as A through L, are depicted by cylinders and β sheets by arrows with labels $\beta 1$ to $\beta 5$. The heme prosthetic group is drawn as a ball-and-stick model.

shape in the heme active site.[20] The β-5 surface loop between helices H and I is often a poorly defined β hairpin, which is absent in CYP119 and P450nor. This type of information is important in preparing the search model for MR, as only those parts of the structure that are expected to be structurally similar are retained. Including regions that are structurally diverse introduces considerable noise into the Patterson function and hence makes the MR process more difficult. Therefore, once a sequence alignment between the model and target P450s is established, a decision must be made on which of these surface variable regions need to be trimmed from the search model.

With many P450 structures in the data bank, the structure-based sequence alignment becomes quite informative. Table II provides the pairwise sequence identity and the overall RMSD in α-carbon positions of all P450s with a known structure calculated after superimposition of each pair of structures. The information in Table II provides a qualitative measure on how closely two P450s resemble

TABLE II
SEQUENCE IDENTITY AND RMSD BETWEEN P450s WITH KNOWN STRUCTURE[a]

	BM3[b] (455)[c]	TERP (412)	CAM (406)	ERYF (403)	NOR (399)	CYP119 (367)	CYP51 (449)	2C5 (449)
BM3		16.5 (350)[d]	11.6 (315)	12.9 (337)	10.3 (339)	17.7 (307)	14.0 (353)	15.4 (381)
TERP	2.14		20.4 (359)	20.3 (354)	24.6 (355)	22.6 (333)	13.1 (302)	13.1 (338)
CAM	2.40	1.95		20.3 (361)	18.5 (339)	19.1 (319)	9.1 (294)	12.6 (309)
ERYF	2.34	1.81	2.33		26.8 (349)	27.2 (329)	14.4 (305)	13.6 (321)
NOR	2.14	1.86	2.03	1.87		23.4 (318)	13.8 (309)	13.5 (334)
CYP119	2.18	1.80	1.93	1.69	1.69		13.6 (278)	13.4 (307)
CYP51	2.19	2.23	2.21	2.21	2.29	2.26		12.9 (350)
2C5	2.22	2.33	2.42	2.41	2.24	2.33	2.29	

[a] The sequence identity (%) and the overall root mean-squared deviation (RMSD, Å) in Cα positions between each pair of P450s are calculated following a structure-based sequence alignment using program LSQMAN Version 8.0 (http://xray.bmc.uu.se/usf). The pairwise sequence identities (%) are listed in the upper right half of the table whereas RMSDs (Å) are listed the lower left half.

[b] Abbreviations and PDB entry codes for P450s: BM3, P450BM-3 or CYP102 (1BU7); TERP, P450terp or CYP106 (1CPT); CAM, P450cam or CYP101 (5CP4); ERYF, P450eryF or CYP107A 1(1OX8); NOR, P450nor or CYP55A1 (1ROM); CYP119 (1F4T); CYP51 (1E9X); 2C5, P4502C5, or CYP2C5 (1DT6).

[c] Total number of residues observed in the crystal structure.

[d] Number of residues matched within 5 Å in the least-square superimposition.

each other. Table II can be used as a guide when the MR search model(s) needs to be chosen for an unknown P450 structure.

Failures and Successes

Solving a new P450 crystal structure with the MR method has been a challenge. In the early days, attempts to solve the P450BM-3 structure using P450cam as a search model failed.[15] MR calculations on P450eryF using a P450cam search model provided only a correct molecular orientation, but the translation search failed to determine the location of the molecule in the unit cell.[17] Given the structure-based sequence identity and RMSD listed in Table II, it is understandable why P450cam/P450BM-3 as a model/unknown MR trial failed to give solutions. The sequence identity of this pair is below 11%, whereas the RMSD is well above 2.0 Å. In the P450cam/P450eryF case, the sequence identity is at 20.4%, close enough to find the rotation solution, but the RMSD is 2.33 Å mainly due to the drastic differences in the B' helix and the F/G loop regions between these two P450 structures.

The first success in solving a P450 structure utilizing the MR technique was determination of the CYP119 structure using P450eryF as a search model.[20] P450eryF was chosen as a model based on its high sequence identity with CYP119. The CYP119–imidazole complex crystallized in space group $P2_1$. Two independent molecules were expected in the asymmetric unit according to an estimation of the Matthew's coefficient.[41] When other available MR programs failed to give any solutions, an early trial version of BEAST was used, which provided a correct rotation solution for the first molecule. AMoRe was used to find a translation solution with this known orientation. Orientation of the second molecule was still difficult to determine and was eventually generated by applying the matrix of the self-rotation solution onto the orientation of the first molecule. The sequence identity between CYP119 and P450eryF is 27.2%, the highest among structure-known P450s. These two P450 structures show a low RMSD of 1.69 Å. However, this is still a marginal case for MR calculation. The improved signal-to-noise ratio using LLG as a scoring criterion in BEAST was a key to this success.

Our laboratory has solved another thermophilic P450 crystal structure by the MR technique using P450BM-3 as a search model (unpublished results). Again, the selection of the model among eight known P450 structures was based on the sequence identity between the model P450 and the unknown structure. The crystal of the new P450 with a 4-phenylimidazole bound belongs to space group $P2_1$ with two molecules per asymmetric unit. Both AMoRe and BEAST found the same rotation solution for the first molecule. The BEAST solution LLG score of 20.07 contrasted with 13.15 for the highest scoring incorrect solution [mean 7.50 and sigma (σ) 5.08]. The top rotation solution in AMoRe had a CC value of 30.7 against 30.0 for the first incorrect solution. In BEAST the rotation solution was optimized further using a fine rotation of $1°$ around the correct solution to produce a LLG score of 22.07 (a feature that is not present in AMoRe). AMoRe seemed to perform better than BEAST in finding the correct translation solutions, although the accurate orientation determined with BEAST made the translational search in AMoRe much easier, producing a higher CC score. The rigid body fitting in AMoRe proved to be a powerful tool for confirming the final answers. MR solutions were verified further by the fact that both the anomalous difference Fourier map calculated with the MR model phases and the anomalous difference Patterson map found the same heme iron positions. This presents the first MR case where both P450 molecules in the asymmetric unit were located by cross rotation and translation function searches.

---

[41] B. W. Matthews, *J. Mol. Biol.* **33,** 491 (1968).

A Case Study

CYP119 is the first P450 structure solved by the MR technique. Here we use another CYP119 crystal form, space group $C222_1$ with 4-phenylimidazole bound to the heme, as a test case to illustrate some technical points involved in a MR calculation. For comparison, calculations with both AMoRe and BEAST are performed at a resolution range of 20.0–5.0 Å. Unfortunately, although there are two molecules per asymmetric unit, neither program was able to find a solution for the second molecule. However, with two different strategies of choosing the search model, trimmed, or composite model, the search results of locating the first molecule demonstrate the strengths and weaknesses of each program.

Trimmed P450eryF Search Models

While we can expect the core P450 structure to remain relatively unchanged from one P450 to the next, surface loops and helices can exhibit substantial differences. These differences can lead to an increase in noise when comparing Patterson vector maps or structure factors. Therefore, crystallographers will often remove those regions from the known structure that are expected to deviate from the unknown. In our test case, the P450eryF structure first was converted to a polyalanine search model followed by the removal of surface regions in a stepwise manner. With each trimmed model the same rotation/translation function searches were carried out. The AMoRe outputs are tabulated in Table III. The rotation solution did not improve using a trimmed model. However, the translation solution was not located until the F/G loop (model 4) and β-5 loop (model 5) were trimmed. The contrast in the CC (and R factor) value was sharpest when all the surface variable regions were removed from the P450eryF search model (model 5). Given the fact that CYP119 does not bear an N-terminal tail like the P450eryF, it is puzzling why the full-length P450eryF polyalanine search model (model 1) gave a correct translation solution while the N-terminal trimmed model (model 2) failed to do so.

Composite Search Models

When more than one sequence-homologous structures are available, constructing a composite search model by superimposition of those structures has been demonstrated in the literature[34,35] to be a superior method when compared to a single search model. Tests were performed for CYP119 using the composite search model consisting of up to four P450 structures: P450eryF, P450cam, P450nor, and P450terp. The test results of MR searches with various composite models using both AMoRe and BEAST are presented in Table IV. The composite model, including all four P450 structures (Comp 1), gave the best solution with a sharp signal-to-noise ratio. Removing P450eryF from the composite model (Comp 2), a solution could still be found. However, a composite model with P450nor left out

TABLE III

AMoRe MR Search on Target Protein CYP119 Using P450eryF as a Model[a]

		Euler angles			Translation				
		$\theta 1$	$\theta 2$	$\theta 3$	x	y	z	CC	R
Model 1. P450eryF Polyalanine full-length with residues 2–404									
SOLUTIONTF1	1	131.44	64.63	307.81	0.1787	0.1789	0.0882	37.8	55.8
SOLUTIONTF1	1	16.02	70.11	47.23	0.1946	0.4832	0.4934	33.7	58.3
SOLUTIONTF1	1	144.54	90.00	244.17	0.0337	0.4494	0.4385	33.5	57.6
Model 2. P450eryF poly(A) with N terminus (2–15) trimmed off									
SOLUTIONTF1	1	131.97	65.92	308.50	0.4463	0.4332	0.0226	34.9	56.9
SOLUTIONTF1	1	67.15	75.16	17.58	0.4466	0.4346	0.1947	33.5	57.1
SOLUTIONTF1	1	148.71	60.44	135.71	0.2555	0.2466	0.2384	33.1	58.4
Model 3. P450eryF poly(A) with N terminus (2–15) and B' helix (56–69) trimmed off									
SOLUTIONTF1	1	131.70	65.53	308.52	0.3508	0.2718	0.1707	34.5	57.4
SOLUTIONTF1	1	143.50	90.00	246.92	0.0220	0.4568	0.4450	33.1	58.1
SOLUTIONTF1	1	35.88	90.00	69.00	0.4811	0.4566	0.0543	33.0	58.0
Model 4. P450eryF poly(A) with N terminus (2–15), B' helix (56–69), and F/G loop (219–227) trimmed off									
SOLUTIONTF1	1	131.58	65.20	308.31	0.1781	0.1911	0.0904	38.0	56.2
SOLUTIONTF1	1	148.44	61.31	137.36	0.2412	0.2526	0.2420	33.4	58.1
SOLUTIONTF1	1	109.73	90.00	267.73	0.1272	0.0284	0.1659	33.4	57.1
Model 5. P450eryF poly(A) with N terminus (2–15), B' helix (56–69), β-5 loop (176–184), and F/G loop (219–227) trimmed off									
SOLUTIONTF1	1	130.72	64.63	307.59	0.1803	0.1947	0.0918	39.8	55.1
SOLUTIONTF1	1	137.20	88.74	249.86	0.2183	0.2611	0.1640	33.6	57.6
SOLUTIONTF1	1	109.64	90.00	267.00	0.1299	0.0287	0.1683	33.3	57.5

[a] The first three solutions from each search are listed. Each solution consists of three Euler angles and three translations with CC and R factor as score. A correct MR solution was found as the top solution when models 1, 4, and 5 were used, respectively.

(Comp 3) failed to find a solution. In contrast, with the same composite model (Comp 3), BEAST was able to find the right translational solution, illustrating its higher discriminating power by using the LLG score versus CC used in AMoRe. Utilizing the P450cam model alone in both AMoRe and BEAST failed to give the right translations, although a molecular orientation was still roughly defined. The last two examples (Comp 3 and 4) in Table IV demonstrate that a composite model consisting of two not so good search models can outperform each model alone.

Molecular Replacement and Beyond

Determining a new P450 crystal structure by the MR technique has become more practical because of the increased size of the P450 structure data base. Much

TABLE IV

AMoRe and BEAST MR Search on Target Protein CYP119 Using Composite P450 Models[a]

			Euler angles			Translation				
		$\theta1$	$\theta2$	$\theta3$	x	y	z	CC	R

Comp 1. ERYF, CAM, NOR, and TERP composite model
AMoRe

SOLUTIONTF1	1	132.67	66.71	307.60	0.1759	0.1834	0.0932	33.5	56.9
SOLUTIONTF1	1	109.90	71.57	212.94	0.4064	0.9962	0.0089	30.8	57.1
SOLUTIONTF1	1	149.11	90.00	245.37	0.2051	0.4582	0.4702	30.2	57.5

BEAST
For 304405 trials, best LLG = 35.083 with mean = −19.748 and σ = 8.422
Euler angles: 132.558 66.730 305.730   Translation: 0.3229 0.1640 0.0860

Comp 2. CAM, NOR, and TERP composite model
AMoRe

SOLUTIONTF1	1	132.97	65.62	306.85	0.1749	0.1847	0.0935	32.0	57.0
SOLUTIONTF1	1	149.11	90.00	245.83	0.2014	0.4582	0.4704	28.7	57.8
SOLUTIONTF1	1	38.14	75.08	209.09	0.1001	0.0544	0.4298	28.7	57.3

BEAST
For 304405 trials, best LLG = 27.975 with mean = −14.377 and σ = 7.496
Euler angles: 132.750 66.730 308.649   Translation: 0.3294 0.1615 0.0816

Comp 3. CAM and TERP composite model
AMoRe

SOLUTIONTF1	1	12.72	61.89	121.25	0.2802	0.1050	0.2157	28.7	57.5
SOLUTIONTF1	1	53.50	80.01	185.77	0.0314	0.3890	0.2647	28.6	57.5
SOLUTIONTF1	1	133.50	65.82	306.39	0.2796	0.1253	0.2868	28.3	57.9

BEAST
For 304405 trials, best LLG = 28.596 with mean = −5.809 and σ = 6.540
Euler angles: 133.366 66.930 305.445   Translation: 0.3229 0.1640 0.0860

Comp 4. CAM model only
AMoRe

SOLUTIONTF1	1	130.65	65.45	308.41	0.4922	0.4608	0.4934	26.5	56.4
SOLUTIONTF1	1	158.24	53.50	101.00	0.0218	0.1200	0.3068	26.3	56.2
SOLUTIONTF1	1	134.36	59.71	302.18	0.1301	0.1171	0.1943	25.7	56.3

BEAST
For 304405 trials, best LLG = 20.492 with mean = −0.454 and σ = 4.820
Euler angles: 136.573 66.930 302.757   Translation: 0.3294 0.1565 0.4299

[a] The first three solutions from each search are listed. Each solution consists of three Euler angles and three translations with CC and R factor as score. Note that the translational components found in BEAST are offset by 0.5 in x from that in AMoRe.

more accurate structure-based sequence alignments are now feasible, which provides more checkpoints along the sequence vital to a reliable sequence analysis between the model and the target P450s. Choosing the best search model can be accomplished in two ways: (1) a single model according to the sequence identity between the model and the unknown P450 or (2) a composite model with all the homologous P450 structures superimposed. MR search programs have been advanced to include more sophisticated algorithms and scoring methods. To tackle the difficult MR cases such as P450s, one needs to explore the strength of more than one program package so that the correct solutions can be cross-checked and confirmed.

In addition to MR techniques, any strategy for solving a new P450 structure should also include the use of anomalous scattering by the heme iron. It now should be routinely possible to locate the heme iron atom by taking advantage of the fact that iron has an anomalous dispersion signal at its absorption edge, 1.74 Å. This wavelength is easily reached at one of the several synchrotron facilities. In fact, good anomalous signals can often be obtained using in-house Cu Kα X-ray sources at 1.54 Å. It is critically important to have as high a redundancy as possible such that each reflection is measured many times. As with any experiment, repeated measurements increase the signal-to-noise ratio. Because the anomalous signal from iron is relatively weak compared to other heavy atoms, high redundancy is important. This is now possible using a single crystal, as cryogenic techniques are routinely used. With good anomalous data in hand, it is quite straightforward to determine the iron position by interpreting an anomalous difference Patterson using programs such as SOLVE[42] or SHELXD.[43] This effectively solves the translation problem. As discussed, the rotation solution usually works easier than the translation search. Hence, once the rotation problem is solved, the oriented molecule can be translated such that the heme iron is placed at the position derived from the anomalous difference Patterson. The problem is more complex when there is more than one P450 molecule per asymmetric unit. The trick then is to match the rotation solution with one of the iron positions. This should be achievable by a few trial-and-error MR runs. In addition to locating the heme, the iron anomalous signal can also be used for obtaining experimental phase information. Therefore, a coupling of iron anomalous dispersion information with MR should circumvent the need for the use of heavy atom methods in future P450 structure determinations.

Acknowledgments

We thank David Schuller (now at CHESS, Cornell University) for introducing the maximum likelihood MR programs to our laboratory and Randy Read (University of Cambridge) for providing preliminary versions of BRUTE and BEAST. Work in our laboratory was supported by NIH Grant GM33688 (TLP).

[42] T. C. Terwilliger and J. Berendzen, *Acta Cryst.* **D55**, 849 (1999).
[43] G. M. Sheldrick and T. R. Schneider, *Methods Enzymol.* **277**, 319 (1997).

[10] Optical Biosensor and Scanning Probe Microscopy Studies of Cytochrome P450 Interactions with Redox Partners and Phospholipid Layers

By ALEXANDER I. ARCHAKOV and YURI D. IVANOV

Principles

Nanotechnological methods, such as the optical biosensor technique and scanning probe microscopy (SPM), enable the effective analysis of the structure–function properties of cytochromes P450 and their redox partners. The combination of the two techniques makes it possible to reveal protein–protein complexes, to measure the kinetic parameters of their formation and decay, and to visualize their spatial structures. In the optical resonant mirror biosensor technique, the nanolayer of protein or lipid under study is deposited onto the surface of the optical resonant structure (prisma-waveguide), which, at the same time, functions as the measuring cuvette bottom. Interaction of immobilized molecules with their respective partners on complex formation brings about changes in the refractive index that are monitored by the resonant angle shift of the probe laser beam propagating along the waveguide.[1] The SPM method enables the visualization of protein molecules and their complexes by monitoring (a) an interaction force between the probe tip of the microscope and the molecular surface as the probe tip is scanned across the support with immobilized molecules (atomic force microscopy, AFM) and (b) the tunneling current between the probe tip and the adsorbed molecule (scanning tunneling microscopy, STM).[2]

Optical Biosensor Technique

Apparatus. An IAsys optical biosensor (Affinity Sensor, Division of Labsystems) is used throughout.

Chemicals. 1-Ethyl-3-(3-dimethylaminopropyl) carbodiimide (EDC), *N*-hydroxysuccinimide (NHS), and ethanolamine (Affinity Sensors UK, division of Labsystems); Emulgen 913 (Kao Atlas, Osaka, Japan), and Tween 20 (Ferak, Berlin, Germany).

Lipids. Dilauroylphosphatidylethanolamine (DLPE) and distearoylphosphatidylethanolamine (DSPE) are saturated phospholipids differing in acyl chain length (Sigma, St. Louis, MO).

[1] D. Yeung, A. Gill, C. Maule, and R. Davies, *Trends Anal. Chem.* **14,** 49 (1995).
[2] H. G. Hansma and J. H. Hoh, *Annu. Rev. Biomol. Struct.* **23,** 115 (1994).

Proteins. Water-soluble proteins cytochrome P450cam (P450cam), putidaredoxin (Pd), and putidaredoxin reductase (PdR) from the P450-containing system (a gift from Dr. Hui Bon Hoa) are expressed in *Escherichia coli,* as described elsewhere.[3,4] Water-soluble proteins adrenodoxin (Ad) and adrenodoxin reductase (AdR) and the membrane protein cytochrome P450scc (P450scc) from the cytochrome P450scc system (a gift from Dr. S. A. Usanov) are purified from bovine adrenocortical mitochondria as described elsewhere.[5,6] The oligomeric full-length membrane proteins cytochrome P4502B4 (d-2B4), cytochrome b_5 (d-b5), and cytochrome P450 reductase (d-Fp) from the membrane-bound P4502B4 system are prepared from the microsomal fraction of phenobarbital-treated rabbit liver as described.[7,8] Purified tryptic fragments of t-b5 and t-Fp are obtained from rabbit liver microsomes following their treatment with trypsin.[9] Truncated 2B4 (t-2B4) lacking the NH_2-terminal residues 2–27 is expressed in *E. coli* as described previously.[10]

Problems

There are difficulties associated with biosensor monitoring of such a complex system as membrane-bound cytochrome P4502B4. To obviate these difficulties, the system is reconstituted from the monomeric proteins, d-2B4, d-Fp, and d-b5, by use of Emulgen 913 without phospholipid. For this purpose, to 30 μl of 500 mM potassium phosphate buffer (KP), containing 5 nmol of protein, is added 13 μl of 2% Emulgen 913 (w/v). After incubation for 10 min at room temperature, the concentrations of protein and Emulgen 913 are brought to 5–10 μM and 0.25 g/liter, respectively, by use of 500 mM KP, pH 7.4 (KP/E), and the mixture is incubated at 4° for 24 hr.[11]

Procedure for Immobilization of Protein Nanolayers in an Optical Biosensor

Immobilization of proteins on the carboxylated dextran surface of the optical cuvette of the biosensor is carried out at a constant temperature of 25°. If the temperature changes, the resonant angle position of the laser beam propagating

[3] I. C. Gunsalus and G. C. Wagner, *Methods Enzymol.* **52,** 166 (1978).

[4] C. Jung, G. Hui Bon Hoa, K.-L. Schroeder, M. Simon, and J. P. Doucet, *Biochemistry* **31,** 12855 (1992).

[5] S. A. Usanov, I. A. Pikuleva, V. L. Chashchin, and A. A. Akhrem, *Bioorg. Khim.* **10,** 32 (1984).

[6] S. A. Usanov, V. L. Chashchin, and A. A. Akhrem, *Biokhimiya* **54,** 472 (1989).

[7] I. I. Karuzina, V. G. Zgoda, G. P. Kuznetsova, N. F. Samenkova, and A. I. Archakov, *Free Radic. Biol. Med.* **26,** 620 (1999).

[8] L. Spatz and P. Strittmatter, *Proc. Natl. Acad. Sci. U.S.A.* **68,** 1042 (1971).

[9] T. Omura and S. Takesue, *J. Biochem.* **67,** 249 (1970).

[10] S. J. Pernecky, N. M. Olken, L. L. Bestervelt, and M. J. Coon, *Arch. Biochem. Biophys.* **318,** 446 (1995).

[11] Yu. D. Ivanov, I. P. Kanaeva, and A. I. Archakov, *Biochem. Biophys. Res. Commun.* **273,** 750 (2000).

along the waveguide of the biosensor is also changed (and so is the baseline) making calculation of the surface concentration of immobilized protein after the immobilization is completed impossible. Prior to immobilization, the cuvette of the biosensor containing 200 μl PBS/t buffer (20 mM sodium phosphate buffer, pH 7.4, 138 mM NaCl, 2.7 mM KCl, 0.05% Tween 20) is incubated at 25° for 30 min, which makes it possible to attain thermic equilibrium. Besides that, prolonged (30 min) incubation in detergent-containing buffer allows the dextran layer to swell. The cuvette is then washed twice with 200 μl of PBS/t (here and later the repeat washing of the cuvette with this volume of PBS/t is required), and the baseline of the optical biosensor (R_o) is recorded for 20 min; the baseline position is subsequently taken into account in calculating the surface concentration of immobilized protein. The chemistry of the method involves covalent coupling of protein amino groups with carboxy groups of the dextran layer on the surface of the cuvette by use of EDC/NHS.[12] For this purpose, PBS/t in the cuvette is changed to the solution containing the activating EDC/NHS mixture (0.4 M/0.1 M). Activation of the carboxymethylated dextran surface of the cuvette of the biosensor is achieved by its incubation in 200 μl of the activating mixture for 7 min. Then the cuvette is washed twice with 200 μl of PBS/t, and immobilization is carried out in 10 mM of immobilizing buffer (either acetate, at pH 4.0–6.0, or KP, at p.H 7.0) and 10 μl of protein solution (5–50 μM) in KP (50–500 mM, pH 7.4 were added to 180 μl of immobilizing buffer. The final pH values of the incubation mixture are 6.6, 5.0, and 5.3 for PdR, Pd, and P450cam; 7.2, 5.9, and 7.2 for AdR, Ad, and P450scc; and 7.0, 7.2, 6.1, 5.0, and 5.9 for d-, t-2B4, d-, t-Fp, and d-, t-b5. Binding of ligand to activated support is monitored by changes in the resonant angle position (i.e., by the R value increase). Uncoupled ligand is removed with (twice-repeated) washing of the cuvette with PBS/t. After ligand removal, the signal amplitude level of R is compared to that of R_0 to estimate prospects for possible protein immobilization. If the signal amplitude level corresponding to the resonant angle of the probe laser beam is higher than the baseline level, immobilization may occur. If R is lower than R_o, the pH of the immobilizing buffer should be decreased by 0.1–0.4 and the procedure repeated anew. The repeat procedure is required mainly in cases of Ad and b5 immobilization because these two proteins are most sensitive to the influence of immobilization conditions. Protein-uncoupled carboxy groups of the matrix are blocked with 1 M ethanolamine, pH 8.5. To attain this, the cuvette is washed twice with PBS/t, then (also twice) with 1 M ethanolamine, pH 8.5, and incubated in 200 μl of the latter for a period of 2 min. The cuvette is again washed twice with PSB/t, and the position of the final resonant angle R_f is registered. The surface concentration of the protein is calculated from the calibration curve of the device

$$I_{im} = (R_f - R_o)6.02 \times 10^{14}/(M \times 200) \text{ (in molecules/mm}^2)$$

[12] User Guide, "IASys." Human-Computer Interface Limited, Cambridge, England, 1993.

(where M is the molecular mass of protein and is estimated to be 10^{10}–10^{11} molecules/mm^2.

Analysis of Interactions of Cytochromes P450 with Their Redox Partners

The binding of proteins is attained as follows. To 180 μl of buffer a ligand is added in a large excess over immobilized ligate (ligand concentration ranging from 10^{-7} to 10^{-5} M, its volume being 5 to 20 μl). The association (k_{on}) and dissociation (k_{off}) rate constants are determined by fitting the experimental binding curves (based on the FASTfit program) to equations describing the reaction kinetics as[12]

$$R = R_0 + R_\infty\{1 - \exp[-(k_{on}C + k_{off})t)]\} \qquad (1)$$

where R, R_0, R_∞ are responses of the device at times t, 0, and ∞ (at the equilibrium state), respectively, and C is ligand concentration. To calculate k_{off} more exactly, the process of complex dissociation is presented as

$$R = R_d \exp(-k_{off}t) \qquad (2)$$

where R_d is the difference in R values at the initial time point of the dissociation process and after its completion. The complex lifetime is determined as $\tau = 1/k_{off}$. The equilibrium constant (K_{eq}) is calculated as $K_{eq} = k_{on}/k_{off}$.

Comments

1. The possibility of participation of the protein amino groups in complex formation is verified by monitoring the interactions of protein partners on their alternative immobilization through appropriate amino groups. The observation of complex formation parameters' dependence on the immobilized partner indicates that the amino groups of a given protein are involved in the complex formation process. If such dependence is lacking, the participation of amino groups in complex formation is insignificant. The role of hydrophobic membranous fragments in complex formation may be estimated from the comparative analysis of interactions of the full-length proteins and proteins devoid of their membranous fragments. The complex formation parameters for the full-length protein partners, d-2B4/d-Fp and d-2B4/d-b5, from the membrane-bound system appear to be independent of whichever protein is immobilized and hence are presented as one line in Table I. In all other cases, such dependence is clearly seen.

2. The optical biosensor method reveals ternary complexes as well. The following scheme for their identification is proposed. First, one of the partners (e.g., AdR in the case of the P450scc-containing system) is immobilized. Then two other

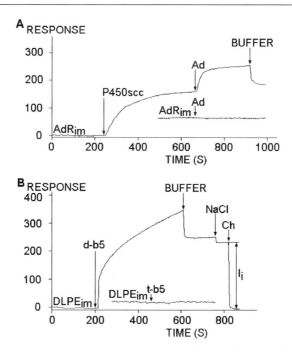

FIG. 1. Interaction of redox partners with one another (A) and with phospholipid layer DLPE (B). The incubation mixture contains (A) 175 μl of 50 mM KP buffer, pH 7.4; arrows indicate the addition of 20 μl of 50 μM P450scc and 5 μl of 250 μM Ad; T, 25° and (B) 180 μl of 500 mM KP/E buffer, pH 7.4; arrows indicate the addition of 20 μl of 0.25 μM d,t-b5; T, 36°. NaCl is 200 μl of 50 mM KP, pH 7.4, containing 1 M NaCl; Ch is 200 μl of 50 mM KP, pH 7.4, containing 20% sodium cholate and 1 M NaCl; I_i, phospholipid incorporated protein.

partners, P450scc and Ad, are added successively (Fig. 1A). Addition of the third partner, Ad, to the preliminarily formed binary complexes leads to formation of the ternary AdR_{im}/P450scc/Ad complex. AdR_{im} cannot bind Ad directly. The lifetime of a ternary complex is determined from its dissociation curve, the dissociation being brought about by the replacement of protein solution with appropriate buffer. The lifetime of the ternary complex thus formed (about 15 sec) is several times longer than the time required for realization of the cholesterol-to-pregnenolone conversion cycle[13] $(\tau_e) = 3$ sec and is sufficient to permit completion of up to five catalytic cycles (Table I). The P450cam and P4502B4 systems are also able to form ternary complexes, the lifetimes of which are sufficient for realization of 60 and 1 substrate hydroxylation cycles, respectively (Table I).

[13] I. Hanukoglu, V. Spitsberg, J. A. Bumpus, K. M. Dus, and C. R. Jefcoate, *J. Biol. Chem.* **256**, 4321 (1981).

TABLE I
KINETIC CONSTANTS OF REDOX PARTNERS COMPLEX FORMATION

Pairs	$k_{on} \times 10^5 M^{-1} s^{-1}$	$k_{off}(s^{-1})$	$\tau(s)$	$K_{eq} \times 10^6 M^{-1}$	$\tau_e(S)$	Turnover
d-2B4 + d-Fp[a]	4.5 ± 2.0	0.20 ± 0.1	5.0 ± 2.5	2.2 ± 1.0		
d-2B4$_{im}$ + t-Fp[a]	0					
t-Fp$_{im}$ + d-2B4[a]	5.0 ± 2.0	0.30 ± 0.1	3.3 ± 1.1	1.67 ± 0.8		
d-2B4 + d-b5[a]	12.0 ± 5.0	0.2 ± 0.1	5.0 ± 2.5	6.0 ± 4.0		
t-2B4$_{im}$ + d-b5[a]	0					
d-b5$_{im}$ + t-2B4[a]	24.0 ± 12.0	0.5 ± 0.1	2.0 ± 0.4	4.8 ± 2.6		
(d-2B4$_{im}$ + d-Fp) + d-b5[b]			7.0 ± 2.0		8.3 ± 1.4	1
(AdR$_{im}$ + P450scc) + Ad			15 ± 0.5[c]		3[d]	5[c]
(Pd$_{im}$ + PdR) + P450cam			1.8 ± 1.3[e]		0.03[f]	60[e]

[a] Yu. D. Ivanov, I. P. Kanaeva, V. Yu. Kuznetsov, M. Lehnerer, J. Schulze, P. Hlavica, and A. I. Archakov, *Arch. Biochem. Biophys.* **362**, 87 (1999). Conditions: 500 mM KP, pH 7.4.

[b] Yu. D. Ivanov, I. P. Kanaeva, and A. I. Archakov, *Biochem. Biophys. Res. Commun.* **273**, 750 (2000). Conditions: 50 mM Tris–HCl buffer, 200 mM KCl, 0.4 mM camphor, pH 7.4. T, 25°.

[c] Yu. D. Ivanov, S. A. Usanov, and A. I. Archakov, *Bioch. Mol. Biol. Int.* **47**, 327 (1999). Conditions: 50 mM KP, pH 7.4.

[d] I. Hanukoglu, V. Spitsberg, J. A. Bumpus, K. M. Dus, and C. R. Jefcoate, *J. Biol. Chem.* **256**, 4321 (1981).

[e] Yu. D. Ivanov, I. P. Kanaeva, I. I. Karuzina, A. I. Archakov, G. Hui Bon Hoa, and S. G. Sligar, *Arch. Biochem. Biophys.* (2001).

[f] E. J. Mueller, P. J. Loida, and S. G. Sligar, *in* "Cytochrome P450: Structure, Mechanism, and Biochemistry" (P. R. Ortiz de Montellano, ed.), p. 103. Plenum Press, New York, 1995.

Procedure for Immobilization of Phospholipid Nanolayers in an Optical Biosensor

Synthetic phospholipids DLPE and DSPE have an advantage over natural non-saturated ones of being unable to form radicals and hence possessing greater stability. Phospholipids are covalently immobilized on the aminosilane surface (AS) of the measuring cuvette by the use of succinic anhydride. Prior to immobilization, the AS cuvette is incubated for 10 min with 200 μl of 10 mM KP, pH 7.6, at 25°. The buffer is then replaced with 200 μl succinic anhydride solution (310 g/liter in a mixture of acetonitrile/10 mM KP, pH 7.6, 1 : 1), and the AS cuvette is incubated with this mixture for 1 hr, which leads to carboxylation of the cuvette surface. After washing the cuvette with 10 mM KP, pH 7.0, DLPE or DSPE can be covalently coupled through their respective amino groups to this carboxylated layer using a standard procedure. The carboxylated surface is activated with EDC/NHS as described earlier, which leads to the formation of succinimide on the surface of the cuvette. After 7 min activation, DLPE or DSPE (20 g/liter in dioxane, 200 μl by volume) can be immobilized for 10 min. In the course of the nucleophilic substitution reaction (with amino groups of phospholipid acting as nucleophil), phospholipid is covalently bound with the surface of the cuvette through the amide bond. The uncoupled ligand is removed with PBS/T. Unbound succinimide ester on the surface of the cuvette is blocked by the addition of 1 M ethanolamine, pH 8.5. The cuvette is then washed with 50 mM KP, pH 7.4, containing 1 M NaCl and 20% sodium cholate.

Problems

Phospholipid immobilization on the surface of the biosensor cuvette is verified based on two facts: (1) the observation of essential distinctions in the binding of membrane protein to the lipid-modified and unmodified AS cuvettes. In the former case, the protein would be more tightly bound to the immobilized lipid layer due to hydrophobic interactions between membranous protein fragments and the phospholipid layer. (2) The reproducibility of the binding curves for protein with immobilized lipid layer on repetitive addition of protein to the cuvette, regenerated with 50 mM KP, pH 7.4, containing 1 M NaCl and 20% sodium-cholate (20% Ch); in these conditions, the noncovalently adsorbed phospholipid must be desorbed, whereas the lipid covalently bound to the surface of the cuvette must be retained, ensuring stable reproducibility of protein-binding curves.

Analysis of Protein Incorporation into the Phospholipid Layer

Lipid–protein binding is followed by adding 10–40 μl of protein (2–5 μM) to 160–190 μl of buffer in the measuring cuvette. The surface concentration of bound protein (I, mol/cm^2) is calculated as $I = R(5 \times 10^{-10})/M$, where R is the response of the device and M is the molecular weight of the protein. The total amount of coupled protein (I_t) involves protein adsorbed to the surface (I_a) and protein incorporated into the phospholipid layer (I_i). Protein incorporation is stopped by replacing the protein solution with the appropriate buffer; this causes desorption of slightly adsorbed protein from the phospholipid surface (Fig. 1B). Removal (under the action of electrostatic forces) of tightly adsorbed protein molecules is attained by the addition of 50 mM KP (200 μl) containing 1 M NaCl, pH 7.4 (NaCl). Protein remaining on the surface is considered as incorporated, through hydrophobic interactions, into the phospholipid layer (here, I_i). This portion of protein could only be removed from the biochip surface by treatment with 200 μl of 0.3–20% Ch (the chosen concentration level being dependent on the protein type). The incorporation of protein is modeled as

$$dI_a/dt = k_{on}{}^a f C - k_{off} I_a \qquad (3)$$
$$dI_i/dt = k_{on}{}^i f C \qquad (4)$$

where $I_t = I_a + I_i$ is the concentration of protein binding to the phospholipid layer (in moles adsorbed protein/cm^2); $k_{on}{}^a$ and $k_{on}{}^i$ are the association rate constants for the adsorption and incorporation processes; k_{off} is the dissociation rate constant for protein–lipid complexes, and C is the concentration of protein added to the measuring cuvette; $f = S/S_0$; S_0 and S are the surfaces accessible for protein adsorption at the initial time point and at the time of measurement; f is the ratio of these surfaces. Constants k_{on}^a, k_{on}^i, k_{off} are calculated by solving the reverse kinetic problem for the registered binding curve $I_t(t)$.[14]

Comments

Data on the interactions of d-b5 and t-b5 with the DLPE surface in 500 KP/E are presented in Fig. 1B. It can be seen that d-b5 is able to incorporate into DLPE with the efficiency $k_{on}^i = 1.3 \times 10^{-4}$ cm sec^{-1}.[15] The lack of interaction of t-b5 with DLPE indicates that the hydrophobic membranous fragment of this protein has a crucial role in its incorporation.

Atomic Force Microscopy: A Technique for Visualization of Free Cytochromes P450 and Cytochrome P450 Complexes with Their Redox Partners

Apparatus. An atomic force (Nanoscope IIIa) microscope (AFM; Digital Instruments, Santa Barbara, CA)

Proteins. Monomers of 2B4 and Fp are prepared as described earlier.

Procedure for Preparation of Cytochromes P450 and Their Redox Partners for Measurements

Experiments are carried out using direct surface adsorption. As supports, hydrophobic high-oriented pyrolytic graphite (HOPG) and hydrophilic charged mica are used. As reported earlier, in air humidity of 45% and higher, the mica surface is covered with a water layer.[16] There is reason to suppose that the protein molecules under study remain hydrated. To present formation of protein multilayers on the support and ensure its adsorption as separate molecules, the solution of 2B4 or Fp monomers (5 μM) in 100 mM KP/E is diluted up to 0.2 μM with the same buffer. Protein samples are deposited on the HOPG or mica surface and left for 2 min. Then each sample is rinsed with distilled water and dried in an air flow. 2B4/Fp complexes are obtained by mixing the monomeric proteins in solution (the concentration, for either protein, is 5 μM) with 100 mM KP/E buffer, pH 7.4.

[14] I. S. Zaslonko, V. N. Smirnov, and A. M. Teresa, *Kinet. Catal.* **34,** 599 (1993).

[15] Yu. D. Ivanov, I. P. Kanaeva, O. V. Gnedenko, V. F. Pozdnev, A. M. Tereza, R. D. Schmid, and A. I. Archakov, *J. Mol. Recognit.* **14,** 1 (2001).

[16] R. Guckenberger, M. Heim, G. Cevc, H. F. Knapp, W. Wiegrabe, and A. Hillerbrand, *Science* **266,** 1538 (1994).

Then the mixture is incubated for 10 min, diluted 50 times in the same buffer, and immediately placed onto the support.

To determine the average heights and diameters of objects, no less then 50 objects of each type (monomers, oligomers, complexes) should be measured. Commercially available Nanoprobe cantilevers for the tapping mode (Digital Instruments, Santa Barbara, CA) are used. The length of the cantilevers is 125 μm, and the resonant frequency is 308–340 kHz.

Problems

There are difficulties associated with insufficient adhesion of samples onto the support during the direct adsorption of molecules. The monomeric proteins d-2B4 and d-Fp are adsorbed poorly to mica and adsorbed readily to HOPG. Aggregates of these proteins are adsorbed to HOPG as well. In studying Fp/2B4 complexes, it is better to use mica supports; the molecules of these complexes are virtually unable to be adsorbed on to the hydrophobic HOPG.

Analysis of AFM Images

Because AFM offers an ability to measure the exact height of observed rigid objects, while the lateral dimensions are broadened due to a tip-related effect,[2] the main criterion for distinguishing between monomeric and oligomeric protein forms is their height. The images of objects have bell-like profiles. The diameter of an object is determined as the width of its bell-shaped image measured at

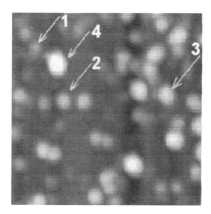

FIG. 2. AFM images of 2B4/Fp complexes on mica surface. Experimental conditions: 0.2 μM of each protein in 100 mM KP buffer, pH 7.4, containing 0.25 g/liter Emulgen 913. The image area is 0.54 × 0.54 μm. Arrows 1–4 indicate the images of 2B4 monomers, Fp monomers, binary complexes of 2B4/Fp monomers, and aggregates of higher orders, respectively.

half-height. For control experiments, the appropriate buffer mixture without proteins has to be applied to the support and then imaged. Randomly distributed contaminations in control measurements must not be higher than 1 nm. The measured dimensions of HOPG-adsorbed d-2B4 monomers are approximately 2.5–3 nm high and 15–18 nm wide and those of d-Fp monomers are 4–5 nm high and 20–22 nm wide. d-2B4 oligomers are visualized as globules with unresolved structure having a height of 10 nm and lateral dimensions of 50 nm. The average height and diameter of d-Fp oligomers are 7–8 and 40–60 nm, respectively. The images of d-2B4/d-Fp on mica are presented in Fig. 2. In the order of increasing height, the images are sized as follows: 2.5–3, 4–5, 6–8, 10, and those higher than 10 nm. The images of 10 nm and higher apparently correspond to highly aggregated proteins. A group of imaged objects with a height of about 6–8 nm is assigned to the binary d-2B4/d-Fp complexes, taking into account that (a) the height of these complexes is greater than the height of the individual monomeric proteins d-2B4 (2.5–3 nm) and d-Fp (4–5 nm) and (b) the height of the binary complexes is roughly equal to the sum of heights of the monomeric proteins, d-2B4 and d-Fp.

Thus the atomic force microscopy technique offers an ability to visualize both the individual cytochrome molecules and the cytochrome complexes with their redox partners.

[11] Cryoradiolysis for the Study of P450 Reaction Intermediates

By ILIA G. DENISOV, THOMAS M. MAKRIS, and STEPHEN G. SLIGAR

Introduction

A detailed understanding of chemical mechanism implies knowledge of the most important steps of the reaction, including the structure and main features of the intermediate species on the path from reagents to products. Information about the reaction intermediates may be obtained through a variety of kinetic methods. However, information from direct observation of evolution of the system with time is limited to a few (very often only one) most stable intermediates, which are accumulated during the reaction because of the presence of a sufficient activation barrier. Although other intermediate species on a reaction pathway may be present as a very minor fraction at any moment, their importance for the reaction mechanism is by no means diminished. The limited ability to isolate and study the properties of such transient species is essential for the relaxational methods;

the only possibility to prove the existence of such species is through indirect methods, e.g., measuring the kinetic isotope effect of the reaction. Such situations cannot be improved even with an increase in the time resolution of experimental techniques.

For direct studies, alternative approaches to stabilization of the transient intermediates are based on entrapment of the species of interest in order to increase their fraction by changing the experimental conditions. Cryogenic entrapment is generally used in matrix isolation chemistry for structural and spectroscopic studies.[1] A similar method has been adapted for biochemical needs in mechanistic studies of cytochrome P450cam[2-4] and heme oxygenase.[5] The method is based on preparing the precursor (oxyferrous heme complex) at ambient conditions, freezing it at 77 K, and irradiating it with high-energy photons (X-rays or gamma rays from a ^{60}Co source). Radiolysis of the solvent molecules (water and glycerol or ethylene glycol) results in the generation of high concentrations of free electrons, which perform the reduction of the oxyferrous heme complex 4 (Scheme 1). The reduced oxyferrous complex 5a of cytochrome P450cam, which is protonated easily (5b) and undergoes subsequent chemical transformations, could not be isolated and studied at ambient or moderately low temperatures, as several previous attempts have proven inconclusive.[6-8] At 77 K, however, it is stable for months and can be studied by electron paramagnetic resonance (EPR)[3,4,9] and UV–VIS spectroscope.[10] The additional advantage of cryogenic radiolytic reduction is the fact that all side products of nonspecific radiolysis of the solvent and protein are immobilized in the solvent matrix as well and do not react with the species of interest. Because of the absence of translational diffusion at the glass transition temperature, it is possible to anneal the active intermediate and follow its subsequent transformations, including conformational relaxation, protonation via the nearest water molecule, and even quantitative formation of the product through the suggested proton abstraction–oxygen rebound mechanism involving

[1] B. Meyer, "Low Temperature Spectroscopy." American Elsevier, New York, 1971.

[2] I. Schlichting, J. Berendzen, K. Chu, A. M. Stock, S. A. Maves, D. E. Benson, R. M. Sweet, D. Ringe, G. A. Petsko, and S. G. Sligar, *Science* **287**, 1615 (2000).

[3] R. Davydov, I. D. G. Macdonald, T. M. Makris, S. G. Sligar, and B. M. Hoffman, *J. Am. Chem. Soc.* **121**, 10654 (1999).

[4] R. Davydov, T. M. Makris, V. Kofman, D. E. Werst, S. G. Sligar, and B. M. Hoffman, *J. Am. Chem. Soc.* **123**, 1403 (2001).

[5] R. M. Davydov, T. Yoshida, M. Ikeda-Saito, and B. M. Hoffman, *J. Am. Chem. Soc.* **121**, 10656 (1999).

[6] P. Debey, E. J. Land, R. Santus, and J. Swallow, *Biochem. Biophys. Res. Commun.* **86**, 953 (1979).

[7] D. E. Benson, K. S. Suslick, and S. G. Sligar, *Biochemistry* **36**, 5104 (1997).

[8] K. Kobayashi, M. Amano, Y. Kanbara, and K. Hayashi, *J. Biol. Chem.* **262**, 5445 (1987).

[9] R. Davydov, R. Kappl, R. Hutterman, and J. Peterson, *FEBS Lett.* **295**, 113 (1991).

[10] I. G. Denisov, T. M. Makris, and S. G. Sligar, *J. Biol. Chem.* **276**, 11648 (2001).

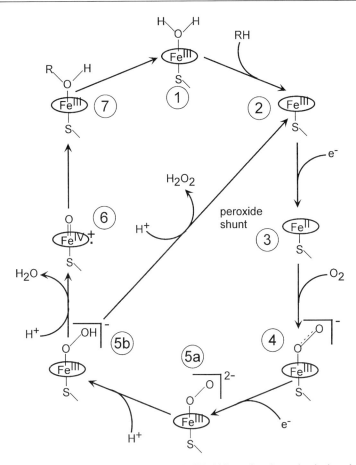

SCHEME 1. The mechanism of cytochrome P450 (CYP101)-catalyzed camphor hydroxylation.

ferryl-oxo compounds.[2–4] These processes proceed without any apparent indication of the side reactions with other nonspecific radicals generated in the sample by radiolysis.

Radiolysis

Extensive discussions of various aspects of radiation chemistry are available in the literature.[11–16] The fundamental bibliographic database (more than 4000 references) maintained by the Notre Dame Radiation Laboratory (Notre Dame University, IN) provides a much needed resource in searching the literature for this complex multidisciplinary field (http:\\allen.rad.nd.edu) (see also Schuler[17] for a description).

Radiolysis of the solvent water and glycerol provides electrons that reduce the oxyferrous heme complex and turn it into the active peroxo-heme or hydroperoxo-heme complex. At 77 K, well below the glass transition temperature of the solvent, the peroxo species are stable for months because of the lack of conformational relaxation and mobility. The same is true for the side products of radiolysis, which are also immobilized in the frozen solvent matrix and do not interfere with the target compound. Working at cryogenic temperature in the frozen solution incorporates the advantages of radiolysis and avoids its drawbacks, the most important advantage being the multiple side reactions with different radical products of radiolysis.

The majority of results on radiolysis of simple substances were obtained using gamma rays. Cobalt-60 sources provide photons with energies of 1.13 and 1.3 MeV that interact with elements having a low atomic number Z (all organic substances and water) mainly through Compton scattering with generation of a lower energy photon and the primary product of radiolysis, usually an electron and a cation radical. The scattered photon and electron generate more products of radiolysis. The final radiolytic yield is usually several particles (electrons and radicals) per 100 eV of absorbed energy (references), i.e., about 10^4 radiolytic electrons per one fully absorbed 1 MeV gamma photon. From the aforementioned discussion, it is clear that all products of radiolysis except electrons, and to a certain extent protons and hydrogen atoms, are frozen in the solvent matrix and exhibit minimal reactivity only with radiolytic electrons.[18] The latter reactions result in the saturation behavior of the radiolytic yield at a higher irradiation dose.[19,20] For gamma radiolysis of alcohols, the main products of radiolysis are hydrogen atoms, CO, aldehydes, alkanes, alkenes, ketones, and glycols, the amount of the products strongly dependent on the conditions and the history of the sample before analysis.[12]

[11] J. W. T. Spinks and R. J. Woods, "An Introduction to Radiation Chemistry." Wiley, New York, 1990.

[12] R. J. Woods and A. K. Pikaev, "Applied Radiation Chemistry: Radiation Processing." Wiley, New York, 1994.

[13] H. C. Box, "Radiation Effects. ESR and ENDOR Analysis." Academic Press, New York, 1977.

[14] E. N. Jenkins, "Radioactivity: A Science in Its Historical and Social Context" (N. Mott and G. R. Noakes, eds.), The Wykeham Science Series, Wykeham Publication, London, 1979.

[15] E. L. Alpen, "Radiation Biophysics." Academic Press, San Diego, 1998.

[16] J. Belloni, M. O. Delcourt, C. Houee-Levin, and M. Mostafavi, *Annu. Rep. Progr. Chem. Sec. C* **96**, 225 (2000).

[17] R. H. Schuler, *Radiat. Phys. Chem.* **47**, 9 (1996).

[18] M. C. R. Symons, *Progr. React. Kinet. Mechan.* **24**, 139 (1999).

[19] J. E. Willard, *in* "Fundamental Processes in Radiation Chemistry" (P. Ausloos, ed.), p. 599. Wiley, New York, 1968.

[20] R. W. Fessenden and R. H. Schuler, *in* "Advances in Radiation Chemistry" (M. Burton and J. L. Magee, eds.), p. 1. Wiley, New York, 1970.

Radiolytic reduction at cryogenic temperatures has been used in studies of heme proteins since the 1970s by several groups.[21–31] Traditionally, gamma irradiation from ^{60}Co sources has been used. By means of optical spectroscopy, magnetic circular dichroism, and electronic paramagnetic resonance spectroscopy, it has been shown that heme proteins are reduced at 77 K by a reaction with radiolytic electrons. Usually the reduced protein is frozen in a conformationally unrelaxed nonequilibrium state and undergoes relaxation only after annealing at temperatures close to the glass transition temperature of the solvent matrix. After annealing, the protein relaxes and returns to a conformational state observed at ambient conditions.

Using [^{32}P]phosphate as an internal source, the same result can be achieved, and the conversion of the oxy-ferrous complex to the hydroperoxo-heme complex can be observed using UV–VIS spectroscopy at low temperatures.[10] This method provides the tool for the study of unstable one-electron-reduced intermediates of heme enzymes without the need for expensive and exotic equipment. Transient radical intermediates, which do not accumulate in the course of reaction and hence cannot be directly detected even with fast kinetic methods, can also be isolated and characterized.

Sample Preparation

Method 1. Gamma Irradiation

Samples for cryoradiolysis and optical spectroscopy at low temperatures are prepared directly in methacrylate disposable cells [UV-enhanced semimicro cells from Fisher Scientific (Pittsburgh, PA), linear dimensions $10 \times 4.3 \times 25$ mm, total sample volume about 1 ml, used with the optical path 4.3 mm at 15–40 μM final protein concentration]. For preparation of the oxyferrous complex of

[21] R. M. Davydov, S. N. Magonov, A. M. Arutyunyan, and Yu. A. Sharonov, *Mol. Biol.* **12**, 1037 (1978).

[22] S. N. Magonov, R. M. Davydov, L. A. Blyumenfeld, R. O. Vilu, A. M. Arutyunyan, and Yu. A. Sharonov, *Mol. Biol.* **12**, 725 (1978).

[23] S. N. Magonov, R. M. Davydov, L. A. Blyumenfeld, R. O. Vilu, A. M. Arutyunyan, and Yu. A. Sharonov, *Mol. Biol.* **12**, 913 (1978).

[24] S. N. Magonov, R. M. Davydov, L. A. Blyumenfeld, A. M. Arutyunyan, and Yu. A. Sharonov, *Molek. Biol.* **12**, 919 (1978).

[25] R. M. Davydov, *Mol. Biol.* **14**, 211 (1980).

[26] R. M. Davydov, A. V. Karyakin, and E. Greschner, *Biofizika* **25**, 404 (1980).

[27] L. A. Blumenfeld and R. M. Davydov, *Biochim. Biophys. Acta* **549**, 255 (1979).

[28] M. C. R. Symons and R. L. Petersen, *Proc. R. Soc. Lond. B* **201**, 285 (1978).

[29] Z. Gasyna, *FEBS Lett.* **106**, 213 (1979).

[30] Z. Gasyna, *Biochim. Biophys. Acta* **577**, 207 (1979).

[31] R. Kappl, M. Hohn-Berlage, J. Huttermann, N. Bartlett, and M. C. R. Symons, *Biochim. Biophys. Acta* **827**, 327 (1985).

cytochrome P450, the deaerated solution of the ferric protein is transferred to the anaerobic chamber and reduced using solid dithionite. The unreacted dithionite and the products of dithionite decomposition are removed by gel filtration chromatography on a small (4 ml) sephadex G-25 column equilibrated with anaerobic phosphate buffer containing 0.6 mM camphor (substrate of interest). The resulting solution of the ferrous protein is concentrated if necessary and then mixed with the appropriate amount of the phosphate buffer and glycerol to give the ferrous cytochrome P450 dissolved in 75% glycerol, 0.1 M phosphate, 0.2 M potassium, and 0.6 mM camphor. The high glycerol concentration is necessary to maintain the optical transparency of the sample at any temperature between 77 and 300 K.[32] This solution in the methacrylate cells is sealed with Parafilm and, for anaerobic experiments, the cells are placed into a second container to ensure the anaerobicity. When removed from the anaerobic chamber, the sample is cooled in the refrigerator to −20° before opening the external container. These conditions are sufficient to maintain the enzyme in a deoxygenated state due to the high viscosity of 75% glycerol at subzero temperatures and the extremely slow diffusion of oxygen into the bulk of the enzyme solution. After the cell is opened, the enzyme is oxygenated by bubbling with pure oxygen gas with simultaneous stirring for 1–2 min.

Oxygenation is done at temperatures near 0° because of the relatively fast autoxidation of oxyferrous P450.[33] After oxygenation, the cell is fixed in the sample holder and placed into an optical cryostat. The cryostat consists of an unsilvered Pyrex glass dewar flask transparent at $\lambda > 300$ nm. The sample is mounted in the holder fixed on the top cover so it is always over the surface of liquid nitrogen; however, the bottom of the brass holder can be immersed into liquid nitrogen to reach the 77 K temperature. The sample is then cooled by adding liquid nitrogen into the dewar through a funnel. Direct contact of the coolant with the sample should be avoided to prevent cracking. If the dewar flask is precooled to 100 K, the cooling time is less than 10 min. Temperature is controlled by a calibrated thermocouple fixed in a thermal contact with the sample close to a measuring light beam position. UV–VIS absorption spectra of the sample are measured while cooling, and no autoxidation is detected at temperatures below 220 K.

The frozen solutions of oxyferrous cytochrome P450 are irradiated using gamma rays from a ^{60}Co source at the Notre Dame Radiation Laboratory (dose rate 1.2 megarad/hr) for 3–4 hr. During irradiation, samples are kept in the glassy silvered dewar flask filled with liquid nitrogen. It is necessary to add liquid nitrogen every 2 hr because of evaporation. After irradiation, the samples are transferred quickly into a larger liquid nitrogen storage dewar flask. For UV–VIS spectral measurements, irradiated samples are mounted on the holder while being fully

---

[32] Y. Shibata, A. Kurita, and T. Kushida, *Biochemistry* **38,** 1789 (1999).

[33] L. Eisenstein, P. Debey, and P. Douzou, *Biochem. Biophys. Res. Commun.* **77,** 1377 (1977).

immersed in liquid nitrogen and then moved to a cryostat at 77–80 K. Thus the samples are always kept at 77 K from the moment they are frozen to stabilize the active compounds generated as a result of radiolytic reduction.

Method 2. ^{32}P Irradiation

Despite the convenient and relatively fast radiolytic reduction using gamma irradiation from the ^{60}Co source, method 1 has certain limitations. The main limitations are the inability to monitor the gradual conversion of the oxyferrous complex to the product during irradiation and the need to find a source convenient to the laboratory. We have therefore developed an alternative method for irradiation using ^{32}P-enriched phosphate as an internal source of high-energy electrons. The nuclide ^{32}P has a half-life of 14.26 days and decays to ^{32}S by pure β emission. It emits an electron with a mean energy of 0.695 MeV and a maximum energy of 1.71 MeV. The aqueous solution of orthophosphoric acid with an activity of 50 mCi/ml can be purchased from Amersham Pharmacia (Piscataway, NJ). The activity of 1 Ci is by definition equivalent to 3.7×10^{10} disintegrations per second (Bq/sec). The dose rate for a sample containing 10 mCi/ml ^{32}P (mean energy of 0.7 MeV) is 2.3×10^8 MeV/sec, or 9×10^{-6} cal/sec per 1 ml. Because 1 rad $= 6.24 \times 10^{13}$ eV/g $= 10^{-5}$ J/g, there are 13 krad/hour at the initial stage of experiment. Integration over time gives the total dose:

$$D = R[(t_{1/2})/\ln 2]\{1 - 6.5[1 - \exp(-t \ln 2/t_{1/2})]\}$$
$$= 6.5[1 - \exp(-t/494)] \text{ megarad}$$

where t is expressed in hours of incubation with $[^{32}P]$phosphate and $t_{1/2}$ is 342 hr, the half-life of the ^{32}P isotope. However, the sample loses a fraction of this energy because some of the electrons originating at the surface layer of the sample escape. The mean distance of 0.7 MeV electrons from ^{32}P in water is 2.6 mm, and in 75% glycerol it is approximately 2.1 mm. Taking this into account, the corrected dose absorbed by the rectangular sample $4.3 \times 10 \times 25$ mm containing ^{32}P-enriched phosphate will be approximately 0.75 of a total emitted by the radioactive nuclei, with the rest being absorbed by the cell walls and liquid nitrogen surrounding the sample. The maximum distance of ^{32}P electrons in water is less than 1 cm, so being immersed into a liquid nitrogen dewar (stopping power of liquid nitrogen is approximately equal to that of water), the sample is perfectly shielded from the environment and will not cause any radiation damage to the operator. It is, however, necessary to maintain precautions while moving the sample between the storage dewar and the measuring dewar and while taking the spectra because the sample is not shielded by the layer of liquid during this time. In this case, 0.75 ml of 90% glycerol/phosphate buffer containing camphor (approximately 1 mM) is mixed with 0.25 ml of ^{32}P-enriched orthophosphoric acid (50 mCi/ml) at $4°$, and 60 μl of bovine catalase solution (final activity 1200 units) is added to consume hydrogen peroxide produced by the radiolysis of water during transportation and storage of the radioactive solution (note that the accumulation of hydrogen peroxide due to the autoradiolysis of radioactive isotopes must be taken into account while using samples with high radioactivity). After a 45-min incubation at 4°, concentrated, deoxygenated, reduced cytochrome P450 is added and stirred for 1 min. The final concentration of the enzyme is 25–35 μM, and it is completely oxygenated in the aerobic aqueous/glycerol solution. The sample is then transferred into a precooled dewar flask and cooled to temperatures below 200 K for 3–4 min. No sign of the ferric enzyme is detected, formed either by autoxidation or through the reaction with hydrogen peroxide, if the reaction with catalase is run for sufficient time. The concentration of catalase is approximately 25 times lower than that of P450 (on the heme molar basis), and spectra of this enzyme do not interfere with spectra of P450.

The dose absorbed by the sample with ^{32}P irradiation is measured at room temperature using a Fricke dosimeter.[34] We confirmed the general observations that the radiolysis of aqueous solutions and organic substances can be done using the β decay of ^{32}P, ^{35}S, ^{3}H, and other β-active isotopes, and the products of radiolysis are the same as for the gamma radiolysis from ^{60}Co.[35–38]

Products of Radiolysis

Absorption of high-energy (from 10 keV to 10 MeV) photons and electrons is essentially nonspecific with respect to the chemical structure of the media. The stopping power of the material is determined by the mean volume density of electrons and is similar for water, glycerol, liquid nitrogen, and proteins—the materials and compounds that contain only the low atomic number elements.[12] Thus, for dilute protein solutions, almost all gamma photons and radiolytic electrons will be absorbed by the frozen solvent matrix because the volume occupied by protein is small. For example, for a 30-μM solution of a protein with a molecular mass of 46 kDa, the weight concentration is 1.4 mg/ml. As the density of the protein is about 0.77 g/ml, the fraction of the total volume it occupies will be less than 0.2%, which means that approximately the same fraction of high-energy particles (0.2%) will encounter the protein molecules and cause primary radiolytic damage. In other words, the dose absorbed by protein will be 0.2% of the total dose. The main effect on the protein is produced by "free" (thermalyzed) electrons, which will reach

[34] R. H. Schuler and A. O. Allen, *J. Chem. Phys.* **24**, 56 (1956).
[35] T. J. Hardwick, *Can. J. Chem.* **30**, 39 (1952).
[36] M. Peisach and J. Steyn, *Nature* **187**, 58 (1960).
[37] I. V. Berezin and K. Martinek, *Russ. J. Phys. Chem.* **38**, 1468 (1964).
[38] O. Yamamoto, *Int. J. Radiat. Biol.* **42**, 661 (1982).

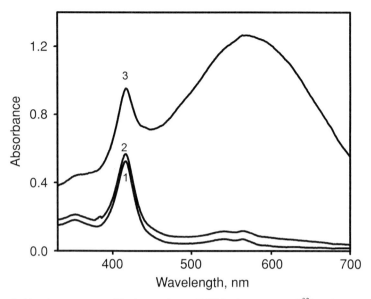

FIG. 1. Absorbance spectra of ferric cytochrome P450 in the presence of ^{32}P at the temperature 80 K. Spectrum 1: 10 min after mixing with ^{32}P-enriched phosphate; spectrum 2: 1 hr after mixing; and spectrum 3: 20 hr after mixing.

the electron-affinic groups and reduce them.[18] All other products of radiolysis are much more bulky and do not diffuse in the frozen solution below the glass transition temperature. The radiolytic electrons that escape geminate recombination with the trapped positive radicals become trapped themselves at the defects of the glassy structure, and thus generate color centers. As a result, we observe a very broad absorbance band with a maximum at 560 nm, a typical spectrum for an electron trapped in a glycerol matrix.[39] This absorption can be bleached by illumination with a visible light, which confirms the origin of this spectrum as a result of the accumulation of trapped electrons. With samples containing ^{32}P, it was possible to observe the gradual increase of this band with an initial rate at maximum of ~0.07/hr (Fig. 1). This rate is in agreement with the reported values of molar absorption of an electron solvated in water and in alcohols, and with the estimated radiolytic yield of solvated electrons at the given dose rate. After illuminating, the absorption band at 560 nm disappears. However, the multiple broad absorption bands due to radicals produced by the radiolysis of glycerol still contribute a suffi-cient background in the spectrum. Blank spectra of the 75% glycerol/buffer glass, which was irradiated at the same dose, were measured at temperatures between

[39] M. Kolodziejski and H. Abramczyk, *J. Mol. Struct.* **436–437,** 543 (1997).

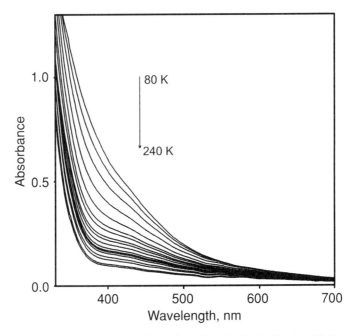

FIG. 2. Absorbance spectra of 75% glycerol/phosphate buffer irradiated at 77 K, total dose 5 megarad. Spectra were taken at different temperatures from 80 to 240 K at 4 K intervals. Not all spectra are shown. The arrow shows the decrease in absorbance during annealing.

80 and 240 K (Fig. 2). This spectral array was used for background subtraction; the absorption spectrum of the irradiated solvent at an arbitrary temperature was calculated by linear interpolation of the spectra shown in Fig. 2.

The irradiation of ferric cytochrome P450 with a 5 megarad dose results in a reduction of approximately 65% of the hemes (Fig. 3). The reduced ferrous P450 has very distinct Soret and visible spectra characteristic for the low-spin hexacoordinated ferrous heme proteins such as ferrocytochrome c, in agreement with previously published results.[40] It differs from the cytochrome P450 reduced in solution at ambient temperature, which shows high-spin spectra. The spectrum of the protein, which was reduced at room temperature and then cooled to 77 K, undergoes only minor changes; the protein remains in a high-spin state. Thus the radiolytic reduction of ferric cytochrome P450 at 77 K clearly results in formation of the nonequilibrium ferrous protein, which is stabilized in the nonequilibrium low-spin state. After annealing at higher temperatures, the protein undergoes structural

[40] S. Greschner, R. M. Davydov, G.-R. Janig, K. Ruckpaul, and L. A. Blumenfeld, *Acta Biol. Med. Germ.* **38**, 443 (1979).

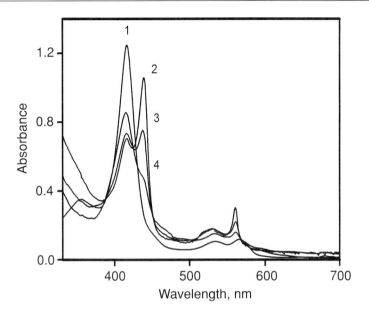

FIG. 3. Absorbance spectra of ferric cytochrome P450 before and after irradiation at 77 K. Spectrum 1: ferric cytochrome P450 before irradiation, 77 K; spectrum 2: same sample after irradiation at 5 megarad dose, 80 K; spectrum 3: same sample after annealing at 160 K; and spectrum 4: same sample after annealing at 180 K.

relaxation and returns to an equilibrium high-spin state (Fig. 3, spectra 2–4). The same relaxation processes were also observed by ultrafast optical methods in solution[41,42] that monitor the transition of the ferrous heme from low-spin to high-spin equilibrium conformation after the photolysis of ligated myoglobin and hemoglobin. At ambient temperatures, this conformational relaxation of the heme (iron out-of-plane movement and heme doming) is completed in several picoseconds. However, at 77 K in the aqueous/glycerol glass, conformational relaxation of the protein is frozen so that the protein is entrapped in the nonequilibrium low-spin state by the solvent matrix. The technique of cryogenic entrapment thus allows unprecedented detailed investigations of the elementary steps in heme protein reactions, e.g., spectroscopic observation of photodissociation of the water molecule from radiolytically reduced 12 K metmyoglobin.[43,44]

[41] S. Franzen, B. Bohn, C. Poyart, G. DePhillis, S. G. Boxer, and J.-L. Martin, *J. Biol. Chem.* **270,** 1718 (1995).
[42] S. Franzen, L. Kiger, C. Poyart, and J.-L. Martin, *Biophys. J.* **80,** 2372 (2001).
[43] D. C. Lamb, A. Osterman, V. E. Prusakov, and F. G. Parak, *Eur. Biophys. J.* **27,** 113 (1998).
[44] D. C. Lamb, J. Kriegl, K. Kastens, and G. U. Nienhaus, *J. Phys. Org. Chem.* **13,** 659 (2000).

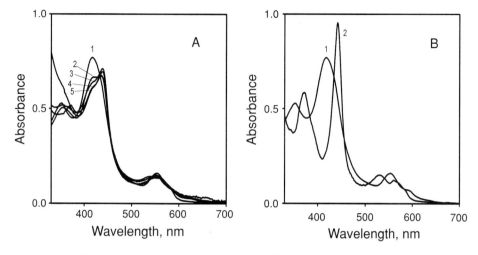

FIG. 4. (A) Absorbance spectra of oxycytochrome P450 at 77 K before irradiation (1) and after incubation with ^{32}P-enriched phosphate to a total dose of 0.9 megarad (2), 2.3 megarad (3), 3.9 megarad (4), and 5.1 megarad (5). (B) Absorbance spectra of oxycytochrome P450 (1) and pure reduced oxy-P450 (2) calculated from spectra in (A). Part of the figure is reproduced by permission from the American Society for Biochemistry and Molecular Biology [from I. G. Denisov, T. M. Makris, and S. G. Sligar, *J. Biol. Chem.* **276**, 11648 (2001)].

We have utilized the cryoreduction technique for studies of the active intermediates of P450-catalyzed camphor hydroxylation. Figure 4 shows spectra of the initial oxyferrous P450 measured in 75% glycerol/phosphate buffer in the presence of the substrate camphor and the gradual formation of the reduced oxy complex, the hydroperoxo-ferric heme complex.[10] The assignment of this transient compound was made by Davydov *et al.*[3,4,9] using EPR spectroscopy on the basis of earlier works on oxyhemoglobin and oxymyoglobin.[28,29,31] The calculated spectrum of the pure reduced oxycytochrome P450 intermediate averaged from several independent experiments is shown at Fig. 4B. The spectrum has the split Soret band at 441 and 370 nm, and two maxima in the visible region centered at 530 and 561 nm. This is the first reported UV–VIS spectrum of the hydroperoxo-ferric heme protein complex with a thiolate as a proximal ligand. There was a report on the kinetically resolved optical spectrum of the hydroperoxo-ferric complex, isolated as an intermediate in the reaction of the His-64 mutant of sperm whale myoglobin (Mb) with H_2O_2.[45] The assignment was confirmed with EPR spectra of freeze-quenched samples, which is in agreement with other EPR results of peroxo- and hydroperoxo-heme complexes obtained by cryoradiolytic reduction.[28,29,31] In the

---

[45] T. Brittain, A. R. Baker, C. S. Butler, R. H. Little, D. J. Lowe, C. Greenwood, and N. J. Watmough, *Biochem. J.* **326**, 109 (1997).

hydroperoxo-ferric Mb complex, the red-shifted Soret band and two maxima in the visible region were also observed. However, the proximal ligand in Mb is histidine, so direct comparison of these results is impossible. Harris *et al.*[46] calculated the absorbance spectrum for the thiolate-ligated hydroperoxo-heme model complex as a suggested intermediate in the cytochrome P450 reaction cycle. Our results are in a good agreement with their calculated spectrum, which shows the pronounced split Soret band and the 30-nm red shift. Similar results were also reported in pulse radiolysis experiments with oxy-P450 reconstituted with a deuteroheme,[8] although the authors suggested the formation of compound II.

As expected, UV–VIS spectra of the reduced oxy complex obtained by radiolysis with ^{32}P electrons and with ^{60}Co gamma photons are the same. Early studies comparing the radiochemistry of water and aqueous solutions by β particles from ^{3}H, ^{32}P, and ^{35}S with ^{60}Co gamma irradiation revealed the close similarity of the yield and products of radiolysis.[35–38]

The method of cryogenic radiolytic reduction of heme proteins, as well as non-heme enzymes containing metals,[47–50] provides new insights into the mechanisms of enzymatic catalysis due to information obtained directly from the studies of stabilized reactive intermediates. Radiolysis at 77 K may be accomplished using not only the traditional ^{60}Co gamma sources, but also ^{32}P-enriched buffer systems as an internal source. Other isotopes, such as ^{35}S, may also be used. The two irradiation methods have their own advantages, as well as drawbacks, so the combination of both may provide the investigator with different options.

[46] D. Harris, G. Loew, and L. Waskell, *J. Am. Chem. Soc.* **120,** 4308 (1998).

[47] M. Brugna, S. Rodgers, A. Schricker, G. Montoya, M. Kazmeier, W. Nitschke, and I. Sinning, *Proc. Nat. Acad. Sci. U.S.A.* **97,** 2069 (2000).

[48] R. Davydov, G. Kuprin, A. Graslund, and A. Ehrenberg, *J. Am. Chem. Soc.* **1994,** 11120 (1994).

[49] A. Karlsson, J. V. Parales, R. E. Parales, D. T. Gibson, H. Eklund, and S. Ramaswamy, *J. Inorg. Biochem.* **78,** 83 (2000).

[50] J. Telser, R. Davydov, Y.-C. Horng, S. W. Ragsdale, and B. M. Hoffman, *J. Am. Chem. Soc.* **123,** 5853 (2001).

[12] Analyzing Binding of N-Terminal Truncated, Microsomal Cytochrome P450s to Membranes

By Jose Cosme *and* Eric F. Johnson

Introduction

Identification of the structural features that underlie the association of P450 enzymes with biological membranes will contribute to understanding the role of the membrane in redox partner interactions and substrate binding. Modification of the surfaces of the enzyme that contribute to membrane binding and aggregation when dissociated from the membrane are also likely to contribute to the crystallization of P450s for structural characterization.[1,2] A hydrophobic region, which encompasses at least 20 consecutive residues near the N terminus, targets microsomal P450s for insertion into the endoplasmic reticulum and anchors the proteins to the membrane.[3] In general, expression of microsomal P450s lacking the N-terminal leader sequence in yeast or *Escherichia coli* results in significant association of the truncated protein with cellular membranes.[4,5] Anisotropic absorption decay studies following flash photolysis suggest that truncated and full-length P450 1A1 interact similarly with yeast membranes.[6] These and similar studies indicate that, in addition to the N-terminal leader sequence, other regions of microsomal P450s are likely to be involved in membrane association as is seen for mitochondrial P450s. Mutations that alter the interactions of mitochondrial or truncated microsomal P450s with phospholipid membranes will contribute to the identification of structural features that contribute to membrane binding and facilitate crystallization and structure determination.[1,2] This chapter describes two approaches for analysis of the effects of mutations on the binding of N-terminal-truncated microsomal P450s to phospholipid membranes.

Membrane Binding Determined by Subcellular Fractionation

When expressed in *E. coli,* truncated microsomal P450s are generally associated with membrane-containing subcellular fractions. Differential sedimentation of membrane fractions prepared in detergent- and salt-containing buffers provides

[1] J. Cosme and E. F. Johnson, *J. Biol. Chem.* **275,** 2545 (2000).

[2] P. A. Williams, J. Cosme, V. Sridhar, E. F. Johnson, and D. E. McRee, *Mol. Cell* **5,** 121 (2000).

[3] S. D. Black, *FASEB J.* **6,** 680 (1992).

[4] Y. Sagara, H. J. Barnes, and M. R. Waterman, *Arch. Biochem. Biophys.* **304,** 272 (1993).

[5] S. J. Pernecky, J. R. Larson, R. M. Philpot, and M. J. Coon, *Proc. Natl. Acad. Sci. U.S.A.* **90,** 2651 (1993).

[6] Y. Ohta, T. Sakaki, Y. Yabusaki, H. Ohkawa, and S. Kawato, *J. Biol. Chem.* **269,** 15597 (1994).

a means of estimating the degree of membrane association and the potential for disruption of this association by detergents or high salt buffers.

The following procedure is used to characterize the association of full-length and modified, truncated forms of microsomal P450 2C5.[1–7] cDNAs encoding the enzymes are cloned into the expression vector pCW ori+ and expressed in *E. coli.* Cultures are grown as described previously by von Wachenfeldt *et al.*[7] The cells are harvested by centrifugation at 5000g for 10 min, and the cells are suspended in 10% of the original culture volume of 20 mM KP$_i$, pH 7.4, containing 20% (v/v) glycerol, 1 mM phenylmethylsulfonyl fluoride (PMSF), and 10 mM 2-mercaptoethanol. The cell suspension is incubated with lysozyme (0.2 mg/ml) for 30 min, an equal volume of water is added, and the incubation is continued for an additional 10 min. The resulting spheroplasts are pelleted by centrifugation at 5000g at 4° for 10 min, suspended in 10–500 mM KP$_i$, pH 7.4, containing 20% glycerol, 10 mM 2-mercaptoethanol, and 1 mM PMSF and are then disrupted by sonication by three repetitive 45-sec bursts with cooling in an ice bath. Incompletely disrupted spheroplasts are discarded by centrifugation at 5000g at 4° for 10 min, and the resulting supernatant is used for membrane association assays.

The lysate is centrifuged at 123,000g at 4° for 90 min, and the supernatant is removed carefully from the pelleted membranes. The membranes are then suspended in 500 mM KP$_i$, pH 7.4, containing 20% glycerol and 1 mM PMSF. The amount of P450 present in the soluble and membrane fractions is determined spectrophotometrically from difference spectra determined for the formation of the carbon monoxide complex with the protein after reduction with dithionite following the method described by Omura and Sato.[8] Solubilization of full-length and truncated enzymes is relatively complete using 0.3% (v/v) Nonidet P-40 (NP-40), a nonionic detergent, and the yield of the enzyme in the soluble fraction under different buffer conditions can be expressed as a percentage of the yield obtained in the presence of detergent. In general, increasing the concentration of the KP$_i$ buffer increases the fraction of truncated P450 recovered in the soluble fraction. For P450 2C5dH, roughly 90% of the P450 is recovered in the soluble fraction using the 500 mM KP$_i$ buffer, whereas this amount is reduced to 35–40% in 50 mM KP$_i$.

In order to better characterize the distribution of the enzyme in membrane fractions, the lysates can be fractionated using a discontinuous sucrose gradient[9] consisting of 1 ml of 70% (w/v) sucrose, 4 ml of 30% sucrose, and 3.5 ml of 15% sucrose prepared in 10–500 mM KP$_i$, pH 7.4, containing 10 mM 2-mercaptoethanol and 1 mM PMSF. Aliquots of lysate (1.5 ml) that contain 5 to 10 nmol of P450 are layered at the top of the sucrose gradient and are centrifuged at 150,000g for 4 hr at 4° using a Beckman SW41 rotor. Fractions (1 ml) are collected from the

[7] C. von Wachenfeldt, T. H. Richardson, J. Cosme, and E. F. Johnson, *Arch. Biochem. Biophys.* **339,** 107 (1997).

[8] T. Omura and R. Sato, *J. Biol. Chem.* **239,** 2379 (1964).

[9] T. D. Porter and J. R. Larson, *Methods Enzymol.* **206,** 108 (1991).

bottom of the tube after puncturing the tube with a needle and are analyzed for the distribution of cytochrome P450, NADH oxidase activity, and phospholipids. This procedure provides a means of more clearly separating inner membranes from soluble proteins than the assay described earlier.

NADH oxidase activity is used to follow the distribution of inner membranes. The activity is determined for 20–100 μl of each fraction in a final volume of 1 ml of 50 mM KP$_i$, pH 7.4, containing 0.1 mM EDTA. The reaction is initiated by the addition of 100 μM NADH, and the rate of NADH depletion is monitored spectrophotometrically at 340 nm using an extinction coefficient of 6250 cm$^{-1}M^{-1}$.

Phospholipid concentrations can be determined by a direct colorimetric method described by Yoshida et al.[10] Fractions (0.1 ml) collected from the sucrose gradient are mixed with 1 volume of ethanol and are heated for 5 min at 60°. Then, 0.5 ml thiocyanatoiron reagent (0.97 g ferric nitrate and 15.2 g ammonium thiocyanate in 100 ml of water) and 0.3 ml 0.17 N HCl are added, and the mixture is incubated for 5 min at 35°. The thiocyanatoiron–phospholipid complex is extracted with 1.5 ml of 1,2-dichloroethane. Following vigorous mixing, the mixture is centrifuged for 2 min in a clinical centrifuge, and the absorbance of the lower layer is measured at 470 nm against a blank without phospholipids. The amount of phospholipids is determined using a calibration curve established with standard solutions of phospholipids.

P450 Binding to Phospholipid Vesicles

P450 binding to vesicles prepared from dilauroyl-L-α-phosphatidylcholine (DLPC) can also be used to characterize membrane binding.[1] The vesicles are prepared by suspending lyophilized DLPC in 50 mM KP$_i$, pH 7.4, 20% glycerol, 0.2 mM EDTA, and 0.1 mM dithiothreitol (DTT) at a concentration of 16 mM. The DLPC suspension is sonicated with a Branson sonifier (Bransonic 220) at 25° for 2 hr, and the clarified suspension is then extruded through a 0.1-mm Whatman (Clifton, NJ) filter. The filtrate can be used directly for the binding assay. DLPC vesicles (0–13 mM phospholipid) are mixed with 2 nmol of purified P450 in 10–500 mM KP$_i$, pH 7.4, 20% glycerol, 0.2 mM EDTA, and 0.1 mM DTT in a final volume of 0.6 ml and incubated for 10 min at 37° followed by an incubation at 4° for 12 hr. The phospholipid–P450 mixture is then centrifuged at 120,000g for 30 min at 4° in a Beckman TLA 100.3 rotor using a Model TL100 centrifuge (Beckman Instruments). Solutions omitting the phospholipid or the P450 that have been subjected to the same procedure are used as controls. The supernatant is isolated carefully, and the pellet, if present, is suspended in 500 mM KP$_i$, pH 7.4, 20% glycerol, 0.2 mM EDTA, and 0.1 mM DTT. This assay can be used to determine the salt dependence of binding and the optimum ratio of phospholipid to P450. Alternatively, a glycerol gradient formed by 2 ml of 20% glycerol, 3 ml of

---

[10] Y. Yoshida, E. Furuya, and K. Tagawa, *J. Biochem. (Tokyo)* **88,** 463 (1980).

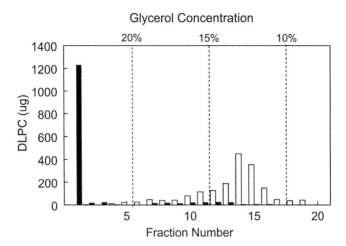

FIG. 1. Distribution of DLPC vesicles on a discontinuous glycerol gradient. In the presence of P450 2C5dH (filled bars), both the P450 and the lipid form a pellet at the bottom of the gradient. In the absence of P450 (open bars), vesicles remain at the top of the gradient. In the absence of DLPC, P450 is also found at the top of the gradient.

15% glycerol, and 3 ml of 10% glycerol in the same buffer can be used to provide a greater degree of separation. For this purpose, the glycerol content of the incubation buffer is reduced to 10%, and the amount of P450 is increased to 3 nmol. Before application to the gradient, a 0.5-ml sample is diluted with an equal volume of buffer without glycerol. Following centrifugation at 150,000g for 150 min at 4° in a Beckman SW41 rotor, 0.5-ml fractions are collected and analyzed for their content of P450 and phospholipid.

In the absence of the other component, the P450 or the DLPC vesicles remain at the top of the gradient as shown for the DLPC vesicles in Fig. 1. However, when the P450 binds to the phospholipid vesicle, the protein-containing lipid vesicles form a pellet after centrifugation. The ability of high salt buffers to dissociate P450 from the vesicles can be judged by repeating the centrifugation of the sample following suspension of the pellet in the high ionic strength buffer.

Sedimentation of the vesicles with or without incorporation of the protein is dependent on the fatty acyl chain length of the phospholipid. If longer chain length phospholipids are used, the vesicles will be found at the air–liquid interface following centrifugation. In order to incorporate a full-length, microsomal P450 into the vesicle, P450 should be incubated with the vesicle in the presence of a detergent such as cholate that can be removed slowly by dialysis prior to analysis.[11,12]

[11] M. Ingelman-Sundberg and I. Johansson, *Biochemistry* **19**, 4004 (1980).
[12] B. Bosterling, A. Stier, A. G. Hildebrandt, J. H. Dawson, and J. R. Trudell, *Mol. Pharmacol.* **16**, 332 (1979).

Conclusion

Detailed knowledge of the membrane association of cytochrome P450s is important for understanding conformational changes, dynamic interactions with redox partners, substrate accessibility, heme topology, and aggregation. The assays described here provide a basis for ascertaining the effects of protein modification on membrane binding and, by inference, the role of specific regions of the protein in membrane interactions. These assays have implicated the region between F and G helices of P450 2C5 in membrane interactions and in the aggregation of the protein when dissociated from membranes.[1]

Acknowledgment

This work is supported in part by USPHS Grant GM31001 (E.F.J.).

[13] Sensitizer-Linked Substrates and Ligands: Ruthenium Probes of Cytochrome P450 Structure and Mechanism

By IVAN J. DMOCHOWSKI, ALEXANDER R. DUNN, JONATHAN J. WILKER,
BRIAN R. CRANE, MICHAEL T. GREEN, JOHN H. DAWSON,
STEPHEN G. SLIGAR, JAY R. WINKLER, and HARRY B. GRAY

Introduction

Covalent attachment of the photosensitizer $[Ru(bpy)_3]^{2+}$ (bpy is 2,2'-bipyridine) to a substrate or ligand (Fig. 1) constitutes a powerful method for probing the steric and electronic properties of buried cytochrome P450 active sites. Researchers have exploited the versatile laser-triggered photoredox chemistry of $[Ru(bpy)_3]^{2+}$ for the rapid modulation of enzyme oxidation states.[1-7] Relative to conventional

[1] D. W. Low, J. R. Winkler, and H. B. Gray, *J. Am. Chem. Soc.* **118,** 117 (1996).

[2] M. J. Bjerrum, D. R. Casimiro, I.-J. Chang, A. J. DiBilio, H. B. Gray, M. G. Hill, R. Langen, G. A. Mines, L. K. Skov, J. R. Winkler, and D. S. Wuttke, *J. Bioenerget. Biomembr.* **27,** 295 (1995).

[3] J. Berglund, T. Pascher, J. R. Winkler, and H. B. Gray, *J. Am. Chem. Soc.* **119,** 2464 (1997).

[4] J. J. Wilker, I. J. Dmochowski, J. H. Dawson, J. R. Winkler, and H. B. Gray, *Angew. Chem. Int. Ed.* **38,** 90 (1999).

[5] I. F. Sevrioukova, C. E. Immoos, T. L. Poulos, and P. Farmer, *Israel J. Chem.* **40,** 47 (2000).

[6] I. Hamachi, S. Tanaka, S. Tsukiji, S. Shinkai, M. Shimizu, and T. Nagamune, *Chem. Commun.* 1735 (1997).

[7] I. Hamachi, S. Tanaka, S. Tsukiji, S. Shinkai, and S. Oishi, *Inorg. Chem.* **37,** 4380 (1998).

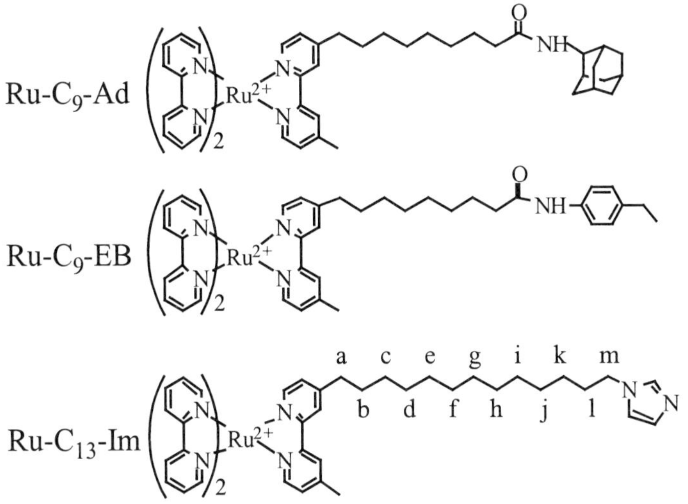

FIG. 1. Sensitizer-linked substrates (Ru-Ad, Ru-EB) and ligands (Ru-Im) for cytochromes P450. Methylenes in the Ru-C_{13}-Im alkyl chain are labeled (a–m) for purposes of NMR characterization (see Synthesis of Ru Probes).

stopped-flow methods, this photochemical approach affects oxidation and reduction over a wide range of time scales (nanosecond–millisecond) with varying thermodynamic driving forces. Ruthenium (Ru) probes can be synthesized in three to four steps from commercially available photosensitizers, linkers, and substrates; varying each component "tunes" specific binding and electronic interactions with P450.[4,8,9] Due to their roles in xenobiotic metabolism and steroid biosynthesis, P450s are important pharmaceutical targets. Active site-directed photosensitizers can bind both bacterial and human P450s with high affinity ($K_D \sim 1\ \mu M$).[10] Sensitizer-linked substrates target cytochromes P450 in a background of other heme enzymes, and their demonstrated biosensing and substrate screening capabilities may facilitate future drug design efforts.[8,9]

Unlike many structural probes of the P450 hydrophobic pocket,[11] sensitizer-linked substrates bind reversibly and may undergo modest turnover during biological catalysis. Resonance Raman spectroscopy interrogates the heme environment in P450 Fe^{2+}–CO Ru–substrate complexes, as was shown previously with other

[8] I. J. Dmochowski, B. R. Crane, J. J. Wilker, J. R. Winkler, and H. B. Gray, *Proc. Natl. Acad. Sci. U.S.A.* **96**, 12987 (1999).

[9] I. J. Dmochowski, J. R. Winkler, and H. B. Gray, *J. Inorg. Biochem.* **81**, 221 (2000).

[10] A. R. Dunn, I. J. Dmochowski, C. D. Stout, J. R. Winkler, and H. B. Gray, manuscript in preparation (2002).

[11] P. R. Ortiz de Montellano and M. A. Correia, *in* "Cytochrome P450: Structure, Mechanism, and Biochemistry" (P. R. Ortiz de Montellano, ed.), 2nd Ed., p. 305. Plenum Press, New York, 1995.

A) <u>Förster Energy Transfer:</u> $[Ru(bpy)_3]^{2+*}:Fe^{3+}Por \longrightarrow [Ru(bpy)_3]^{2+}:Fe^{3+}Por^*$

B) <u>Flash-Quench ET:</u> $[Ru(bpy)_3]^{2+*}:Fe^{3+} \overset{Q \quad Q^+}{\longrightarrow} [Ru(bpy)_3]^{+}:Fe^{3+} \longrightarrow [Ru(bpy)_3]^{2+}:Fe^{2+}$

C) <u>Photoinduced ET:</u> $[Ru(bpy)_3]^{2+*}:Fe^{3+} \longrightarrow [Ru(bpy)_3]^{3+}:Fe^{2+}$

FIG. 2. Interaction of Ru^{2+*} with the P450 heme. The quencher (Q) is p-methoxydimethylaniline.

P450 substrates.[12–15] X-ray structure determination of cytochrome P450cam cocrystallized with Ru-C_9-Ad identified an "open conformation" by which loop and helix movements accommodate substrates in the channel.[8,16]

Photoexcitation of Ru–substrate:P450 complexes, either in steady-state or in time-resolved experiments, generates the excited state, Ru^{2+*}, which can interact with the heme by a variety of energy transfer and electron transfer (ET) processes (Fig. 2). The fraction of Ru^{2+*} luminescence quenched by the enzyme shows binding constants for this class of molecules that agree with traditional measurements of spin shifts at the ferric heme. Furthermore, energy transfer kinetics from Ru^{2+*} to the heme provide a "molecular ruler" for measuring the depth of P450 channels. Indeed, Ru–Fe distances calculated by Förster analysis[17] are entirely consistent with crystallographic data for two different Ru–adamantyl:P450 complexes.[8,16] Thus, luminescence data provide detailed structural information not easily obtained by other spectroscopic methods.

Finally, Ru–substrates complement existing mechanistic probes by providing an efficient ET pathway to the P450 heme through the hydrocarbon linker. Due to rate-limiting ET ($k \approx 50 \text{ sec}^{-1}$) in the natural enzymatic system (e.g., via NADH, putidaredoxin reductase, and putidaredoxin), reactive Fe-peroxy and ferryl intermediates in P450 catalytic cycles are not readily observable.[18] These reductant cofactors are also economically infeasible for industrial biosynthetic applications

[12] A. V. Wells, P. Li, and P. M. Champion, *Biochemistry* **31**, 4384 (1992).

[13] C. Jung, G. H. B. Hoa, K.-L. Schroeder, M. Simon, and J. P. Doucet, *Biochemistry* **31**, 12855 (1992).

[14] O. Bangcharoenpaurpong, P. M. Champion, S. A. Martinis, and S. G. Sligar, *J. Chem. Phys.* **87**, 4273 (1987).

[15] T. Uno, Y. Nishimura, R. Makino, T. Iizuka, and Y. I. Tsuboi, *J. Biol. Chem.* **260**, 2023 (1985).

[16] A. R. Dunn, I. J. Dmochowski, A. M. Bilwes, H. B. Gray, and B. R. Crane, *Proc. Natl. Acad. Sci. U.S.A.* **98**, 12420 (2001).

[17] T. Förster, *Discussions Faraday Soc.* **27**, 7 (1959).

[18] M. J. Hintz, D. M. Mock, L. L. Peterson, K. Tuttle, and J. A. Peterson, *J. Biol. Chem.* **257**, 14324 (1982).

and remain unidentified for certain P450s.[19,20] The ongoing search for electrochemical and photochemical alternatives[4,5,21–23] has led us to develop conjugated Ru probes that mediate photoinduced ET on a nanosecond time scale (Fig. 2C).[10] This article describes methodologies for employing sensitizer-linked substrates and ligands as probes of P450 structure and mechanism.

Methods

Enzymes, Reagents, and Ruthenium Probes

P450, putidaredoxin reductase (PdR), and putidaredoxin (Pd) are expressed and purified with minor modifications to literature procedures,[18,24–26] stored at $-70°$, and thawed just before use. P450 as described here is camphor-free, unless stated otherwise. Camphor is removed by eluting the enzyme at $4°$ on a desalting column (PD-10, 50 mM Tris–HCl, pH 7.4), followed by a second PD-10 column equilibrated with buffer A (50 mM potassium phosphate, 100 mM potassium chloride, pH 7.4). NADH (Sigma, St. Louis, MO), substrate amines (Aldrich, Milwaukee, WI), bromoalkyl linkers (Aldrich), and cis-[Ru(bpy)$_2$Cl$_2$] (Strem, Newburyport, MA) are purchased and used without further purification. The reductive quencher, para-methoxy-N,N-dimethylaniline (p-MDMA), is synthesized following established procedures.[27,28]

[Ru-C$_9$-Ad]Cl$_2$ (Fig. 1) is synthesized as described previously,[9] where "C$_n$" indicates the number of methylene units separating 2,2′-bipyridine (bpy) from adamantylamine (Ad). Ru-C$_9$-EB (EB refers to ethyl benzene, Fig. 1) is synthesized analogously by linking 8-bromooctanoic acid to 4-ethylaniline via an amide bond.[28] Imidazole-terminated Ru probes (Ru-Im), through their covalent linkage to the heme, are better coupled electronically and generally bind more selectively than their substrate analogs.[8] We have described the synthesis of Ru-C$_n$-Ad/EB/Im with various alkyl linkers ($n = 7$–13).[28] The synthesis and binding of Ru-C$_{13}$-Im (Fig. 1) are described herein.

[19] M. A. McLean, S. A. Maves, K. E. Weiss, S. Krepich, and S. G. Sligar, *Biochem. Biophys. Res. Commun.* **252**, 166 (1998).

[20] J. K. Yano, L. S. Koo, D. J. Schuller, H. Y. Li, P. R. Ortiz de Montellano, and T. L. Poulos, *J. Biol. Chem.* **275**, 31086 (2000).

[21] U. Schwaneberg, D. Appel, J. Schmitt, and R. D. Schmid, *J. Biotechn.* **84**, 249 (2000).

[22] K. K. Lo, L.-L. Wong, and H. A. O. Hill, *FEBS Lett.* **451**, 342 (1999).

[23] N. Sugihara, Y. Ogoma, K. Abe, Y. Kondo, and T. Akaike, *Polymers Adv. Tech.* **9**, 307 (1998).

[24] C. A. Tyson, J. D. Lipscomb, and I. C. Gunsalus, *J. Biol. Chem.* **247**, 5777 (1972).

[25] P. W. Roome, J. C. Philley, and J. A. Peterson, *J. Biol. Chem.* **258**, 2593 (1983).

[26] J. A. Peterson, M. C. Lorence, and B. Amarneh, *J. Biol. Chem.* **265**, 6066 (1989).

[27] N. Sekiya, M. Tomie, and N. J. Leonard, *J. Org. Chem.* **33**, 318 (1968).

[28] I. J. Dmochowski, "Probing Cytochrome P450 with Sensitizer-Linked Substrates." California Institute of Technology, Pasadena, 2000.

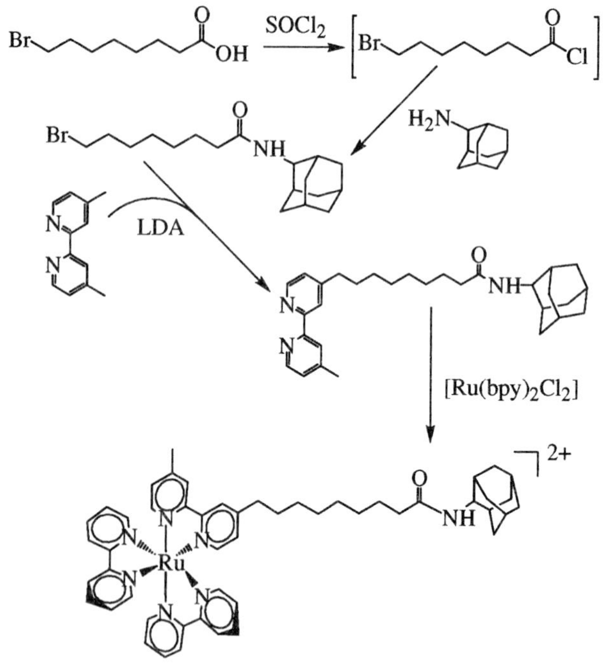

FIG. 3. Synthetic scheme for a prototypal sensitizer-linked substrate, [Ru-C$_9$-Ad]Cl$_2$.

Synthesis of Ru Probes

A general procedure for generating sensitizer-linked substrates is outlined in Fig. 3. This protocol may be generalized to include any combination of photosensitizer, linker, and substrate/ligand that targets the desired P450 active site. We have synthesized numerous compounds: Ru and Os photosensitizers with substituted bipyridine, terpyridine, phenanthroline, and imidazole ligands; alkyl, polyethylene glycol, perfluorobiphenyl, and polyxylyl linkers; attached via amide, ether, or amino bonds to substrates (adamantane, ethyl benzene, borneol, norbornane, thioanisole styrene) or imidazole ligands. All reactions are performed under an inert atmosphere with dried and freshly distilled solvents. Nonfluorescing silica thin-layer chromatography (TLC) plates are used for monitoring bpy ligand syntheses, as bpy coordination of the metal in fluorescing plates renders the compounds stationary. TLC plates are imaged with an aqueous ferric salt solution, which turns bpy-containing spots red.

Synthesis of [Ru-C$_{13}$-Im]Cl$_2$

4-(13-Bromotridecyl)-4′-methyl-[2,2′]bipyridinyl(bpy-C$_{13}$-Br). Diisopropylamine (0.770 g, 7.61 mmol), *n*-butyllithium (7.60 mmol in hexanes), and tetrahydrofuran (THF, 10 ml) are combined in a Schlenk flask chilled over an ice bath. A cold solution of 4,4'-dimethyl-2,2'-bipyridine (1.40 g, 7.60 mmol) in 40 ml of THF is added by cannula over 15 min. To this solution is added 1,12-dibromododecane (25.0 g, 76.2 mmol) in THF (20 ml). The reaction is stirred on ice for 3 hr and is then allowed to warm to room temperature for further stirring overnight. The solution is transferred to a separatory funnel to which water (15 ml) and ether (15 ml) are added. The organic layer is washed with saturated $NaHCO_3$, dried with $MgSO_4$, and evaporated to a beige solid under vacuum. Silica gel column chromatography with $CHCl_3$ as the eluent yields 1.31 g (40.0%) of a white solid. ^{1}H NMR ($CDCl_3$): δ ~1.7 (m, CH_{2c-k}), 1.75 (p, CH_{2b}), 1.85 (p, CH_{2l}), 2.55 (s, bpy-CH_3), 2.76 (t, bpy-CH_2), 3.48 (t, CH_2-Br), 7.26 (d, bpy 5 and 5'), 8.40 (d, bpy 3 and 3'), 8.65 (t, bpy 6 and 6').

4-(13-Imidazol-1-yl-tridecyl)-4'-methyl-[2,2']bipyridinyl (bpy-C_{13}-Im). Imidazole (1.0 g, 15 mmol) and bpy-C_{13}-Br (0.30 g, 0.70 mmol) are combined in a flask with THF (50 ml) and refluxed for 4 days. The solvent is removed under vacuum, and the resulting solid is dissolved in $CHCl_3$ for washing by saturated $NaHCO_3$, water, and NaCl. The product is purified by silica gel column chromatography using ethyl acetate as the eluent to yield 0.26 g (90%) of a white solid. ^{1}H NMR ($CDCl_3$): δ ~1.3 (m, CH_{2c-k}), 1.71 (p, CH_{2b}), 1.76 (p, CH_{2l}), 2.42 (s, bpy-CH_3), 2.73 (t, bpy-CH_2), 3.95 (t, CH_{2m}), 6.90 (s, imid H-5), 7.10 (s, imid H-4), 7.19 (d, bpy 5 and 5'), 7.68 (s, imid H-2), 8.24 (s, bpy 3 and 3'), 8.60 (t, bpy 6 and 6').

[Ru(bpy)$_2$(bpy-C_{13}-Im)]Cl$_2$. The ligand bpy-C_{13}-Im (460 mg, 1.10 mmol) and *cis*-[Ru(bpy)$_2$Cl$_2$] (538.6 mg, 1.04 mmol) are combined with 5 : 1 (v/v) water/ethanol (18 ml) and refluxed for 12 hr. The solvent is removed under vacuum, and the dark red solid is dissolved in water (60 ml). This aqueous solution is combined with a solution of NH_4PF_6 (1.20 g, 7.36 mmol) in water (20 ml) to yield an orange precipitate. The aqueous slurry is extracted with CH_2Cl_2 (75 ml); the organic layer is washed with 0.01 M HCl (2 × 50 ml), 0.01 M NaOH (2 × 50 ml), and water (2 × 75 ml) prior to rotary evaporation. The PF_6^- salt of this ruthenium complex is purified by silica gel flash chromatography employing an eluent of 3% (v/v) methanol in CH_2Cl_2. Pure fractions are combined and dried by rotary evaporation. In order to metathesize the ruthenium complex to the Cl^- salt, the purified PF_6^- salt is dissolved in methanol (10 ml) and loaded onto a CM Sepharose cation-exchange column (2 × 13 cm). The column is washed with water (600 ml), and the ruthenium complex is then eluted with 500 mM NaCl (300 ml) and dried by vacuum. The desired [Ru(bpy)$_2$(bpy-C_{13}-Im)]Cl$_2$ is isolated from the NaCl-containing solid by repeated extractions with CH_2Cl_2, followed by filtering and drying under vacuum. Yield of the dark red solid is 188 mg (20.0%). ^{1}H NMR (CD_2Cl_2): δ ~1.3 (m, CH_{2c-k}), 1.72 ($CH_{2b,l}$), 2.60 (s, bpy'-CH_3), 2.85 (t, bpy'-CH_2), 3.99 (t, CH_{2m}), 7.04 (m), 7.26 (m), 7.48 (m), 7.71 (m), 8.13 (m), 8.63 (s), 8.70 (s), 9.07 (m). LRMS (electrospray, positive ion) calcd for $C_{47}H_{54}N_8Ru$ (M + H$^+$) m/z 833, found 833.

Transient Absorbance and Time-Resolved Luminescence Spectroscopic Methods

Solution experiments are performed in sealed cuvettes with equimolar P450 (camphor-free) and Ru-probe (5–10 μM) in buffer A. Flash-quench experiments also include the reductive quencher p-MDMA at concentrations (5–10 mM) sufficient for efficient Ru^{2+*} quenching. Cuvettes are fitted with a magnetic stir bar and are deoxygenated by repeated evacuations on a vacuum line, followed by backfilling with purified argon (three rounds of 10 cycles).

All samples are excited with a tunable (220–2000 nm, excitation at 480 nm) optical parametric oscillator (Spectra Physics, Mountain View, CA, MOPO) pumped by a frequency-tripled Q-switched Nd:YAG laser (Spectra Physics, 355 nm, 350 mJ/pulse, 8 ns FWHM). The YAG fires continuously at 10 Hz; thus, for longer time base experiments (>50 ms), software was written to select pulses with a shutter. The laser output is attenuated with a polarizer to give a 1- to 2-mJ/pulse at the sample. Laser shots with energies differing by more than 10% from the mean value (laser pulses detected by a photodiode and selected by a discriminator, Phillips Scientific Model 6930) are rejected.

Kinetics data are collected as averages of at least 250 laser shots.[28] Transient absorption traces are typically fit to mono- or biexponential functions ($y = c_0 + c_1 e^{-(k_{en} + k_0)t} + c_2 e^{-k_0 t}$) using a least-squares fitting routine. Luminescence decay data are fit similarly, but must be deconvolved from the instrument response function (~10 ns FWHM) in cases where Ru^{2+*} luminescence is highly quenched by the protein (τ <50 ns).

Resonance Raman Spectroscopy of P450 Fe^{2+}–CO Substrate Complexes

Samples (200 μl, 100 μM P450 in buffer A, substrate concentration, 1 mM) are prepared in quartz NMR tubes sealed with a rubber septum. The solutions are bubbled gently with carbon monoxide for several minutes before adding a spatula tip of dithionite (~1 mg). Formation of the Fe^{2+}–CO complex is verified by UV–Vis (λ_{max} = 446 nm) and reconfirmed after each experiment. Samples are excited at 441 nm with a HeCd laser (Liconix (Santa Clara, CA) Model 4240NB, 40 mW), and Raman scatter is focused using longitudinal and transverse collection optics onto a double spectrometer (SPEX 1403, 0.85 m) interfaced to a PC via the SPEX MSD2 module. The signal is collected using a PMT (Hamamatsu R955, Hamamatsu City, Japan) biased at −1100 V (Pacific Precision Instruments, San Diego, CA). The sample control unit, photon counter, amplifier/discriminator, and buffered interface are all from EG&G Instruments (Gaithersburg, MD).

Catalytic Oxidation of [Ru-C_9-Ad]Cl_2 with NADH/PdR/Pd/P450

All reagents (NADH, PdR, Pd, P450) are tested prior to use by measuring NADH consumption in the presence of camphor: our finding of 575 ± 25 μmol NADH/min/μmol P450 agrees with literature values.[29]

[29] J. J. De Voss and P. R. Ortiz de Montellano, *J. Am. Chem. Soc.* **117,** 4185 (1995).

A 1.5-ml solution in buffer A containing 1 μM P450, 1 μM PdR, 10 μM Pd, 10 μM [Ru-C$_9$-Ad]Cl$_2$, and 200 μM NADH ($\varepsilon = 6.22$ mM^{-1}cm^{-1} at 340 nm)[18] is put in a cuvette. The solution is maintained at 20° and is mixed continuously (500 rpm) with a Hewlett Packard (Palo Alto, CA) 89090A stirrer/temperature controller. The consumption of NADH is monitored at 340 nm for 4 min (at 2-sec intervals) using the kinetics software package on a PC-controlled Hewlett Packard 8452A spectrophotometer.

Aliquots (250 μl) are removed 1, 2, and 3 min after the addition of P450 and are quenched immediately with a 1-M ethanolic camphor solution (final camphor concentration, 2 mM). Camphor displaces Ru substrates from the P450 pocket, and the enzyme consumes leftover NADH rapidly. The remaining reaction volume ($\sim$750 μl) is quenched with camphor after 4 min. Water (3 ml) and CH$_2$Cl$_2$ (3 ml) are added to each aliquot and shaken in a small separatory funnel. Improved separation of organic and aqueous layers is achieved by leaving the mixture at $-20°$ overnight. The CH$_2$Cl$_2$ layer (containing all ruthenated species, and minimal buffer salt) is isolated and evaporated to dryness. The Ru complex is redissolved in minimal methanol, and the concentration is checked by UV–Vis ($\varepsilon = 14.5$ mM^{-1}cm^{-1} at 456 nm).[30]

Electrospray Ionization (ESI) Mass Spectroscopy to Monitor Product Formation

Ru samples ($\sim$10 μM in methanol) are injected with a 250-μl Hamilton syringe into a LcQ (Finnigan Mat, Bremen, Germany) quadrupole ion trap mass spectrometer at a rate of 5 μl/min; typical runs require less than 100 μl per sample, and data sets are averaged for 100 scans. Only positive ions are detected, and sensitivity is highest in the selected ion monitoring (SIM) mode, with a mass range of 420–460. The isotope patterns of both starting and product materials are checked against expected values. The peak spacings ($m/z = 0.5$) are particularly diagnostic for these divalent species. The observation of a product ($m/z = 445$) with a mass 16 units higher ($z = +2$; $m/z = 16/2$) than the starting material ($m/z = 437$) is consistent with one or more singly hydroxylated adamantyl species, [Ru-C$_9$-Ad-OH]Cl$_2$.

Calibration of ESI for Ratiometric Analysis of Product Yields

A stock solution of [Ru-C$_9$-Ad-OH]Cl$_2$ is obtained by scaling the catalytic oxidation procedure 10-fold and continuing until NADH is consumed entirely. The resulting product is diluted in acetonitrile, and its purity is confirmed by ESI mass spectroscopy. Stock solutions (10 μM) of [Ru-C$_9$-Ad]Cl$_2$ and [Ru-C$_9$-Ad-OH]Cl$_2$ in methanol are combined in proportions varying from 10 : 1 to 1 : 4. Each solution is injected three different times, with intermediary blank runs of 1 : 1 (v/v) acetonitrile : methanol. Relative peak intensities are determined using

[30] K. Kalyanasundaram, "Photochemistry of Polypyridine and Porphyrin Complexes." Academic Press, London, 1992.

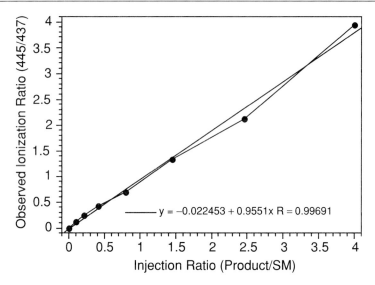

FIG. 4. Calibration line showing the relationship between the injected ratio of [Ru-C$_9$-Ad-OH]$^{2+}$/[Ru-C$_9$-Ad]$^{2+}$ to the relative ionization intensity of product and starting material (SM) cations in the ESI.

standard LcQ analysis software. A calibration graph (Fig. 4) is generated from these data showing that the ionization efficiencies of Ru-C$_9$-Ad and Ru-C$_9$-Ad-OH are indistinguishable within experimental error (slope of 0.96). Thus, for this Ru substrate, turnover efficiency may be analyzed by directly taking the ratio of the peaks, Ru-C$_9$-Ad-OH/Ru-C$_9$-Ad.

Results

Ru Probe Binding Monitored by UV–Vis and Time-Resolved Luminescence Measurements

The stoichiometric addition of Ru-C$_9$-Ad to P450 induces modest low-to-high spin conversion, as shown by a blue-shifted Soret (417 → 414 nm) and increased absorption at 392 nm (Fig. 5). Ru-C$_9$-EB barely perturbs the Soret of six-coordinate Fe^{3+}-OH$_2$ P450, much like the substrate ethyl benzene.[31] Binding of Ru-C$_{13}$-Im to ferric P450 induces a small red shift (417 → 420 nm, Fig. 5). The larger spectral changes observed for the binding of imidazole itself (7 nm red shift) suggest competition between aquo and Ru-Im ligands for Fe^{3+} P450. Full

---

[31] P. J. Loida and S. G. Sligar, *Prot. Eng.* **2,** 207 (1993).

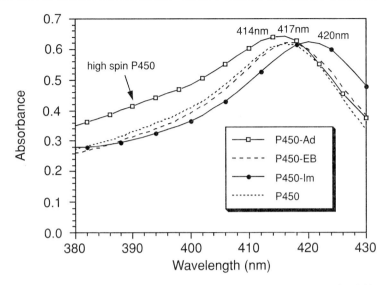

FIG. 5. UV–Vis spectra showing P450 alone (5.3 μM) and in the presence of stoichiometric Ru-C$_9$-Ad, Ru-C$_9$-EB, and Ru-C$_{13}$-Im. Solutions are buffered (50 mM potassium phosphate, 100 mM potassium chloride, pH 7.4).

heme ligation (Ru-Im-Fe^{2+}) accompanies reduction with dithionite (Soret shifts completely to 446 nm, as seen for P450 ferrous imidazole species).[32] P450$_{cam}$ shows tremendous selectivity for Ru-C$_{13}$-Im over Ru-C$_{11}$-Im[8]; thus, by attaching linkers with varying steric and electronic properties to the imidazole ligand,[28] it should be possible to engineer Ru-Im probes with considerable isozyme specificity.

Apparent dissociation constants for Ru probes are determined quickly and accurately from time-resolved luminescence measurements. Laser excitation of Ru-C$_{13}$-Im in the presence of equimolar (10 μM) P450 produces biphasic luminescence (60% of Ru^{2+*} decays by a rapid process, Fig. 6). Ru^{2+*} quenching has been shown by competition experiments with camphor to correlate directly with Ru probe binding at the active site.[9] Because [Ru]$_{total}$ = [Ru]$_{quenched}$ + [Ru]$_{free}$, and [Ru]$_{free}$ = [P450]$_{free}$, the apparent dissociation constant fo Ru-C$_{13}$-Im is

$$K_D(\text{app}) = [\text{Ru}]^2_{\text{free}}/[\text{Ru}]_{\text{bound}} = [4 \ \mu M]^2/[6 \ \mu M] \approx 2.7 \mu M \qquad (1)$$

$K_D(\text{app})$ depends somewhat on the concentrations of Ru and P450 due to competition between Λ and Δ Ru stereoisomers in the racemic mixture.[9]

[32] J. H. Dawson, L. A. Andersson, and M. Sono, *J. Biol. Chem.* **258,** 13637 (1983).

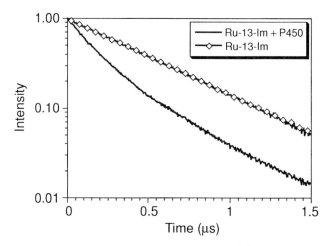

FIG. 6. Time-resolved luminescence traces showing $[Ru-C_{13}-Im]^{2+*}$ alone (diamonds, monophasic decay, $k_0 = 2 \times 10^6$ s^{-1}) and in the presence of stoichiometric (5 μM) P450$_{cam}$ (biphasic decay, $k_{en} = 5 \times 10^6$ s^{-1}, $k_0 = 2 \times 10^6$ s^{-1}).

Rapid Flash-Quench Photoreduction of P450 by Ru-C$_{13}$-Im

Photoexcitation of P450:Ru probe complexes in the presence of p-MDMA leads to [Ru-probe]$^+$ formation, followed by heme reduction on a microsecond time scale. This flash-quench scheme (Fig. 2C) has been applied to Ru-Ad, Ru-EB, and Ru-Im and promotes electron injection ($k \sim 2 \times 10^4$ s^{-1}) at rates three orders of magnitude faster than putidaredoxin. Rapid formation of Ru-C$_{13}$-Im-Fe^{2+} P450 is indicated by an absorbance increase at 445 nm and a decrease of the ferric Soret at 420 nm (Fig. 7), in agreement with the spectra of well-characterized P450 ferrous-imidazole species.[32]

Resonance Raman Spectra of P450 Fe^{2+}-CO:Substrate Complexes

As shown in Table I, and as was reported previously,[12–15] the Fe^{2+}-CO (ν_{Fe-C}) stretching mode in the P450–carbon monoxide complex is quite sensitive to the nature of the substrate. These results suggest a direct interaction between bound CO and the substrate and agree with IR absorption data.[33] Camphor-bound Fe^{2+}-CO P450 has $\nu_{CO} \sim 1940$ cm^{-1}, and camphor-free Fe^{2+}-CO P450 has two bands (~ 1942 and 1963 cm^{-1}) that have been assigned to bent and linear structures, respectively.[33] The ν_{Fe-C} stretching frequency differs by 4 cm^{-1} in the

---

[33] D. H. O'Keefe, R. E. Ebel, J. A. Peterson, J. C. Maxwell, and W. S. Caughey, *Biochemistry* **17,** 5845 (1978).

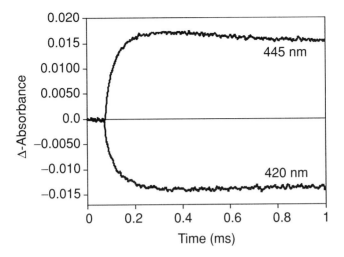

FIG. 7. Single-wavelength transient absorption: Δ-absorbance versus time plots for the reaction of $[Ru\text{-}Im]^+$ with P450. Changes in the Soret region (bleach of Fe^{3+}-Im at 420 nm and increase at 445 nm) are observed after laser excitation of a 10 μM P450, 10 μM Ru-C_{13}-Im, 10 mM p-MDMA sample.

series Ru-C_{11}-Ad, Ru-C_9-Ad, and adamantane, suggesting that these substrates bind with slightly increasing proximity to the heme (Table I). These frequencies (472–476 cm^{-1}) lie between those found for substrate-free and camphor-bound P450 and likely reflect subtle differences in solvation and hydrogen bonding at the active site.[33] The similarity between binding modes for Ru-Ad compounds

TABLE I

SUBSTRATE-INDUCED PERTURBATION OF P450 Fe-CO
($\nu_{\text{Fe-C}}$) STRETCHING FREQUENCY

Substrate	$\nu_{\text{Fe-C}}{}^a$
None	463^b
Ru-C_{11}-Ad	472
Ru-C_9-Ad	474
Adamantane	476
Camphor	482^b

[a] All units are cm^{-1}.
[b] These values are in good agreement with published values for substrate-free (465 cm^{-1}) and camphor-bound P450 (484 cm^{-1}). From A. V. Wells, P. Li, and P. M. Champion, *Biochemistry* **31**, 4384 (1992).

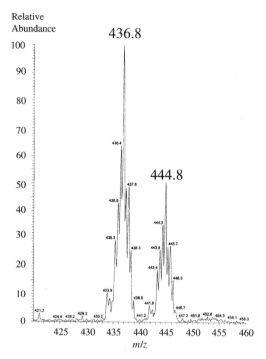

FIG. 8. ESI mass spectrum showing the starting material, [Ru-C$_9$-Ad]$^{2+}$ ($m/z = 436.8$), on the left-hand side and the conversion to products, [Ru-C$_9$-Ad-OH]$^{2+}$ ($m/z = 444.8$), on the right-hand side. This mixture was obtained by quenching the reaction with camphor 1 min after the addition of P450. The yield of product is derived from the relative peak heights (2 : 1).

and adamantane in solution is also reflected in X-ray structure determinations.[8,34]

Kinetics of Ru-C$_9$-Ad Hydroxylation

By supplying electrons to P450 via the biological route (NADH → PdR → Pd → P450), the rate of NADH consumption was found to be 10 ± 1 μmol NADH/min/μmol P450. In the absence of substrate, NADH consumption is only slightly slower (8 ± 1 μmol NADH/min/μmol P450). UV–Vis data and product (ESI) assays of aliquots taken after 1 min (Fig. 8) show that the efficiency of Ru-C$_9$-Ad-OH formation is $30 \pm 5\%$, based on NADH consumption. These data indicate that P450 oxidase chemistry (conversion of O$_2$ into water) competes with the hydroxylation

[34] R. Raag and T. L. Poulos, *Biochemistry* **30**, 2674 (1991).

of Ru-C$_9$-Ad. In a control experiment with NADH/PdR/Pd and no P450, the rate was found to be only 1 μmol NADH/min; thus, most of the observed NADH decay may be attributed to P450-mediated catalysis.

To probe the effect of linker and [Ru(bpy)$_3$]$^{2+}$ moieties on binding and turnover, the rate of 2-adamantylacetamide[28] hydroxylation was measured and found to be 90 ± 10 μmol NADH/min/μmol P450, in good agreement with the literature value for 1-adamantylacetamide ($\sim$110 μmol NADH/min/μmol P450).[29] Slower hydroxylation of Ru-C$_9$-Ad can largely be attributed to differences in water displacement from the heme (95% high-spin conversion for 2-adamantylacetamide versus 25% for Ru-C$_9$-Ad), as the biological delivery of electrons to low-spin P450$_{cam}$ is thermodynamically disfavored.

Acknowledgments

This work was supported by the National Institutes of Health Metalloprotein Program Project Grant P01 GM48495, the National Science Foundation, and NIH GM 54383 (J.H.D.). I.J.D. was a NIH predoctoral trainee (GM08346), A.R.D. is a Fannie and John Hertz Graduate Fellow, B.R.C. was a Helen Hay Whitney Postdoctoral Fellow, and M.T.G. is a Burroughs-Wellcome and NIH Postdoctoral Fellow. Special thanks go to the Beckman Institute Laser and Mass Spectrometry Resource Centers at the California Institute of Technology.

[14] Combining Pharmacophore and Protein Modeling to Predict CYP450 Inhibitors and Substrates

By COLLEN M. MASIMIREMBWA, MARIANNE RIDDERSTRÖM, ISMAEL ZAMORA, and TOMMY B. ANDERSSON

Introduction

The interest in computational methods to understand the properties of cytochrome P450 (CYPs) enzymes combined with factors governing the selectivity of substrates and inhibitors for each enzyme has increased considerably over the past few years due to the development of computational methods, graphical representation, and the crystallization of a mammalian CYP enzyme. CYP enzymes are a major drug-metabolizing system, and the pharmaceutical industry has started to appreciate the potential to be able to predict metabolic properties of drugs, i.e., to virtually screen compound libraries or only to synthesize compounds with desirable metabolic properties. This chapter summarizes our experience in homology modeling of CYPs 2C8, 2C9, 2C18, and CYP2C19 based on the rabbit CYP2C5 crystal structure. A substrate selectivity analysis for the CYP2C subfamily is also shown highlighting the amino acids responsible for the selectivity. Furthermore, generation of a 3D-QSAR model for a diverse set of CYP2C9 inhibitors taking into account important parameters, such as mechanism of inhibition and stereochemistry, is described.

Challenges in Modeling CYP Structure, Inhibitor, and Substrate Selectivity, and Potency of Interaction

CYP Protein Structure

Whereas most human receptors and enzymes have very strict ligand/substrate specificity, drug-metabolizing CYPs exhibit a broad and overlapping specificity. This has fundamental implications in modeling protein–ligand interaction. Major drug-metabolizing CYPs include CYP1A2, 2A6, 2C8, 2C9, 2C19, 2D6, 2E1, and 3A4. Members of the different families share <45% amino acid similarity (e.g., CYP1 and CYP2), and members of different subfamilies share <60% amino acid similarity (e.g., CYP2A and CYP2D). CYPs of the same subfamily therefore share a high amino acid similarity. For example, CYPs 2C8, 2C9, 2C18, and 2C19 have between 77 and 93% sequence identity with each other. This implies that differences between their protein structure and active site chemistry are on one hand minor, thus explaining substrate specificity overlap and, on the other, subtle in some cases to explain the selectivity of interaction observed with some compounds. Until 2000, there was no crystal structure of a membrane-bound CYP, and homology models of human CYPs were based on the crystal structures of soluble bacterial CYPs (CYP101, 102, 107A, and 108). These bacterial CYPs share 15–20% amino acid sequence similarity with human CYPs. Such low levels of similarity have been shown to generate unreliable homology models with root mean square (rms) deviations of the backbone atoms in the core regions as a function of sequence identity of 3 Å for globins[1] and 1.5–2.0 Å for CYP450s.[2] Greater deviations would therefore be expected in surface loops. Whereas the overall helical topography of CYPs might be similar,[3] packing of the helical elements might exhibit wide diverges. The situation has since significantly improved with the resolution of a mammalian membrane-bound CYP, a rabbit CYP2C5.[4] Homology models of human CYPs, the CYP2C family in particular, has benefited from this development because they have up to 83% amino acid sequence similarity. Homology models will, however, remain limiting in understanding details of

[1] C. Chothia and A. M. Lesk, *J. Mol. Biol.* **186,** 225 (1980).
[2] T. L. Poulos, *Methods Enzymol.* **206,** 11 (1991).
[3] O. Gotoh, *J. Biol. Chem.* **267,** 83 (1992).
[4] P. A. Williams, J. Cosme, J. V. Sridhar, E. F. Johonson, and D. E. McRee, *Mol. Cell* **5,** 121 (2000).

enzyme–ligand interactions until human CYPs are crystallized with and without substrates/inhibitors.

Chemical Diversity

Most receptor–ligand modeling in drug design is performed using compound sets with a common core structure. However, CYPs have the capacity to interact with compounds with extremely diverse chemical features. Ligands interacting with a specific CYP can include chemical classes such as acids, bases, and heterocyclic amines and compound classes including flavonoids, steroids, and coumarins. The compounds can range in molecular mass from around 200 to over 800 kDa. Substrates and inhibitors interact with CYPs in a stereospecific manner, and most of the compounds are very flexible (with at least three rotatable bonds), which increases the number of possible conformers from which one has to select the one likely to interact with a specific CYP. These features make it difficult to generate models based on alignment rules such as comparative molecular field analysis (CoMFA)[5] for a diverse set of compounds.

CYP Inhibitors

Inhibitors can interact with the CYP protein in many ways. These are described as reversible competitive, noncompetitive, and uncompetitive inhibitors. Inhibitors could also be metabolism dependent and result in both reversible and nonreversible inhibition. Even if inhibitors are described as, e.g., competitive, the binding in the CYP protein may be inhibitor dependent. One inhibitor could also act through a combination of several mechanisms of inhibition. The mode of interaction can thus be very complex, and common molecular descriptors to explain the biological behavior for a set of inhibitors may not exist. These facts mean that models should be generated using data that at least are associated with a specific mode of inhibition, and general parameters such as percent inhibition or IC_{50} are not sufficient. Furthermore, laboratory–laboratory variation due to the source of enzymes, types of marker substrates, and so on necessitate the use of a clearly defined set of data and disables the possibility of using data compiled from literature to derive models.

CYP Substrates

It is very common that a single compound is metabolized by one or more CYPs at one or more sites. This implies the possibility of many substrate-docking alternatives for CYP–substrate interaction determined by different sets of molecular descriptors. The rates of metabolism at different sites and by the different CYPs

---

[5] R. D. Cramer, D. E. Paterson, and J. D. Bunce, *J. Am. Chem. Soc.* **110,** 5959 (1988).

are different and seem to imply that metabolic stability can be driven predominantly by the reactivity of the compound, by the enzyme–compound affinity, or by a mixture of both factors. Prediction of site of metabolism and rate of metabolism therefore become a complex challenge and, as for inhibition, data used to arrive at any model also become important. Caution must be exercised in using K_m as a measure of CYP–substrate affinity and V_{max}/K_m as the major driving factor for the rate of metabolism and isoform selectivity. This is because, experimentally, the K_m is just a mathematical expression of the concentration that gives half the maximal velocity (V_{max}), which in some cases approximates the binding constant, K_s.

Integrated Approaches to Generate Homology, QSAR, and Selectivity Models for CYP Substrates and Inhibitors

Although the fundamental challenges outlined earlier imply that understanding of CYP–protein structure and active site chemistry; qualitative and quantitative models for substrates, inhibitors, and activators would be very difficult, a few approaches show promising results. With the basic understanding that "inappropriate data produces bad models and that inappropriate computational techniques misinterpret good data," we have come to the conclusion that only an integrated approach of expertise in computational chemistry, molecular biology, and enzymology, can tackle the challenge to generate reliable models in CYP metabolism research. Experience gained in the authors' laboratory indicates that three-dimensional descriptors adequately capture the CYP interactions both qualitatively and quantitatively. Use of the currently available CYP2C5 coordinates has enabled us to produce better quality models for CYP2C forms. The substrate selectivity in the closely related CYP2C isoforms was analyzed by consensus principal component analysis (CPCA).[6] Figure 1 summarizes the integrated approach for the generation of *in vitro* data, homology modeling, 3D-QSAR, and substrate selectivity analysis modeling used in the authors' laboratory. The procedures and software used in the computational aspects of this approach are now described briefly.

Homology Modeling

The crystal structure of CYP2C5 is used as the template in the modeling of human CYPs 2C8, 2C9, 2C18, and 2C19. The sequence identities between CYP2C5 and the human forms are 79, 83, 77, and 83%, respectively. The amino acid sequences of the five CYPs are imported to ClustalW, a program used for sequence alignment.[7] The crystal structure of CYP2C5 lacks the first 29 N-terminal

[6] M. A. Kastenholz, M. Pastor, G. Cruciani, E. E. J. Haaksma, and T. Fox, *J. Med. Chem.* **43**, 3033 (2000).

[7] J. D. Thompson, D. G. Higgins, and T. J. Gibson, *Nucleic Acids Res.* **22**, 4673 (1994).

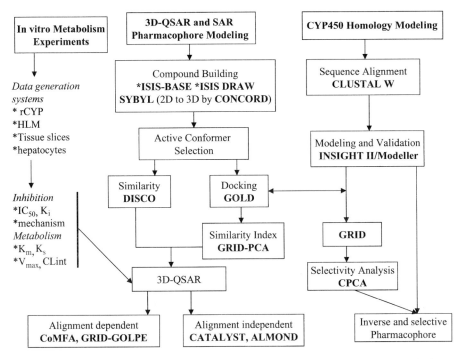

FIG. 1. Flow chart of the integrated approaches used to model for metabolic properties in the authors' laboratory.

amino acids and has five mutations compared with the wild-type enzyme, changes that are made to solubilize the enzyme for crystallization.[4] The corresponding amino acids in the human CYP2C isoforms are therefore also removed before amino acid sequence alignments were done. The aligned amino acid sequences are imported into the Insight II 98.0 Homology Module (Molecular Simulations, Inc., San Diego, CA) or Look v3 (Molecular Application Group, Palo Alto, CA) for modeling. The models are analyzed using Prostat in the Insight II Homology module and the SYBYL ProTable module. Cα atoms of the homology models have a rmsd between 0.18 and 0.23 compared to the CYP2C5 crystal structure. This is a significant improvement compared to the rmsd of 1.5–2 Å obtained when modeling human CYPs against crystal structures of bacterial CYPs. $\Phi–\psi$ angles in the core region of the models present in the allowed region (of the Ramachandran plot) are higher in the models (79–85%) compared to the template (67%) due to the optimization of the angles by MODELER. The MatchMaker average energy score for the models varies between −0.13 and −0.10 kJ (the CYP2C5 crystal structure is −0.13).

Active Site Docking and Conformer Selection

Traditional 3D-QSAR modeling using CoMFA (Comparative Molecular Field Analysis)[5] and DISCO (DIStance Comparisons)[8] is based on conformer selection by similarity search between low-energy conformers calculated *in vacuo* using a set of pharmacophoric features as alignment rules. When we use a chemically diverse set of CYP2C9 inhibitors, this conformer selection technique does not yield any alignment rule useful for generating a QSAR model (unpublished results). de Groot *et al*[9] showed that although a qualitative prediction model for the site of oxidation for CYP2D6 substrates could be generated by using conformers generated *in vacuo,* the model improves significantly when the CYP2D6 substrates are energy minimized within the homology model active site. We therefore decided to choose conformers from the solutions obtained in docking experiments with the CYP2C9 homology model.

The compound structures are imported from ISIS Base and converted into three-dimensional structures using the CONCORD program. Each structure is energy minimized using the Tripos force field with the Gesteiger–Marssili atom charges *in vacuo* conditions. With the CYP450 heme catalytic center as a reference point and a radius of 10 Å, an active site volume is defined into which substrates/inhibitors are docked by the program GOLD.[10] GOLD is a docking procedure based on a genetic algorithm that considers the ligand flexible but not the protein. In order to find a conformer on which to base the analysis of each compound, a template molecule(s) approach is used. First, some compounds with known binding modes are docked into the CYP2C9-binding site using information from the site of metabolism or known coordination of the nitrogen to the iron of the heme. A maximum of 10 conformers found in the docking are chosen for each molecule. The grid interaction fields with the −OH and DRY probes[11] for these conformers are subjected to a principal component analysis (PCA) using the GOLPE program.[12] Alignment of the selected conformers of each template molecule is done using the CYP2C9 protein structure (Fig 2). Second, for the other compounds in the training data set, the distance in the score space from each of the conformers to the template molecules is used as a similarity criteria for active conformer selection.[13]

[8] Y. C. Martin, M. G. Bures, E. A. Danaher, J. DeLazer, I. Lico, and P. A. Pavlik, *J. Comp.-Aided Mol. Design* **7,** 83 (1993).

[9] M. J. de Groot, M. J. Ackland, V. A. Horne, A. A. Alex, and B. C. Jones, *J. Med. Chem.* **42,** 1515 (1999).

[10] G. Jones, P. Willettt, R. C. Glen, A. R. Leach, and R. Tayor, *J. Mol. Biol.* **267,** 727 (1997).

[11] P. J. Goodford, *J. Med. Chem.* **28,** 849 (1985).

[12] B. Massimo, G. Cruciani, G. Costantino, D. Riganelli, R. Valigi, and S. Clementi, *Quant. Struct.-Activity Relat.* **12,** 9 (1993).

[13] L. Afzelius, I. Zamora, M. Ridderström, T. B. Andersson, A. Karlen, and C. M. Masimirembwa, *Mol. Pharmacol.* **59,** 909 (2001).

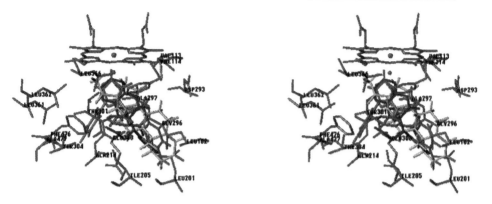

FIG. 2. Alignment of (S)-wafarin (blue), phenytoin (green), progesterone (yellow), and sulfaphen-azole (red) into the active site of CYP2C9. Reprinted from L. Afzelius, I. Zamora, M. Ridderström, T. B. Andersson, A. Karlén, and C. M. Masimirembwa, *Mol. Pharmacol.* **59**, 909 (2001).

Molecular Interaction Fields

Molecular interaction fields (MIF) for compounds or amino acid in a protein can be calculated using a program called GRID.[11] These fields describe the energy of interaction between the structure of the compounds and a small chemical group called a probe. In order to do this, the molecule is surrounded by a specified grid size. In each cross point of the grid, the chemical probes interact with the molecule to generate associated descriptors. Some of the probes used are water (for hydrophilic interactions, both hydrogen bond donor and acceptor properties), DRY (for hydrophobicity interactions), NH_3 (for hydrogen bond acceptor features), $C=O$ (for hydrogen bond donor properties), or OH (for hydrogen bond donor properties) probes (Fig 3).

The interactions are calculated by summing up the steric (Leonard–Jones potential), electrostatic, and hydrogen-bonding potentials. This generates large amounts of molecular descriptors, which are then related to biological parameters using multivariate analysis for the generation and validation of QSAR models. A new program for generating 3D-QSAR based on a nonalignment dependent approach has been proposed: ALMOND.[14] The procedure involves calculation of a set of molecular descriptors termed grid-independent descriptors (GRIND), which are obtained starting from a set of molecular interaction fields computed by GRID or other programs (Fig. 4), which are then simplified and encoded as alignment-independent variables. These can then be represented as "correlograms" depicting the positive interaction associated with the different pharmacophoric features.

[14] M. Pastor, G. Cruciani, I. McLay, S. Pickett, and S. Clementi, *J. Med. Chem.* **17**, 3233 (2000).

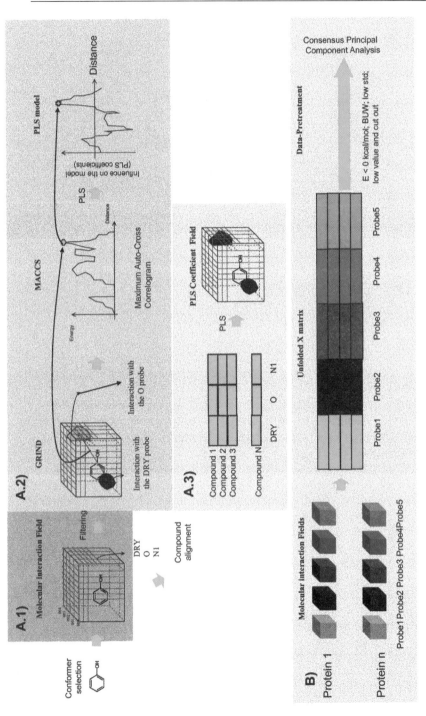

FIG. 3. Depiction of how molecular interaction fields for GRID are derived and used in three-dimensional QSAR modeling using GOLPE or ALMOND and in selectivity analysis and reverse pharmacophore modeling. (A. 1–3) shows how GRID molecular interaction fields are converted to GRIND descriptors which can be used to generate alignment-independent QSAR models using ALMOND. (B) shows how molecular probes can be used to derive molecular interaction fields in the active sites of different CYPs, hence mapping their differences.

CYP2C8   **CYP2C9**

CYP2C18   **CYP2C19**

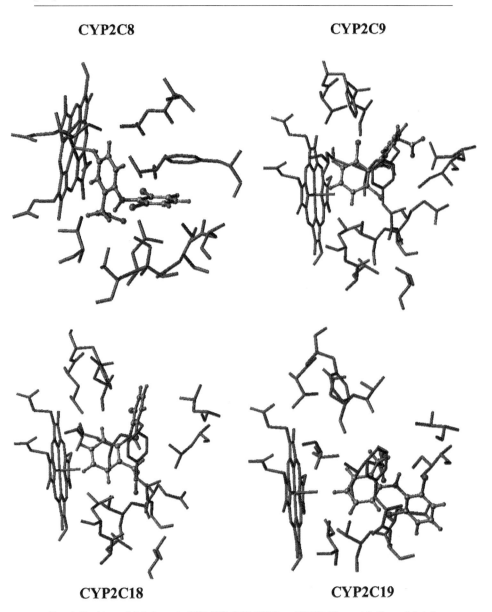

FIG. 4. Docking of diclofenac to CYP 2C8, 2C9, 2C18, and 2C19. The metabolism of diclofenac by CYP2C members is 4′-hydroxylation by CYP2C9, 5-hydroxylation by CYP2C8 and CYP2C18, and no metabolism by CYP2C19.

QSAR Model Generation Using PCA and PLS

The many descriptors generated by GRID or ALMOND represent a multi-dimensional problem that cannot be handled by multiple linear regression, but demand multivariate chemometric analyses, such as PCA (principal component analysis) and PLS (partial least squares). The large amount of variables can be handled by GOLPE (Generating Optimal Linear PLS Estimations) or an algorithm installed in ALMOND.

Three-Dimensional QSAR Modeling and Selectivity Analysis

Three-Dimensional QSAR Models for CYP2C9 Inhibitors

For 3D-QSAR modeling, data sets are divided into a training (to generate the model) and a test set (to validate the model). In this application, experimental data are K_i values for competitive inhibitors of CYP2C9. For alignment-based modeling, "active conformers" are chosen by their similarity to known conformers of the template molecules as described earlier. GRID interaction energies of these molecules are calculated using OH (hydrogen bond donor or acceptor) and DRY (hydrophobic) probes, and GOLPE is used to relate these descriptors to K_i values. After a variable selective procedure, a model with r^2 of 0.947, a cross-validation of $q^2 = 0.730$, and prediction of the external data set of $q^2 = 0.560$ was obtained.[13] For an alignment-independent model using ALMOND, GRIND descriptors are then related to K_i by PLS analysis and produced a model with $r^2 = 0.75$, a cross-validation of $q^2 = 0.58$, and prediction of the external data set of $q^2 = 0.73$ (Afzelius *et al.*, submitted).

Selectivity Analysis (GRID/CPCA)

To explore the basis of substrate-or inhibitor-binding specificities associated with the very similar CYP2C isoforms (2C8, 2C9, 2C18, and 2C19), a GRID/PCA analysis is made to define the selectivity of each active site (Ridderström *et al.*[15]). From the analysis it is possible to generate selective pharmacophores in order to anticipate likely substrates or to explain why substrates dock in a particular mode in the active site of CYPs. The approaches used have been published by Kastenholz *et al.*[6] using consensus principal component analysis (CPCA) to analyze GRID molecular field interactions. GRID descriptors are derived from using a variety of probes with different chemical properties on the active site of the CYP.

Cα atoms of the protein structures are aligned using homology alignment in SYBYL 6.5.3 (Tripos, Inc., St. Louis, MO). PDB files are adjusted to be acceptable by GRIN (version 17). Proteins are neutralized by the addition of counterions in

---

[15] M. Ridderström, I. Zamora, O. Fjellström, and T. B. Andersson, *J. Med. Chem.* **44,** 4072 (2001).

the positions suggested by the program Mimin and Filmap. Only points outside the grid box considered in the selectivity analysis are neutralized. GRIN produces the output files used in GRID calculation. The setup for the GRID calculation defines a box of $18.5 \times 20 \times 15$ Å, including the active site for each of the CYP models. Two GRID calculations are made: one where the amino acids in the active sites are considered rigid and another where they are considered flexible. In the flexible calculation, the MOVE directive is set to 1 (MOVE $= 0$ in the rigid calculation) in the GRID command file to allow for flexible side chains to respond to the interactions with the probe. Hydrophobic interactions are calculated with the DRY probe. C3 and NM3 probes both describe steric interactions. The $N1^+$ and the COO^- probe are charged. The polar probes consist of N1, N:, NH$=$, O, and OH.

Molecular interaction fields (MIFs) from GRID calculations are imported into GOLPE version 4.5.11. Files from the calculation with the single and multi-probe files are merged. Four different types of analyses are performed based on the rigid or flexible GRID calculations, with or without using the cutout tool in the pretreatment that defines the 4-Å radius around the docked diclofenac molecule. Pretreated data are used in CPCA modeling as described by Kastenholz et al.[6]

Selectivity analysis shows that the most important determinants of selectivity among CYP2C models are the geometrical features of the active sites and the hydrophobic interactions (Ridderström et al.[15]). Selectivity analysis singles out CYP2C8 as the most different of the four CYP2C enzymes with amino acids having distinct properties in positions 114, 205, and 476 (Ser, Phe, and Ile, respectively) compared to the other enzymes. An inverse selective pharmacophore model for CYP2C9 is constructed from the selective regions, and the model agrees with the docking of diclofenac in which the properties of the ligand overlap with the pharmacophoric points in the model.

Validation and Verification of Models

Basic validation of the QSAR models involves cross validation using the "leave one out" (L.O.O.) technique or different percentages of elements of the original training set and trying to predict their biological effect by the model generated with the remaining compounds. This method evaluates the predictive power of the model inside the set defined to build it but it could give an overly optimistic view of the performance of the model. A second level of validation involves the prediction of a test set of compounds that have "not seen" the model before, thus mimic a real application situation. The same strategy could be followed on qualitative models to predict likely sites of oxidation by specific CYPs and determination of isoform specificity. The next level of verification that has direct applications will be to introduce chemical changes based on the qualitative (e.g., for sites of oxidations or isoform identification) or QSAR (e.g., for inhibitors and rates of reaction) models to a compound with a predicted and desired metabolic property. It must be noted that many currently available models are either retrospective or explanatory of a fairly familiar chemical space for which there is extensive direct and complementary enzymological data. This has resulted in very impressive models but with a rather narrow range of applicability. These models probably depend on local and sometimes questionable enzymological data assumptions/generation and computational manipulations. It is therefore hoped that models produced in the near future will be truly predictive and based on sound knowledge of CYP–ligand interactions.

Application of Models

Speculative applications of metabolism computational models include the potential to screen company libraries or evaluate external libraries on sale for favorable metabolic properties. With the current knowledge, however, more realistic applications might be in the lead optimization during the drug discovery process in fostering a better understanding of the chemistry of CYP–substrate/inhibitor interactions. Here project-specific models for CYP inhibitors, substrates, or CYP isoform identification might be easier to derive, as the compounds usually belong to related series. If *in vitro* data show undesirable metabolic properties, knowledge gained from the model for that property could be used to assist chemical modification in solving the problem. It should be noted that a panel of models for the major CYPs and for the major metabolic properties would be ideal, as a chemical change (even small) in response to a specific problem for a particular CYP can give birth to new problems with another CYP. This application can be directly linked to the screening of virtual libraries generated by medicinal chemists around the compound of interest and only those promising better metabolic properties are synthesized and tested *in vitro* systems.

Future Perspective in Computational Approaches in Metabolism Research

Whereas the conformer selection approach using the active site represents a significant improvement in understanding CYP–ligand interactions, the fact that the protein part is static is not a true reflection of such interactions is still a limiting factor. In our approach, substrate/inhibitors were docked into the active site with water as the sixth ligand binding to the iron atom, whereas it is known that water is displaced when CYPs interact with substrates or inhibitors. We therefore took the compromise assumption of modeling interactions just before this water is removed. Our success based on this assumption, however, does not necessarily make it valid. This is particularly true for ligand-type inhibitors whose potency of interaction is partly due to the strong nitrogen–iron ligation.

[15] High Pressure: A New Tool to Study P450 Structure and Function

By FRÉDÉRIC BANCEL, GASTON HUI BON HOA, PAVEL ANZENBACHER, CLAUDE BALNY, and REINHARD LANGE

Introduction

Despite several decades of intensive research, many structural and functional properties of P450 in solution are still poorly defined. Some progress has been made by studies conducted under suboptimal conditions, e.g., use of chemical denaturants, detergents, or at temperatures far from physiological conditions. The use of subzero temperatures has allowed the study of several elementary reaction steps in more detail.[1,2] However, the information obtained remained limited, and alternative methods have been sought. The use of high pressure appears to be a promising new approach.[3] Like other perturbants, high pressure affects both the protein conformation and the elementary steps of the P450 reaction cycle due to its specific effects on different types of chemical interactions and reaction rates. However, these effects are not the same as those exerted by chemical reagents or temperature. In many cases the effects are reversible and perturb only the volume of the P450 system. The main contribution to these volume changes comes not only from a protein structural change, but also—and even to a larger extent—from interaction of the different chemical groups on the protein surface with water molecules. Indeed, the hydration shell reflects the heterogeneous density of water molecules around ionized and apolar domains. Changes in conformation and interactions of P450 proteins imply reorganization of these associated water molecules, making them quite sensitive to the effects of pressure. Hydrostatic pressure appears therefore as a useful tool for a better understanding of the structure and function of proteins.

The use of high pressure is not very complicated. It requires, however, a basic understanding of its effects on chemical equilibria and reaction rates and it necessitates some technical adaptation of analytical instruments, e.g., spectroscopic and kinetic instruments. The purpose of this chapter is to familiarize the reader with these techniques and to show some recent examples of applications to P450 research. Examples are concerned with the structural basis of P450 stability with P450cam, P450 from *Sulfolobus solfataricus,* and liver microsomal P450, as well as the perturbation of elementary steps (spin state transition, electron flow).

[1] R. Lange, G. Hui Bon Hoa, C. Larroque, and I. C. Gunsalus, *in* "Cytochrome P-450: Biochemistry and Biophysics" (I. Schuster, ed.), p. 272. Taylor and Francis, London, 1989.

[2] N. Bec, A. C. F. Gorren, C. Voelker, B. Mayer, and R. Lange, *J. Biol. Chem.* **273,** 13502 (1998).

[3] F. Bancel, N. Bec, C. Ebel, and R. Lange, *Eur. J. Biochem.* **250,** 276 (1997); C. Jung, *Biochim. Biophys. Acta* **309,** 1595 (2002); C. Jung, N. Bec, and R. Lange, *Eur. J. Biochem.* **269,** 2989 (2002).

Methods

Physical Background of High-Pressure Effects on Enzyme Reactions

Effect of Pressure on Chemical Equilibria and Reactions. Any chemical reaction that is accompanied by a change in volume is pressure sensitive due to the relation

$$d\mu/dp = V \qquad (1)$$

relating the chemical potential μ and pressure p to the volume V of the reactants. The reaction volume ΔV, i.e., the change in volume of the reactants, can thus be determined experimentally using Eq. (2), where K is the equilibrium constant and R is the universal gas constant:

$$d(\ln K)/dp = -\Delta V^0/RT \qquad (2)$$

For kinetics, the same formalism holds:

$$d(\ln K^{\ddagger})/dp = -\Delta V^{\ddagger}/RT \qquad (3)$$

The unit for pressure is the megapascal (1 MPa = 10 bar).

Effect of Pressure on Proteins. Within the most frequently used pressure range (below 700 MPa), pressure does not affect covalent bonds, but affects the secondary, tertiary, and quaternary structure of proteins. However, it is difficult to predict the effect of pressure on a protein. In fact, pressure acts on several terms: chemical term (for chemical reactions such as reduction and oxidation), conformational term (changes in polypeptide chain conformation), intra- and intermolecular interaction term (at short and long range), and solvation, i.e., interaction with the surrounding medium with other molecules or solvents (water, ions, salts, organic solvents, etc.). The main noncovalent interactions affected are hydration of charged groups (electrostriction), which is accompanied by a volume reduction, and formation of hydrophobic interactions, generally disfavored by pressure. Charge-transfer interactions and stacking of aromatic rings show a small negative volume variation, and hydrogen bonds are almost pressure insensitive.[4]

Effect of Pressure on Enzyme Kinetics. Pressure modifies the rate of enzyme-catalyzed reactions via changes in the structure of an enzyme or changes in the reaction mechanism. In 1950, Laidler formulated the first theoretical basis for explaining the responses of enzymatic activities to high hydrostatic pressures,[5] and a more recent overview concerning this phenomenon was given by Morild.[6] Depending on the reaction under study, a reaction can be accelerated or decelerated by increasing pressure (according to the sign of the activation volume term $\Delta V^{\ddagger}$).

[4] V. V. Mozhaev, K. Heremans, J. Frank, P. Masson, and C. Balny, *Tibtech.* **12,** 493 (1994).

[5] K. J. Laidler, *Arch. Biochem. Biophys.* **30,** 226 (1950).

[6] E. Morild, *Adv. Prot. Chem.* **34,** 93 (1981).

For example, at 20°, if $\Delta V^{\ddagger} = -20$ ml/mol, the rate constant will increase by a factor of 12 when the pressure is raised from atmospheric pressure to 300 MPa. Detailed general descriptions of pressure effects on enzyme reactions have been published elsewhere.[7]

Spectroscopic Techniques Developed to Study P450 under High Pressure

Static Methods. High-pressure spectroscopic experiments can be conducted essentially like experiments under atmospheric pressure. The only difference is that the optical cell containing the enzyme solution is closed with a flexible membrane (polyethylene) and is placed in a thermostatted high-pressure bomb (SDFOP, Rodez, France) made of Marval X12 steel, equipped with sapphire windows (8 mm thick). The cell is surrounded by the pressure vector, which may be water, silicone oil, or pentane. Thus, there is no pressure gradient from inside to outside of the cell. Pressure is applied inside the bomb through a capillary connected to a manual or motor-driven pump, and at any time the pressure of the system can be read by pressure gauges. Due to technical constraints, the most currently used pressure range is from atmospheric pressure to 700 MPa. Typically, the whole pressure cell is placed into the sample compartment area of a spectrophotometer or spectrofluorimeter and spectra are recorded as at atmospheric pressure. However, the choice of the buffer is significant: some anionic buffers, such as phosphate, are highly pressure sensitive, but the pK of other buffers (Tris, MES) is rather insensitive to pressure.

Kinetic Methods. The Montpellier laboratory built a stopped-flow apparatus operational up to 200 MPa between $-40°$ and $40°$. It consists of a stepping motor-driven, classical stopped-flow apparatus, designed to be inserted into a cylindrical high-pressure bomb, which replaces the sample compartment of an Aminco DW2 spectrophotometer. In the Paris as well as in the Montpellier laboratory, a high-pressure jump device has been built that can be used for both fluorescence and absorbance studies. It consists of two high-pressure bombs: one containing the optical cell and the other one serving as a pressure "reservoir." The two bombs are pressurized differently, and the pressure jump is performed on opening a high-pressure pneumatic valve situated between them. For both kinetic devices the experimental dead time is 3–5 ms as measured by a piezoelectric transducer placed inside the optical bomb.

Chemical Methods Developed to Probe Pressure-Induced Active-Site Structure Deformations

Studies of substrate activation by cytochrome P450 have revealed the formation of unusual σ-iron(III)-alkyl (or aryl) P450 complexes. It has been suggested

---

[7] R. Hayashi and C. Balny, eds., "High Pressure Bioscience and Biotechnology." Elsevier, Amsterdam, 1996.

that cytochrome P450, myoglobin, and hemoglobin give rise to such complexes on metabolic oxidation of various monosubstituted hydrazines, with the final products being the corresponding N-alkyl- or N-arylporphyrins.[8] For synthetic iron porphyrins, treatment of the phenyl iron complex with ferricyanide results in phenyl migration from the iron to all the nitrogens of the porphyrin in the same proportion. In the case of cytochrome P450 and other hemoproteins, distribution of the N-arylprotoporphyrin IX regioisomers ($N_B : N_A : N_C : N_D$) as quantitated by high-performance liquid chromatography (HPLC) is not uniform and depends on the presence of protein residues above the heme group and on the height of the aryl probe.[9]

In collaboration with the laboratory of P. R. Ortiz de Montellano, the laboratory in Paris has applied this method to obtain information about changes in the topology (deformation) of the active site, when submitted to high pressure.[10] In the absence of an external oxidant and at atmospheric pressure, the Fe-aryl complex is formed first. The complex shows a typical spectrum with maximum absorbance at 478–480 nm (Fig. 1a) and is stable up to 200 or 300 MPa, depending on the ligand used. At higher pressure, the maximum amplitude at 478–480 nm decreases, and up to 350 MPa a new spectrum appears for the iron-phenyl complex (Fig. 1a). Decompression of the solution is followed by extraction with chloroform ($CHCl_3$). Although the absorbance peak at 478 nm is lost, the chloroform solution shows a visible spectrum with maximum absorbance at 416 nm (Fig. 1b) associated with the formation of N-arylprotoporphyrin IX products, as confirmed by HPLC analysis. The distribution pattern obtained is shown in Fig. 2. The isomer distribution in the pyrrole rings A, B, C, and D obtained at atmospheric pressure (Fig. 2a) when the Fe-phenyl complex is oxidized by potassium ferricyanide was quite different from that obtained at 400 MPa in the absence of external oxidizing agent (Fig. 2b).

Applications

Effect of Pressure on Cytochrome P450 Structure and Stability

Different Structural Organizations of Cytochrome P450s in Monomeric and Oligomeric States. P450cam is a cytosolic bacterial hemoprotein from *Pseudomonas putida* that hydroxylates camphor, its natural substrate, and is the first P450 enzyme for which the crystallographic structure is known.[11,12] The crystal structure shows that the heme (in the absence of substrate) incorporates six molecules of

---

[8] P. R. Ortiz de Montellano and K. L. Kunze, *J. Am. Chem. Soc.* **103**, 6534 (1981).

[9] P. R. Ortiz de Montellano, *Biochimie* **77**, 581 (1995).

[10] R. A. Tschirret-Guth, G. Hui Bon Hoa, and P. R. Ortiz de Montellano, *J. Am. Chem. Soc.* **120**, 3590 (1998).

[11] T. L. Poulos, B. C. Finzel, and A. J. Howard, *Biochemistry* **25**, 5314 (1986).

[12] T. L. Poulos, B. C. Finzel, and A. J. Howard, *J. Mol. Biol.* **195**, 687 (1987).

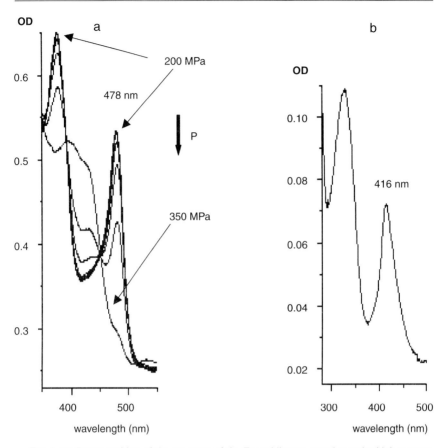

FIG. 1. (a) Decomposition of the spectrum of the Fe-aryldiazene complex under high pressure
between 200 and 350 MPa. At 350 MPa, a new spectrum arises as shown by the loss of the peak centered
at 478 nm. (b) Spectrum of the extracted product in a solution of chloroform showing an absorbing
peak centered at 416 nm. [Adapted with permission from R. A. Tschirret-Guth, G. Hui Bon Hoa, and
P. R. Ortiz de Montellano, *J. Am. Chem. Soc.* **120,** 3590 (1998).]

water, with one water molecule bound as a ligand. Binding of substrate (camphor)
removes the molecules of water, and substrate binding is promoted by the formation of a hydrogen bond between the camphor oxygen and Tyr-96 and hydrophobic
contacts between the substrate and Phe-87, Leu-244, and Val-295. The pressure
behavior of this monomeric protein was compared to that of an oligomeric cytochrome P450 2B4 in order to probe the structure and interactions of the subunit.
P450 2B4 is a microsomal membrane hemoprotein from rabbit liver that hydroxylates benzphetamine and testosterone in position 16α. Electron microscopy studies
show P450 2B4 to be hexamers.

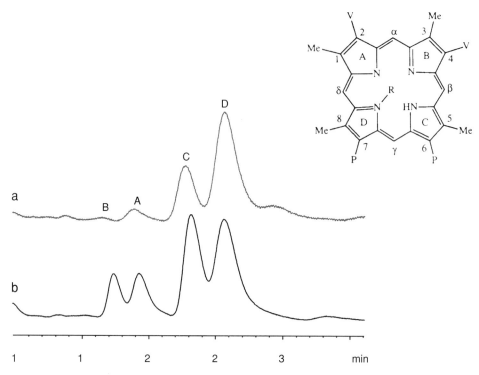

FIG. 2. HPLC pattern of N-arylprotoporphyrin showing different distributions of the N isomers in the four A, B, C, and D pyrrole rings: (a) isomers obtained when the Fe-phenyl complex was oxidized by potassium ferricyanide at atmospheric pressure and (b) isomers obtained when the Fe-phenyl complex was exposed to 400 MPa in the absence of an external oxidizing agent. [Adapted with permission from R. A. Tschirret-Guth, G. Hui Bon Hoa, and P. R. Ortiz de Montellano, *J. Am. Chem. Soc.* **120,** 3590 (1998).]

Application of high pressure perturbed the spectrum of both cytochromes in a complex way. Using factor analysis of the spectral changes in the Soret region allowed the resolution of two pressure-induced processes. The first process was a pressure-induced, high-spin to low-spin shift of the cytochrome substrate-bound P450cam and P450 2B4. Analysis of the pressure-induced spin shift of 2B4 versus benzphetamine concentration showed that both substrate-bound ($\Delta V^\circ_{\text{spin}} = -49$ ml/mol) and substrate-free ($\Delta V^\circ_{\text{spin}} = -21$ ml/mol) species were sensitive to pressure.[13] Although the substrate-dissociation step itself had a very weak pressure dependency ($\Delta V^\circ_{\text{dissoc.}} = -23$ ml/mol) (Table I), for the P450cam-dissociation step it was greater ($\Delta V^\circ_{\text{dissoc.}} = -29$ ml/mol). The origin of these spin volume

[13] D. R. Davydov, E. Deprez, G. Hui Bon Hoa, T. V. Knynushko, G. P. Kuznetsova, Y. M. Koen, and A. I. Archakov, *Arch. Biochem. Biophys.* **320,** 330 (1995).

TABLE I

PRESSURE EFFECT ON SPIN STATE TRANSITION OF CYTOCHROME P450cam AND 2B4

Spin transition	Equilibrium constant at 0.1 MPa		$\Delta V°_{spin}$ (ml/mol)	
	Cam	2B4	Cam	2B4
$(P450)_{hs} \rightarrow (P450)_{ls}$	77	3.21	−20	−21
$(P450\text{-}S)_{hs} \rightarrow (P450\text{-}S)_{ls}$	0.016	0.12	−91	−49
$(P450)_{hs}\text{-}S \rightarrow (P450)_{hs} + S$	0.047 $\mu M (K_{dh})$	36 $\mu M (K_{dh})$	−29	−23

Adapted with permission from R. A. Tschirret-Guth, G. Hui Bon Hoa, and P. R. Ortiz de Montellano, *J. Am. Chem. Soc.* **120**, 3590 (1998).

changes had been established by using new camphor analogs to partially dehy-
drate the heme pocket.[14] A linear correlation between spin volume and the Gibbs-
free energy in the spin transition of the substrate-bound cytochrome P450cam
was obtained. Cytochrome P450 2B4–benzphetamine and BM3–palmitate complexes fit perfectly on the straight line,[15] showing that these proteins exhibit a
similar substrate-binding mechanism but differ in their hydration content in the
heme pocket (an estimate of 16 molecules of water in P450 2B4 and 11 water
in the P450cam active site). Some remarkable barotropic parameters were obtained for the second P450 to P420 process for the oligomeric ferric as well as
reduced carbonyl P450 2B4 (Table II). Only a portion of P420 accounting for
$67 \pm 5\%$ of the total ferric hemoprotein content was reached with corresponding
pressure parameters: $\Delta V°_{inact.} = -32$ ml/mol, $P_{\frac{1}{2}\ inact.} = 156$ MPa.[13] Although
65–70% of the total Fe^{2+}-CO hemoprotein was converted to P420 ($\Delta V°_{inact.} =
-50$ ml/mol, $P_{\frac{1}{2}\ inact.} = 300$ MPa), the remaining portion was stable at pressures
up to 600 MPa.[16] On dissociation of the oligomers by 0.20% Triton N-101, the
inhomogeneity behaviors vanished: 99.9% of the ferric 2B4 was involved in the
spin transition with $\Delta V°_{inact.} = -86$ ml/mol, followed by an intense bleaching
of the hemoprotein. For reduced carbonyl complexes of monomeric P450 2B4,
$95 \pm 10\%$ of the total protein was converted to P420, with pressure characteristics of $\Delta V°_{inact.} = -104$ ml/mol, $P_{1/2\ inact.} = 130$ MPa. Oligomerization results
in a considerable stabilization of the hemoproteins which are less flexible (smaller
inactivation volume). We interpreted the pressure behavior of the oligomer P450
2B4 as evidence of two different conformers of P450 2B4 with different pressure
stabilities. We have proposed that oligomeric P450 2B4 could have some peculiar-
ities in architecture resulting in different conformation and/or orientation of the

[14] V. Helms, E. Deprez, E. Gill, C. Barret, G. Hui Bon Hoa, and R. C. Wade, *Biochemistry* **35**, 1485 (1996).

[15] D. R. Davydov, G. Hui Bon Hoa, and J. A. Peterson, *Biochemistry* **38**, 751 (1999).

[16] D. R. Davydov, T. V. Knyushko, and G. Hui Bon Hoa, *Biochem. Biophys. Res. Commun.* **188**, 216 (1992).

TABLE II
PRESSURE EFFECT ON P450–P420 CONVERSION OF P450cam AND 2B4

P450 → P420 transition	$P_{1/2 \text{ inact.}}$ (MPa)	$\Delta V^{\circ}_{\text{inact.}}$ (ml/mol)	Amplitude (%)
$P450^{3+}_{\text{cam}}$	110	−80	100
$P450^{3+}_{\text{cam}}$-S	170	−160	100
$P450^{3+}_{\text{2B4}}$	156	−32	67
$P450^{2+}_{\text{2B4}}$-CO	300	−50	65–70
$P450^{3+}_{\text{2B4}}$ + 0.2% Triton N-101	120	−86	99.9
$P450^{2+}_{\text{2B4}}$-CO + 0.2% Triton N-101	130	−104	95

Information compiled from Refs. 13 and 16.

subunits. This hypothesis has been confirmed by the study of the effects of pressure on heterooligomer cytochrome P450 2B4-1A2 proteins.[17]

Determination of Liver Microsomal Cytochrome P450 Active-Site Flexibility. Among cytochromes P450 (abbreviated CYP), the form labeled CYP3A4 is important because it is the most abundant CYP in the human liver and it is known to be involved in the biotransformation of most clinically and physiologically active compounds.[18] This is why the flexibility and stability of this enzyme have been examined.[19] Moreover, comparisons were made with other CYP enzymes, namely with CYP1A2 (metabolizing aromatic structures) and microbial soluble flavohemoprotein CYP102 (or BM3, whose crystallographic data are known,[20] and it serves as a good model for membrane CYP enzymes).

Hydrostatic pressure up to 300 MPa was applied to CYP3A4, CYP102 holoenzyme,[19] and CYP1A2[20] both in the presence and in the absence of substrates. Fourth derivative UV spectroscopy showed that there are no significant protein conformational changes present due to the effect of pressure. However, the results have clearly revealed that CYP enzymes, although possessing similar overall structure, differ extensively in the properties of the active site. First, the binding of CO to the reduced heme iron in the active site was followed. Under the same experimental conditions [namely in phosphate buffers containing 20% (v/v) glycerol, which are generally used for P450 studies], CYP1A2 and CYP102 (BM3) were able to withstand pressure up to 300 MPa. On the contrary, CYP3A4 is a highly unstable enzyme, which denatures to the spectrally detectable form P420 (exhibiting absorption maximum of its reduced complex with carbon monoxide at 420 nm

[17] D. R. Davydov, N. A. Petushkova, A. I. Archakov, and G. Hui Bon Hoa, *Biochem. Biophys. Res. Commun.* **276,** 1005 (2000).

[18] F. P. Guengerich, *in* "Cytochrome P450: Structure, Mechanism, and Biochemistry" (P. R. Ortiz de Montellano, ed.), 2nd Ed., p. 473. Plenum Press, New York, 1995.

[19] K. G. Ravichandran, S. S. Boddupalli, C. A. Haseman, J. A. Peterson, and J. Deisenhofer, *Science* **261,** 731 (1993).

[20] P. Anzenbacher, N. Bec, J. Hudecek, R. Lange, and E. Anzenbacherová, *Collect. Czech. Chem. Commun.* **63,** 441 (1998).

TABLE III
ANALYSIS OF SORET BAND RED SHIFT WITH PRESSURE FOR DIFFERENT
CYP ENZYMES[a]

Enzyme	Slope of red shift [cm^{-1}/MPa]	Correlation coefficient
CYP3A4	0.2459 ± 0.0019	0.9881
CYP3A4 + TAO	0.3412 ± 0.0011	0.9978
CYP102	0.3758 ± 0.0021	0.9937
CYP102 + palmitate	0.3853 ± 0.0019	0.9937
CYP1A2	0.2529 ± 0.0013	0.9903
CYP101[b]	0.2724 ± 0.0007	0.9945
CYP101 + S[b]	0.1809 ± 0.0007	0.9836

[a] An equation $S = \alpha^* P + S_0$ has been used [C. Jung, G. Hui Bon Hoa,
D. Davydov, E. Gill, and K. Heremans, *Eur. J. Biochem.* **233**, 600 (1995)]
where S is the position of the Soret band maximum at given pressure P and
S_0 its position at zero pressure, $-\alpha$ is the slope related to compressibility.
[E. Anzenbacherová, N. Bec, P. Anzenbacher, J. Hudecek, P. Soucek, C. Jung,
A. W. Munro, and R. Lange, *Eur. J. Biochem.* **267**, 2916 (2000)].
[b] From C. Jung, G. Hui Bon Hoa, D. Davydov, E. Gill, and K. Heremans, *Eur.
J. Biochem.* **233**, 600 (1995).

differing from the native form absorbing at 450 nm[21]) with denaturation observable
already at 1 MPa.

The second approach applied to studies on the active sites of CYP enzymes
was the calculation of relative compressibility of the heme pocket. According to
theory, information on the compressibility of the hemoprotein active site may be
obtained by following the red shift of the Soret absorption band for the carbon
monoxide complex when hydrostatic pressure is applied. Increasing hydrostatic
pressure may shorten the distance between water molecules or polar groups located
in the vicinity of the heme, thus influencing the microscopic dielectric constant of
the chromophore neighborhood.[22] The slope, α, of the red shift of the Soret band is
related to the compressibility of the heme active site. This (physical) effect has been
used to estimate the compressibility of the heme pocket of another bacterial, soluble
CYP101.[22] Results obtained for CYP3A4, CYP1A2, CYP102, and 101 further
document the differences between active sites of these enzymes (Table III).[23]

Results show that CYP102 (holoenzyme, i.e., containing both the heme and
reductase domain) appears to possess a flexible, compressible structure. Interestingly, this enzyme also does not denature easily with pressure. These properties of
CYP102 make it similar to another CYP enzyme isolated from acidothermophilic

[21] P. R. Ortiz de Montellano, ed., "Cytochrome P450: Structure, Mechanism, and Biochemistry," 2nd
Ed. Plenum Press, New York, 1995.
[22] C. Jung, G. Hui Bon Hoa, D. Davydov, E. Gill, and K. Heremans, *Eur. J. Biochem.* **233**, 600 (1995).
[23] E. Anzenbacherová, N. Bec, P. Anzenbacher, J. Hudecek, P. Soucek, C. Jung, A. W. Munro, and
R. Lange, *Eur. J. Biochem.* **267**, 2916 (2000).

bacteria *Sulfolobus solfataricus.*[24] Denaturation of the CYP102 heme domain under pressure[15] may be explained by two ways. First, the active site of the isolated heme domain apparently exhibits different properties than the active site of the complete enzyme (as evidenced by resonance Raman study[25]). Second, in the experiment discussed,[15] the glycerol (used for its protective effect) was absent from the buffer.

The CYP3A4 molecule is apparently relatively flexible, and binding of the substrate (triacetyloleandomycin) at the active site of this enzyme increases its compressibility. Taken together, CYP3A4, which is known to bind relatively large substrates,[18] has a relatively flexible structure that is vulnerable to denaturation. The flexibility of the active site of this enzyme was also documented by resonance Raman studies.[25] CYP3A4 resembles CYP2B4, which has been shown to denature relatively easily under pressure.[26] Information obtained by the high-pressure technique is hence in line with practical experience: Whereas CYP1A2 is stable during experimental procedures, including isolation, CYP3A4 denatures relatively easily, e.g., during preparation or in the presence of organic solvents.[27]

Pressure-Induced Deformation of Active-Site Structure. Compression of proteins not only favors hydration changes, but changes the interatomic distances within the protein. This effect is related to protein compressibility. Accurate information about how these mechanical forces are transmitted into the structures and how they perturb the secondary structures of the polypeptide chain around the active site can be obtained by using the method of pressure-induced migration of the aryl groups. The iron-aryldiazene complexes of cytochrome P450cam (CYP101), a relatively unstable enzyme, and of a thermophilic cytochrome P450 (CYP119) from the archaebacteria *S. solfataricus,* a hyperstable enzyme, were submitted to hydrostatic pressure. The relative distribution of the four N-arylprotoporphyrin isomers ($N_B : N_A : N_C : N_D$) obtained at 400 MPa was $10 : 14 : 33 : 43$ for CYP101[10] and $23 : 36 : 14 : 27$ for CYP119,[28] whereas at 0.1 MPa, oxidation of the complexes by ferricyanide gave the following results: $00 : 05 : 25 : 70$ for CYP101 and $39 : 47 : 07 : 07$ for CYP119 (Table IV). The rates of complex decomposition at $20°$ were pressure independent for CYP119, but pressure dependent for CYP101. These rates were obtained after pressure jumps from 180 MPa to different higher final pressures. The measured activation volume for the phenyldiazene probe was

[24] M. A. McLean, S. A. Maves, K. E. Weiss, S. Krepich, and S. G. Sligar, *Biochem. Biophys. Res. Commun.* **252,** 166 (1998).

[25] J. Hudecek, E. Anzenbacherová, P. Anzenbacher, A. W. Munro, and P. Hildebrandt, *Arch. Biochem. Biophys.* **383,** 70 (2000).

[26] F. Bancel, N. Bec, and R. Lange, *in* "High Pressure Research in the Biosciences and Biotechnology" (K. Heremans, ed.), p. 71. Leuven University Press, Leuven, Belgium, 1997.

[27] N. Chauret, A. Gauthier, and D. A. Nicol-Griffith, *Drug Metabol. Disposition* **26,** 1 (1998).

[28] R. A. Tschirret-Guth, L. S. Koo, G. Hui Bon Hoa, and P. R. Ortiz de Montellano, *J. Am. Chem. Soc.* **123,** 3412 (2001).

TABLE IV
DISTRIBUTION OF N-ARYLPROTOPORPHYRIN IX ISOMER RATIOS ($N_B : N_A : N_C : N_D$)[a]

Parameter	CYP 101 + phenyldiazene ($N_B : N_A : N_C : N_D$)	CYP 119 + p-Br-phenyldiazene ($N_B : N_A : N_C : N_D$)
Oxidation at 0.1 MPa by ferricyanide at 20°	00 : 05 : 25 : 70	39 : 47 : 07 : 07
Shift obtained by pressure at 400 MPa and 40°	10 : 14 : 33 : 43	23 : 36 : 14 : 27

[a] Obtained by oxidation with ferricyanide at 0.1 MPa and subjected to 400 MPa for P450 Fe-aryl complexes. (Information used from Refs. 10 and 29.)

a negative value (-84.3 ml/mol). The interpretation of results are as follow: the phenyl group, which migrated principally to the pyrrole ring D in CYP102 and to a lesser extent to ring C, encountered some conformational blocking in these pyrrole rings at 400 MPa at room pressure. The decrease in the N_D isomer was compensated for by 10% increases in N_A and N_B isomers. The pressure-induced deformations of the active site were not reversible and showed that the active site of CYP101 was compressed above rings C and D, causing an outward displacement of the I helix. On the contrary, pressure effects on the CYP119–p-Br-phenyldiazene complex were reversible. The topology deduced from these studies indicated that the active site of CYP119 at 0.1 MPa is closed above pyrrole nitrogens A and B but relatively open above pyrrole rings C and D. High pressure induced a small decrease in the relative proportions of the A and B isomers, suggesting that pressure might cause a lateral motion of the I helix (which lies above pyrrole nitrogens A and B) toward the center of the heme. The pressure and thermal stabilities of CYP119 appear to be closely related to its unusual structural plasticity and reversible deformability. Pressure causes subtle and nonisotropic structural changes in the active site of cytochrome P450s.

Effect of Pressure on Elementary Reaction Rates: Electron Flow of Nitric Oxide Synthase

Functionally, NO synthase (NOS) resembles cytochrome P450, due to the thiolate ligand of its heme iron. Its structure is, however, more complex: the reductase and the oxygenase domain are contained in a single peptide chain. In addition, this dimeric enzyme requires tetrahydrobiopterin, Ca^{2+}/calmodulin, and zinc as cofactors. We have studied the electron flow from the flavin (FAD, FMN)-containing reductase domain to cytochrome c, an external electron acceptor. In the absence of Ca^{2+}/calmodulin, the electron transfer rate was about 20 times smaller than in its presence, as expected from the regulatory role of calmodulin. However, under high pressure, the rate increased strongly ($\Delta V^{\ddagger} = -70 \pm 5$ ml/mol), and at

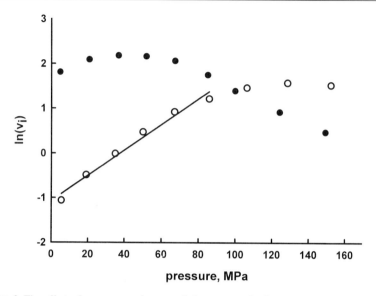

FIG. 3. The effect of pressure on the rate of electron transfer from neuronal NO synthase to cytochrome c. Stopped flow experiments were initiated by mixing a solution containing nNOS (14 to 140 nM) and NADPH (400 μM) with cytochrome c (50 μM) in the presence ($\bullet$) and absence ($\bigcirc$) of $CaCl_2$ (0.5 mM) and calmodulin (20 μg/ml). [Adapted with permission from R. Lange, N. Bec, P. Anzenbacher, A. W. Munro, A. C. F. Gorren, and B. Mayer, *J. Inorg. Biochem.* **87**, 191 (2001).]

150 MPa, a similar value of the initial velocity vi was obtained as the one measured in the presence of Ca^{2+}/calmodulin. The 20-fold increase of the reduction rate was fully reversible after pressure release.[29] These results demonstrate that for this elementary step, high pressure can substitute for Ca^{2+}/calmodulin. Interestingly, a similar accelerating effect has been reported for urea.[30] We could confirm this effect by an experiment carried out under the conditions shown in Fig. 3. The fact that pressure can substitute for Ca^{2+}/calmodulin is intriguing. Generally, pressure acts on proteins as a denaturant, i.e., it favors the unfolding of proteins. In our case, a strong, global protein unfolding effect can be excluded, as this would have resulted in a loss of flavins. The effect of pressure on the NOS structure can therefore be expected to be small, and restricted locally. Because pressure and urea have similar effects in substituting for Ca^{2+}/calmodulin, it appears that one effect of Ca^{2+}/calmodulin can be understood as a weakening or partial unfolding of the reductase domain in the region of the "autoinhibitory loop." The deletion of this peptide sequence had been shown to induce a similar effect.[30]

[29] R. Lange, N. Bec, P. Anzenbacher, A. W. Munro, A. C. F. Gorren, and B. Mayer, *J. Inorg. Biochem.* **87**, 191 (2001).

[30] R. Narayanasami, J. S. Nishimura, K. McMillan, L. J. Roman, T. M. Shea, A. M. Robida, P. M. Horowitz, and B. S. Masters, *Nitric Oxide* **1**, 39 (1997).

Acknowledgments

Support from GACR 203/99/0277 and MSM 15110003 projects is gratefully acknowledged. We gratefully acknowledge the support and advice of Professors Pierre Douzou, Alexander I. Archakov, and Paul R. Ortiz de Montellano. The authors are very grateful to Drs. Eva Anzenbacherová, Nicole Bec, Carmelo Di Primo, Eric Deprez, Dmitri R. Davydov, Richard A. Tschirret-Guth, and Laura S. Koo for their participation in this work. This work was supported by the Institut National de la Santé et de la Recherche Médicale (ex-U310, U128, and U473), Grant Agency of the Czech Republic (Project 203/99/0277), Czech Ministry of Education (MSM 15110003), the Russian Foundation of Fundamental Investigations (Grant 94-06-13764-a), the Institute of Biomedical Chemistry (Moscow), INTAS Grant 96-1314, and the National Institutes of Health (United States) (Grant GM25515).

Section III

Regulation of Expression

[16] Use of *in Vitro* Pregnane X Receptor Assays to Assess CYP3A4 Induction Potential of Drug Candidates

By STACEY A. JONES, LINDA B. MOORE, G. BRUCE WISELY, and STEVEN A. KLIEWER

Introduction

Many hurdles must be overcome before a drug reaches the market. One of the major pitfalls for drug candidates is the induction of liver enzymes, including members of the cytochrome P450 (CYP) family of heme-containing monooxygenases. Induction of the CYP3A4 isoform is particularly relevant, as this enzyme is involved in the metabolism of >50% of all marketed drugs, including chemicals as diverse as macrolide antibiotics, imidazole antimycotics, benzodiazepines, and steroids.[1-3] Notably, transcription of the CYP3A4 gene is increased in response to a remarkable array of drugs and other xenobiotics. The inducibility of CYP3A4 expression coupled with the broad substrate specificity of this enzyme represents the basis for an important class of drug–drug interactions. In this era of polypharmacy, in which many patients are taking multiple drugs, the induction of CYP3A4 by a drug candidate is, in most cases, sufficient reason to discontinue its development. Historically, it has been very difficult to accurately predict the induction potential of drug candidate molecules because of the tremendous differences seen in the induction of CYP3A enzymes across species. Thus, induction studies have generally been performed in primary human hepatocytes, which are difficult to obtain and vary considerably in quality. The availability of robust, high-capacity assays to predict the CYP3A4 induction potential of drug candidates could both expedite the drug discovery process and result in safer medicines.

Nuclear receptors are a family of ligand-activated transcription factors that include steroid, retinoid, and thyroid hormone receptors.[4] In 1998, a new member of this family was reported that is activated by nearly all of the structurally diverse compounds that induce CYP3A gene expression.[5-8] This new receptor was named

[1] P. Maurel, *in* "Cytochromes P450: Metabolic and Toxicological Aspects" (C. Ioannides, ed.), p. 241. CRC Press, Boca Raton, FL.

[2] U. Savas, K. J. Griffin, and E. F. Johnson, *Mol. Pharmacol.* **56**, 851 (1999).

[3] D. J. Waxman, *Arch. Biochem. Biophys.* **369**, 11 (1999).

[4] D. J. Mangelsdorf, C. Thummel, M. Beato, P. Herrlich, G. Schutz, K. Umesono, B. Blumberg, P. Kastner, M. Mark, and P. Chambon, *Cell* **83**, 835 (1995).

[5] S. A. Kliewer, J. T. Moore, L. Wade, J. L. Staudinger, M. A. Watson, S. A. Jones, D. D. McKee, B. B. Oliver, T. M. Willson, R. H. Zetterstrom, T. Perlmann, and J. M. Lehmann, *Cell* **92**, 73 (1998).

[6] J. M. Lehmann, D. D. McKee, M. A. Watson, T. M. Willson, J. T. Moore, and S. A. Kliewer, *J. Clin. Invest.* **102**, 1016 (1998).

the pregnane X receptor (PXR) based on its activation by natural and synthetic C_{21} steroids. PXR binds as a heterodimer with the 9-*cis*-retinoic acid receptor (RXR) to DNA response elements composed of two copies of the consensus sequence AGTTCA organized as either a direct repeat with a three nucleotide spacer (DR3) or an everted repeat with a six nucleotide spacer (ER6). DR3 and ER6 response elements have been identified in the regulatory regions of rat, rabbit, and human CYP3A genes. The binding of ligands to PXR causes a conformational change in the ligand-binding domain (LBD) of the receptor and results in the transcriptional activation of CYP3A4 and other PXR target genes.[5–9] Importantly, it is now established that sequence differences in PXR account for most of the variation in CYP3A induction across species. Compounds that are known to differentially induce CYP3A expression in rat or rabbit or human activate only the corresponding PXR.[10] Moreover, replacement of the mouse PXR with the human PXR results in an animal with a "humanized" CYP3A induction profile.[11] The discovery of PXR and its role in the regulation of CYP3A has provided the insights necessary to develop reliable, cost-effective assays for predicting the induction potential of drug candidates. This article describes two such assays: a cell-based reporter assay and an *in vitro* competition binding assay.

Cell-Based Transfection Assay

The cell-based assay we have developed to measure PXR activity uses transient overexpression of the full-length receptor and a corresponding reporter plasmid, which contains a PXR responsive element in front of the minimal thymidine kinase (tk) promoter and chloramphenicol acetyltransferase (CAT) gene. The throughput for this assay is somewhat lower than for the binding assay (see later), but the ability to profile compounds for their efficacy and potency on PXR in a cellular context is tremendously valuable. Additionally, this assay is a straightforward format for studying PXR activation in different species because only the expression plasmids for PXR from the relevant species are required rather than purified and modified PXR from each species.

---

[7] G. Bertilsson, J. Heidrich, K. Svensson, M. Asman, L. Jendeberg, M. Sydowbackman, R. Ohlsson, H. Postlind, P. Blomquist, and A. Berkenstam, *Proc. Natl. Acad. Sci. U.S.A.* **95**, 12208 (1998).

[8] B. Blumberg, W. Sabbagh, Jr., H. Juguilon, J. Bolado, C. M. Vanmeter, E. Ong, and R. M. Evans, *Genes Dev.* **12**, 3195 (1998).

[9] B. Goodwin, E. Hodgson, and C. Liddle, *Mol. Pharmacol.* **56**, 1329 (1999).

[10] S. A. Jones, L. B. Moore, J. L. Shenk, G. B. Wisely, G. A. Hamilton, D. D. McKee, N. C. O. Tomkinson, E. L. LeCluyse, M. H. Lambert, T. M. Willson, S. A. Kliewer, and J. T. Moore, *Mol. Endocrinol.* **14**, 27 (2000).

[11] W. Xie, J. L. Barwick, M. Downes, B. Blumberg, C. M. Simon, M. C. Nelson, B. A. Neuschwander-Tetri, E. M. Brunt, P. S. Guzelian, and R. M. Evans, *Nature* **406**, 435 (2000).

Plasmids

The human PXR expression vector, pSG5-hPXR△ATG, and the reporter plasmid (CYP3A1 DR3)$_2$-tk-CAT, generated by insertion of two copies of a double-stranded oligonucleotide containing the CYP3A1 DR3 RE (5′ GATCAGACAAG-TTCATGAAGTTCATCTAGATC 3′) into the *Bam*HI site of pBLCAT2, are as described previously.[5,6] pβ-Actin-SPAP, an expression vector containing the human secreted placental alkaline phosphatase (SPAP) cDNA under the control of the β-actin promoter, is used as an internal control in all transfections.

Transient Transfection

1. Maintain CV-1 cells at 37° with 5% (v/v) CO$_2$ in air atmosphere in Dulbecco's modified Eagle's (DME) high glucose HEPES medium (Irvine Scientific, Santa Ana, CA) supplemented with 10% fetal bovine serum (FBS) and 2 mM L-glutamine (Medium #1). Subcultivate twice weekly at a ratio of 1 : 20.

2. Three days prior to plating for transfection, harvest experimental cells with phenol red-free 0.25% trypsin/2 mM EDTA and collect in phenol red-free Dulbecco's modified Eagle's medium (DMEM)–Ham's F12/15 mM HEPES medium (Life Technologies, Inc., Rockville, MD) supplemented with 10% (v/v) charcoal/dextran-treated FBS (HyClone, Logan, UT) and 2 mM L-glutamine (Medium #2). Subcultivate at a ratio of 1 : 10 and maintain in culture for 3 days.

3. Harvest and collect experimental cells as outlined in step 2. Count cells and dilute to a concentration of 200,000 cells/ml with Medium #2. In 96-well plates, seed 20,000 cells/well (100 μl of cell dilution).

4. Return cells to incubator and incubate for 18 to 24 hr at 37°.

5. Prepare transfection mixes to contain 80 ng total DNA and 0.7 μl Lipofect-AMINE (Life Technologies, Inc.)/well as follows. For each transfection, dilute 80 ng total DNA [5 ng pSG5-hPXR△ATG, 12 ng (CYP3A1 DR3)$_2$-tk-CAT reporter plasmid, 8 ng pβ-actin-SPAP as internal control, and 55 ng of carrier plasmid (pBluescript, Stratagene, La Jolla, CA)] into 10 μl of Opti-MEM I (Life Technologies, Inc.). Dilute 0.7 μl LipofectAMINE into 9.3 μl of Opti-MEM for each transfection and then combine with DNA mixture. Mix gently and incubate at room temperature for 30 min. While the DNA and LipofectAMINE mixture incubates, aspirate medium from cells and wash cells once with 100 μl of Opti-MEM I.

6. Dilute the DNA/LipofectAMINE mixture with 80 μl of Opti-MEM I for each transfection.

7. Aspirate Opti-MEM I wash from cells and add 100 μl of diluted transfection mix to each well. Return cells to incubator and incubate for 6 to 18 hr at 37°.

Drug Delivery Protocol

1. Heat inactivate delipidated, charcoal-stripped bovine calf serum (Sigma-Aldrich, St. Louis, MO) to remove endogenous alkaline phosphatase activity.

Allow the serum to warm to room temperature and heat at 62° for 35 min. Store at 4°.

2. Supplement phenol red-free DMEM-F12/15 mM HEPES medium with 10% heat inactivated, delipidated, charcoal-stripped bovine calf serum, 2 mM L-glutamine, 50 units/ml penicillin, and 50 μg/ml streptomycin (Medium #3).

3. Prepare dilutions of drug stocks and vehicle in Medium #3. Aspirate transfection mix from cells and dispense 100 μl of diluted drug or vehicle to each well.

4. Return cells to incubator at 37° for a further 18 hr prior to determination of reporter activity.

Reporter Assays

Alkaline Phosphatase

1. Transfer 40 μl of medium from each well of treated plates to daughter plates to assay for SPAP as an internal control for transfection efficiency. Include a sample of media used for drug delivery as a blank.

2. Prepare 7 mM p-nitrophenyl phosphate in 1 M diethanolamine/0.28 M sodium chloride/0.5 mM magnesium chloride, pH 9.85. Add 200 μl substrate to each well of the daughter plates and the blank.

3. Incubate at room temperature and read absorbance at 405 nm.

Chloramphenicol Acetyltransferase

1. Aspirate remaining media from treated cell plates and add 50 μl 0.1 mg/ml digitonin in 250 mM Tris–HCl, pH 7.8, to each well to lyse the cells. Shake gently and then incubate at room temperature for 30 to 45 min.

2. For each well, prepare substrate containing 13.22 μl 75 μM HCl, 4 μl 250 mM Tris–HCl (pH 7.8), 18 μl 5 mM chloramphenicol in 250 mM Tris–HCl (pH 7.8), 0.514 μl 5 mM acetyl-CoA, and 3.43 μl [^{3}H]acetyl-CoA (specific activity 100 μCi/ml, NEN Life Science Products, Inc., Boston, MA).

3. Add 37 μl of substrate to cell extract and incubate at 37° for 1.5 to 2 hr. Stop the reaction by the addition of 100 μl of 7 M urea. Transfer reaction to a 7-ml scintillation vial containing 750 μl 7 M urea. Add 5 ml toluene containing 8 g/liter 2,5-diphenyloxazole. Mix well and measure counts per minute (cpm).

4. To correct for transfection efficiencies, normalize CAT activity to β-actin SPAP activity. Divide normalized CAT activity of drug-treated cells by normalized CAT activity of vehicle-treated cells to determine fold activation. Plot fold activation vs log[compound concentration] and use nonlinear least-squares curve fitting to generate curves with EC$_{50}$ and V_{max} values (see Fig. 1).

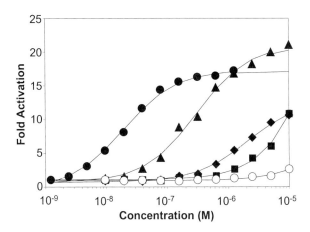

FIG. 1. Activation of human pregnane X receptor (PXR) by hyperforin (●), SR12813 (▲), rifampicin (◆), and troglitazone (■). PCN (○) does not significantly activate human PXR at concentrations up to 10 μM. Data points represent the mean of assays performed in triplicate.

In Vitro Binding Assay

We have developed a high throughput *in vitro* screen in order to rapidly assess the PXR binding profile of large numbers of potential drug candidate compounds. We radiolabeled the bisphosphonate ester SR12813, a high-affinity PXR ligand (K_d = 41 nM),[10] for use in these assays. We chose the scintillation proximity assay (SPA) format because it is easy to automate, has good reproducibility, is cost effective, and is easy to transfer among research sites. Initial studies indicated that gel filtration (Quick Spin Protein columns, Boehringer Mannheim, Indianapolis, IN) could also be used to assess the binding potential of compounds in a low throughput manner. Several of the different SPA bead types were evaluated, including yttrium silicate polylysine, yttrium silicate histidine (His)-tag, poly(vinyltoluene) (PVT) His-tag, and PVT streptavidin beads. PVT beads were selected for high throughput screen development because the PVT bead core offers superior compatibility with liquid handling devices compared with yttrium silicate beads. Additionally, there is very low background binding of the radioligand to the streptavidin PVT bead, and streptavidin–biotin coupling offers an easy and high-affinity method of coupling the receptor to the SPA bead.[10,12]

Protein Expression

Stability and solubility of the ligand binding domain (LBD) of the PXR protein are enhanced greatly by coexpression with a fragment of the nuclear receptor coactivator SRC-1. The PXR LBD expression construct is engineered as an

[12] J. S. Nichols, D. J. Parks, T. G. Consler, and S. G. Blanchard, *Anal. Biochem.* **257**, 112 (1998).

N-terminal polyhistidine-tagged fusion protein with residues 130–434 from the human PXR. The fusion insert is subcloned into the pRSETA expression vector (Invitrogen, Carlsbad, CA). Residues 623–710 of the human SRC-1 gene are subcloned into the bacterial vector pACYC184 along with a T7 promoter.[13] The PXR LBD/pRSETA and the SRC-1/pACYC184 plasmids are cotransformed into the BL21(DE3) *Escherichia coli* strain. One-liter shaker flask liquid cultures containing standard Luria–Bertani (LB) broth with 0.05 mg/ml ampicillin and 0.05 mg/ml chloramphenicol are inoculated and grown at 22° for 20 hr. The cells are harvested by centrifugation (20 min, 3500g, 4°), and the cell pellet is stored at $-80°$.

Protein Purification

The cell pellet is resuspended in 250 ml buffer A [50 mM Tris–HCl, pH 7.8, 250 mM NaCl, 50 mM imidazole, pH 7.5, 5% (v/v) glycerol]. Cells are sonicated for 3–5 min on ice, and cell debris is removed by centrifugation (45 min, 20,000g, 4°). The cleared supernatant is filtered through a 0.45-μm filter and loaded onto a 50-ml ProBond nickel-chelating resin (Invitrogen). After washing to baseline with buffer A, the column is washed with buffer A containing 125 mM imidazole, pH 7.5. The PXR LBD/SRC-1 complex is eluted from the column using buffer A with 250 mM imidazole, pH 7.5. Column fractions are pooled and concentrated using Centri-prep 30K (Amicon, Danvers, MA) units. The protein is subjected to size exclusion, using a column (26 mm × 90 cm) packed with Superdex 75 resin (Pharmacia, Piscataway, NJ) preequilibrated with 20 mM Tris–HCl, pH 7.8, 250 mM NaCl, 5 mM dithiothreitol (DTT), 2.5 mM EDTA, pH 8.0, and 5% glycerol. Column fractions containing the PXR/SRC-1 complex are pooled and diluted fivefold with the dilution buffer (20 mM Tris–HCl, pH 7.8, 5 mM DTT, 2.5 mM EDTA, pH 8.0, 5% glycerol) and loaded onto a 50-ml S Sepharose fast flow column (Pharmacia). The column is washed to baseline with 20 mM Tris–HCl, pH 7.8, 50 mM NaCl, 5 mM DTT, 2.5 mM EDTA, pH 8.0, and 5% glycerol. The protein complex is eluted from the column at 300 mM NaCl in an increasing salt gradient. The appropriate column fractions, which contain the 1 : 1 PXR/SRC-1 complex, are pooled and stored on ice. Purity for the PXR/SRC-1 complex is assessed by SDS–PAGE with silver staining. Protein of >95% purity is biotinylated using a fivefold molar excess of NHS-LC-Biotin (Pierce, Rockford, IL) in a minimum volume of PBS (phosphate-buffered saline). After incubation (60 min at room temperature) with gentle shaking the reaction is stopped by the addition of excess Tris–HCl, pH 8. The biotinylated PXR/SRC-1 complex is dialyzed at 4° against three changes of TBS buffer containing 5 mM DTT, 2 mM EDTA, and 2% sucrose, and is then stored at $-80°$. The extent of biotinylation is determined by mass spectrometric analysis.

[13] R. T. Nolte, G. B. Wisely, S. Westin, J. E. Cobb, M. H. Lambert, R. Kurokawa, M. G. Rosenfeld, T. M. Willson, C. K. Glass, and M. V. Milburn, *Nature* **395,** 137 (1998).

Radioligand Synthesis

[^{3}H]SR12813 is synthesized from 3,5-di-*tert*-butyl-4-hydroxybenzaldehyde by chemists at Amersham International plc (Cardiff, UK).[14] The specific activity is 23 Ci/mmol.

Saturation Binding Assay

Twelve-point saturation curves are typically run in triplicate using clotrimazole to determine nonspecific binding.

1. Streptavidin PVT SPA beads (Amersham Pharmacia Biotech) are resuspended in assay buffer (50 mM Tris, pH 8.0, 50 mM KCl, 1 mM EDTA, 1 mM CHAPS, 1 mM DTT, and 0.1 mg/ml essentially fatty acid free bovine serum albumin) to a concentration of 0.5 mg/ml.

2. The purified, biotinylated PXR ligand-binding domain with coexpressed SRC-1 peptide is added to the beads to a concentration of 75 nM. The receptor is allowed to couple to the SPA beads at room temperature for 30 min.

3. The uncoupled receptor is removed by centrifuging the receptor–bead suspension at 3500g for 15 min at 4° and discarding the supernatant. Receptor-coupled beads are then resuspended in assay buffer to a concentration of 0.5 mg/ml.

4. Fifty microliters of receptor-coupled beads is added to each well of a 96-well Packard OptiPlate.

5. In order to determine nonspecific binding of the radioligand, clotrimazole is added as a 10-μl aliquot of a 100 μM solution in 10% dimethyl sulfoxide (DMSO). Generally, nonspecific binding is determined in triplicate for each radioligand concentration tested. Ten microliters of 10% DMSO in assay buffer is added to the remaining wells (also in triplicate) for determination of total binding at each radioligand concentration.

6. Forty microliters of radioligand is added to each well such that the final radioligand concentration ranges from 0.5 to 1000 nM.

7. Plates are sealed (Packard TopSeal-A), and samples are mixed by momentarily shaking the plates on an orbital shaker.

8. Plates are counted on a Packard TopCount (or equivalent) after incubation for 1 hr at room temperature. In time course studies the signal is determined to be stable for over 15 hr at room temperature.

9. Total and nonspecific data are plotted as counts per minute vs radioligand concentration. Specific binding is calculated by subtracting nonspecific binding from total binding at each concentration of radioligand. Typically, nonspecific samples, which contain clotrimazole, increase linearly with increasing radioligand concentration. The usual hyperbolic relationship is seen when specific binding is

[14] L. M. Nguyen, E. Niesor, H. T. Phan, P. Maechler, and C. L. Bentzen, in U.S. Patent 9011247.

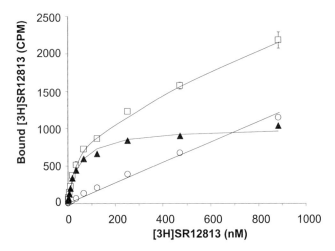

FIG. 2. Purified human PXR LBD immobilized on SPA beads was incubated with concentrations of [³H]SR12813 ranging from 0.5 to 1000 nM in the absence (total binding, □) or presence (specific binding, ▲) of 10 μM clotrimazole to define nonspecific binding (○). Data points represent the mean of assays performed in triplicate. The K_d value for [³H]SR12813 binding to human PXR was 41 nM as calculated by nonlinear regression. Reprinted from Ref. 10 by permission of The Endocrine Society.

plotted vs radioligand concentration (see Fig. 2). Nonlinear least-squares regression curve fitting (e.g., GraphPad Prism software) of the specific binding curve indicates a K_d of around 40 nM.

Competition Binding Assay

Test compounds are generally run across a 96-well plate as 10-point competition curves with the remaining two columns reserved for total binding and nonspecific binding determinations.

1. Threefold serial dilutions of test compounds are made from 10 μM to 0.5 nM in 100% DMSO. These are added to Packard Optiplates as 1-μl aliquots. DMSO concentrations of up to 1% are tolerated in the assay, whereas higher concentrations decrease specific binding.

2. One microliter of DMSO is added to the eight wells in column 11 for the determination of total binding.

3. One microliter of 1 mM clotrimazole in 100% DMSO is added to the eight wells in column 12 for the determination of nonspecific binding.

4. Purified biotinylated PXR/SRC-1 is coupled to streptavidin PVT SPA beads as described earlier. After centrifugation, receptor-coupled beads are resuspended in assay buffer to a concentration of 25–50 μg/100 μl.

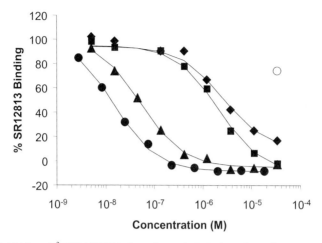

FIG. 3. Inhibition of [³H]SR12813 binding at human PXR by hyperforin (●), SR12813 (▲), trogli-tazone (■), and rifampicin (◆). PCN (○) does not significantly inhibit binding of [³H]SR12813 at the human PXR. Data points represent the mean of triplicate values.

5. Radioligand is added to the receptor–bead suspension to a final concentration of 10 nM.

6. One hundred microliters of the receptor–bead–radioligand suspension is added to the OptiPlates.

7. Plates are sealed and mixed as described earlier.

8. Plates are counted on a Packard TopCount after a 1 to 2-hr room temperature incubation. For competition experiments, the TopCount is programmed to correct for quenching using the 3H SPA PVT color quench and calibration kit (Amersham Pharmacia Biotech).

9. Mean disintegrations per minute (dpm) values for totals (dpm$_T$) and nonspecific (dpm$_{NS}$) are calculated for each plate and used to normalize the dpm values for each well in columns 1–10 (dpm$_{well}$) using Eq. (1). %[³H]SR12813 bound values for each well are then plotted against log[compound concentration], and nonlinear least-squares curve fitting is used to generate the sigmoidal dose–response curves and IC$_{50}$ values (see Fig. 3). K_i values can then be calculated from IC$_{50}$ values using the Cheng–Prosoff equation [Eq. (2)] and the K_d determined from saturation-binding studies.[15]

$$\%[^3H]SR12813 \text{ bound} = 100^*[(dpm_{well} - dpm_{NS})/(dpm_T - dpm_{NS})] \quad (1)$$

$$K_i = IC_{50}/[1 + (\text{radioligand concentration}/K_d)] \quad (2)$$

[15] Y. Cheng and W. H. Prusoff, *Biochem. Pharmacol.* **22,** 3099 (1973).

Conclusions

During the course of drug development, it is critical to assess the potential for a compound to induce levels of liver enzymes, including CYP3A4. With the discovery of the nuclear receptor PXR and the elucidation of its role in regulating CYP3A expression, it is now possible to predict the induction potential of a drug candidate before it is tested *in vivo*. The two assays outlined here provide researchers with the tools needed to profile the potency and efficacy of large numbers of compounds on PXR. The availability of these assays permits the removal of compounds with CYP3A4 induction potential at the very earliest stages of the drug discovery process, thus reducing costs and expediting the development of safer medicines.

Acknowledgment

The authors acknowledge the assistance of Thomas Consler.

[17] Analysis of CYP mRNA Expression by Branched DNA Technology

By Maciej Czerwinski, Peter Opdam, Ajay Madan, Kathy Carroll, Daniel R. Mudra, Lawrence L. Gan, Gang Luo, and Andrew Parkinson

Introduction

The branched DNA (bDNA) signal amplification assay provides a sensitive and relatively straightforward method for quantifying the levels of specific mRNAs. The assay is a nonpolymerase chain reaction method of RNA analysis that is, in principle, similar to an enzyme-linked immunosorbent assay (ELISA) except that oligonucleotide probes serve the function of antibodies. In the final stage of this microtiter plate assay, the mRNA of interest is linked to a branched DNA molecule bound with multiple molecules of alkaline phosphatase. The luminescent product of an alkaline phosphatase-catalyzed reaction is used to quantify the content of mRNA based on a linear relationship between luminescence and mRNA levels. The assay is suitable for the measurement of any mRNA molecule, in cultured cells or tissue, for which a specific probe set can be developed. In our laboratory, bDNA assay probe sets were developed to study the expression and regulation (inducibility) of the major human cytochrome P450 enzymes

(CYPs), UDP-glucuronosyltransferases (UGTs), and sulfonotransferases (SULTs) (www.xenotechllc.com).

Principle of Branched DNA Signal Amplification Assay

The bDNA signal amplification assay is illustrated schematically in Fig. 1. The assay, which is performed in special microtiter plate, requires a probe set for

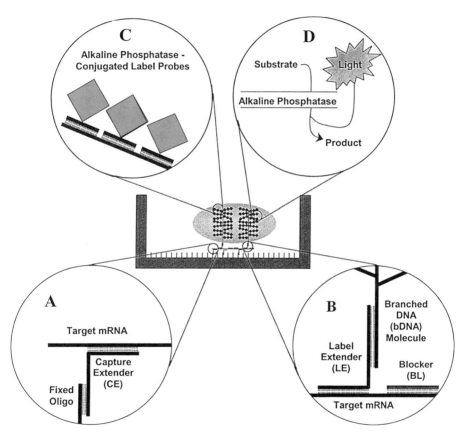

FIG. 1. Schematic representation of the bDNA signal amplification assay. The center depicts a single well containing all key components of the assay. (A) Function of the capture extender (CE) probe; (B) function of the label extender (LE) and blocker (BL) probes; (C) binding of several molecules of oligonucleotide-linked alkaline phosphatase to a single branch of the bDNA molecule; (D) generation of a luminescent signal from the hydrolysis of dioxetane substrate by alkaline phosphatase. Only two bDNA are bound to the mRNA of interest, whereas the average CYP probe set contains 10 or more LEs.

each mRNA of interest. Each probe set consists of three separate types of oligonucleotides: capture extenders (CE), label extenders (LE), and blockers (BL). The
CEs are bifunctional oligonucleotide probes. One end binds to complementary
capture probes that are attached covalently to the special (commercially available)
microtiter plate (Fig. 1A). The other end is designed to bind to the mRNA of
interest. Several CEs are used for each mRNA of interest; each binds to a different region of the mRNA molecule. The LEs are also bifunctional oligonucleotide
probes. One end binds to the mRNA of interest (at different regions to those recognized by the CEs); the other binds to a commercially available branched DNA
molecule (Fig. 1B). Blockers are oligonucleotides that bind to the mRNA of interest at sites that are not recognized by the CEs or LEs. The blockers fill in
sites that would otherwise be susceptible to degradation by RNase. Each branch
of the bDNA molecule can bind multiple molecules of alkaline phosphatase that
are linked covalently to complementary oligonucleotides, as shown in Fig. 1C.
Theoretically, up to 45 molecules of alkaline phosphatase can bind to one bDNA
molecule. Thus, an mRNA molecule bound to six LEs could theoretically bind 270
molecules of alkaline phosphatase, which accounts for the remarkable sensitivity
of the assay. However, steric hindrance likely constrains this process. A chemiluminescent signal is generated by addition of dioxetane, a substrate for alkaline
phosphatase, and plates are read with a 96-well luminometer (Fig. 1D).

Branched DNA Assay Probe Sets for Human CYP Genes

The bDNA probe set for each mRNA of interest is designed with ProbeDesigner
software[1] (Bayer Corp., Emeryville, CA). The probes are synthesized commercially and are stored at $-20°$ either lyophilized or solubilized in $1\times$ TE (10 mM
Tris–HCl, pH 8.0, 1 mM EDTA). Aliquots of CEs, LEs, and BLs are prepared at
50, 200, and 100 fmol/μl, respectively. On average, each CYP probe set consists of
about 30 oligonucleotides that hybridize to a 650-bp portion of the target mRNA.
The average CE is designed to hybridize with 22 bp of the target sequence. CEs are
bifunctional probes responsible for binding the target mRNA to the capture well.
They determine the specificity of the probe set. Table I summarizes the specificity
of human CYP probe sets designed in our laboratory. Individual CEs are identified
by their location within the respective probe set. A CYP probe set is considered
specific for the mRNA of interest if the sequences of CEs are not homologous
to other human CYP mRNAs. The homology is considered insufficient to form a
stable hybrid if three or more mismatched bases are present in a given CE. Although specificity of the majority of the probes for individual mRNAs has been
demonstrated, mRNAs of highly homologous genes in subfamilies 2A, 2C, and 3A
cannot be distinguished by the probe sets. We have not evaluated whether human

[1] S. J. Bushnell, J. Budde, T. Catino, J. Cole, A. Derti, R. Kelso, M. L. Collins, G. Molino, P. Sheridan,
J. Monahan, and M. Urdea, *Bioinformatics* **15,** 348 (1999).

TABLE I
FEATURES AND SPECIFICITY OF bDNA PROBE SETS FOR HUMAN CYP ENZYMES

Probe set designation (GenBank accession number)	Size of probe set's target (bp)	Number of probe set components[a]			Specificity of capture extenders[b]
		CE	LE	BL	
CYP1A1 (K03191)	486	6	12	7	All CEs specific for CYP1A1
CYP1A2 (M31665)	522	6	10	8	All CEs specific for CYP1A2
CYP1B1 (U03688)	632	7	16	11	All CEs specific for CYP1B1
CYP2A (AF182275)	844	7	14	20	CE 63.83: CYP2A6, 2A7, 2A13 CE 282.298: CYP2A6, 2A7, 2A13 CE 468.482: CYP2A6 CE 483.505: CYP2A6, 2A7, 2A13 CE 506.542: CYP2A6, 2A7, 2A13 CE 760.778: CYP2A6, 2A7, 2A13 CE 879.906: CYP2A6
CYP2B6 (AF182277)	649	7	12	12	All CEs specific for CYP2B6
CYP2C (M61856)	735	6	13	14	CE 456.475: CYP2C8, 2C9, 2C18, 2C19 CE 476.496: CYP2C8, 2C9, 2C18, 2C19 CE 678.702: CYP2C8, 2C18 CE 990.1016: CYP2C8, 2C9, 2C18, 2C19 CE 1040.1061: CYP2C9, 2C18, 2C19 CE 1173.1194: CYP2C8, 2C9, 2C18, 2C19
CYP2D6 (X08006)	494	6	14	9	All CEs specific for CYP2D6
CYP2E1 (J02843)	632	9	14	8	All CEs specific for CYP2E1
CYP2F1 (J02906)	700	8	13	13	All CEs specific for CYP2F1
CYP3A (M18907)	592	6	16	4	CE 2.22: CYP3A4 CE 23.42: CYP3A4, 3A43 CE 43.66: CYP3A4, 3A43 CE 115.135: CYP3A4, 3A5, 3A7 CE 458.482: CYP3A4, 3A5, 3A7, 3A43 CE 575.594: CYP3A4, 3A5, 3A7
CYP4A11 (L04751)	832	7	15	12	All CEs specific for CYP4A11
CYP4B1 (J02871)	754	7	14	12	All CEs specific for CYP4B1

[a] CE, Capture extender; LE, label extender; BL, blocker.
[b] Sequence of the probes was analyzed with BLAST algorithm [S. F. Altschul, T. L. Madden, A. A. Schäffer, J. Zhang, Z. Zhang, W. Miller, and D. J. Lipman, *Nucleic Acids Res.* **25,** 3389 (1997)]. Pseudogenes were excluded from the analysis.

TABLE II
SPECIFICITY OF CYP2D6 PROBE SET FOR MAJOR ALLELES OF THE GENE

Allele[a]	Nucleotide changes	Probe set element affected by polymorphism
CYP2D6*1A	None	
CYP2D6*1B	Outside probe-targeted region	
CYP2D6*2	$G_{1749}C$ (exon 3)	BL 433.452
CYP2D6*3	Outside probe-targeted region	
CYP2D6*4A	$C_{188}T$ (exon 1)	LE 118.134
	$C_{1062}A$ (exon 2)	LE 297.313
	$A_{1072}G$ (exon 2)	LE 297.313
	$C_{1085}G$ (exon 2)	LE 314.330
	$G_{1749}C$ (exon 3)	BL 433.452
CYP2D6*4B, *4C	Outside probe-targeted region	
CYP2D6*4D	$C_{1127}T$ (exon 2)	LE 348.366
CYP2D6*5	CYP2D6 deleted	
CYP2D6*6A, *6B	T_{1795} deleted (exon 3)	LE 471.489
CYP2D6*7	Outside probe-targeted region	
CYP2D6*8	$G_{1749}C$ (exon 3)	BL 433.452
	$G_{1846}T$ (exon 3)	BL 523.539
CYP2D6*9	Outside probe-targeted region	
CYP2D6*10A,	$C_{188}T$ (exon 1)	LE 118.134
*10B, *10C	$C_{1127}T$ (exon 2)	LE 348.366
	$G_{1749}C$ (exon 3)	BL 433.452
CYP2D6*11	($G_{971}C$, splicing defect)	
	$G_{1749}C$ (exon 3)	BL 433.452
CYP2D6*12	$G_{212}A$ (exon 1)	BL 135.152
	$G_{1749}C$ (exon 3)	BL 433.452
CYP2D6*13	Hybrid with CYP2D7P exon 1, extra T	CE 153.169
CYP2D6*14	$C_{188}T$ (exon 1)	LE 118.134
	$G_{1846}T$ (exon 3)	BL 523.539
CYP2D6*15	T_{226} insertion (exon 1)	CE 153.169
CYP2D6*16	Hybrid with CYP2D7P	
CYP2D6*17	$C_{1111}T$ (exon 2)	LE 331.347
	$G_{1726}C$ (exon 3)	BL 401.413

[a] Allele nomenclature and numbering of nucleotides according to A. K. Daly, J. Brockmöller, F. Broly, M. Eichelbaum, W. E. Evans, F. J. Gonzalez, J.-D. Huang, J. R. Idle, M. Ingelman-Sundberg, T. Ishizaki, E. Jacqz-Aigrain, U. A. Meyer, D. W. Nebert, V. M. Steen, C. R. Wolf, and U. M. Zanger, *Pharmacogenetics* **6**, 193 (1996). Only nucleotide changes in the probe-targeted region of the mRNA are listed.

CYP probes can be used to examine CYP mRNA expression in other species, although species-specific probe sets can be designed for this purpose. For example, probe sets have been described for rat CYP enzymes.[2]

The specificity of the CYP2D6 probe set was examined for the major alleles of this extensively polymorphic gene. The probe set was developed for the wild-type

[2] D. P. Hartley and C. D. Klaassen, *Drug Metab. Dispos.* **28**, 608 (2000).

allele *CYP2D6*1*. The analysis presented in Table II indicates that polymorphic nucleotides of *CYP2D6*1B, CYP2D6*3, CYP2D6*4B, *4C, CYP2D6*7,* and *CYP2D6*9* are located outside of the probe-targeted region and would not be expected to affect hybridization of the probe set. The polymorphic nucleotides of *CYP2D6*2, CYP2D6*4A, CYP2D6*5, CYP2D6*6A, *6B, CYP2D6*8, CYP2D6*10A, *10B, *10C, CYP2D6*12, CYP2D6*14, CYP2D6*16,* and *CYP2D6*17* are located in the probe-targeted region; however, they affect hybridization with LEs and BLs, but not with CEs. Therefore, they are not expected to change specificity of the probe set. The polymorphism of *CYP2D6*13* and *CYP2D6*15* is expected to reduce the melting temperature of the hybrids between CE153.169 and their mRNAs as compared with *CYP2D6*1* mRNA. Because the mismatch is limited to only 1 bp in one of the six CEs, its effect on the specificity of the probe set is probably minimal.

Branched DNA Assay Conditions

The bDNA assay kit is currently available in two versions, both of which are sold by Bayer Corp. (Emeryville, CA). The QuantiGene bDNA signal amplification kit contains one 96-well capture plate and corresponding amounts of reagents (which do not include the probe set). The QuantiGene high-volume kit contains fifty 96-well capture plates and an appropriate supply of reagents, except for wash buffer, which is prepared by the user. The protocols provided by the manufacturer with the two kits are summarized and compared in Table III. They are intended for assaying expression of one gene per well in cells grown in a 96-well format. In our laboratory, human and rat hepatocytes are often cultured in 60-mm petri dishes, and multiple mRNAs are quantified from a single dish. For mRNA analysis with the QuantiGene high-volume kit, the following protocol, adopted from the manufacturer's recommendations, has been developed and used in our laboratory. The 96-well capture plate and the reagents used in the assay, except the probes and the wash buffer (0.1× SSC, 0.03% lithium lauryl sulfate), are provided in the kit. Freshly isolated hepatocytes are cultured on collagen-coated petri dishes with a Matrigel overlay and are treated with CYP inducers as described previously.[3–6] Cells in a 60-mm dish are lysed at 46° for 2 hr with 1.5 ml of lysis buffer mixed with 3.0 ml of cell culture medium containing 10% fetal bovine serum (FBS). Because cell lysis is essential for blocking the capture wells, the cells are pipetted up and down 10 times with a 10-ml pipette. Upon completion of the lysis process, which

---

[3] E. L. LeCluyse, P. Bullock, A. Parkinson, and J. H. Hochman, *in* "Model Systems for Biopharmaceutical Assessment of Drug Absorption and Metabolism" (R. T. Borchardt, G. Wilson, and P. Smith, eds.), p. 121. Plenum Press, New York, 1996.

[4] B. Quistorff, J. Dich, and N. Grunnet, *in* "Methods in Molecular Biology: Animal Cell Culture" (J. W. Pollard and J. M. Walker, eds.), p. 151. Humana Press, Clifton, NJ, 1989.

[5] P. O. Seglen, *in* "Metabolism in Cell Biology" (D. M. Prescott, ed.), p. 29. Academic Press, New York, 1976.

[6] P. O. Seglen and P. B. Gordon, *Biochim. Biophys. Acta* **630,** 103 (1980).

TABLE III
ASSAY PROTOCOLS FOR QUANTIGENE bDNA SIGNAL AMPLIFICATION AND HIGH-VOLUME KITS

Assay step	QuantiGene bDNA signal amplification kit (volume per capture well)	QuantiGene high-volume kit (volume per capture well)
Cell lysis (cells grown in 96-well plate)	100 μl of lysis buffer per well 30 min at 37°	50 μl of lysis buffer per well 15 min at 37°
Capture of target mRNA		
Hybridization reaction	At least 16 hr at 53°	16–20 hr at 53°
Capture hybridization buffer[a]	100 μl, to block nonspecific binding	Not required
Cell culture media	Not required	
Lysis buffer	100 μl, containing target mRNA	100 μl, containing target mRNA
Capture extenders	240 attomol/μl	50 μl
Label extenders	970 attomol/μl	240 attomol/μl
Blockers	490 attomol/μl	960 attomol/μl
		480 attomol/μl
Posthybridization		
Cooling the plate	Cool plate to room temperature	Cool plate to 46°
Washing the wells	Two washes with 400 μl of wash A[b]	Two washes with 400 μl of wash buffer[c]
Amplification and detection of target mRNA		
Hybridization with amplifier	50 μl of amplifier working reagent 30 min at 53°	100 μl of amplifier working reagent 60 min at 46°
Postamplification wash	Two washes with 400 μl of wash A[b]	Two washes with 400 μl of wash buffer[c]
Hybridization with label probe	50 μl of label probe working reagent 15 min at 53°	100 μl of label probe working reagent 60 min at 46°
Postlabeling wash	Two washes with 400 μl of wash A[b] Three washes with 400 μl of wash D[b]	Two washes with 400 μl of wash buffer[c]
Incubation with substrate	50 μl of substrate mixture 30 min at 37°	100 μl of substrate working reagent 60 min at 46°
Reading the plate	Read immediately after incubation	Read at 45° immediately after incubation

[a] Capture hybridization buffer blocks nonspecific binding of nucleic acids to capture wells. In the high-volume kit, the function is fulfilled by cellular proteins.

[b] Washes A and D are provided with the kit from Bayer Diagnostics. The composition of the buffers is proprietary.

[c] Buffer is prepared by the user.

can be monitored visually with a microscope, the lysate can be stored at $-20°$. On the first day of the analysis, 75 μl of lysate is transferred to individual wells of the capture plate followed by the addition of 25 μl of lysis buffer [diluted 1 : 3 with 10% FBS in phosphate-buffered saline (PBS)] containing the desired probes at concentrations resulting in final concentrations of CEs, LEs, and BLs of 240, 960, and 480 attomol/μl, respectively. Hybridization is carried out for 16–20 hr at 53° with the plate sealed with an aluminum foil plate sealer (provided with the kit). On the second day of the assay, the capture plate is cooled to 46°, the hybridization mixture is aspirated, and the wells are washed two times with approximately 400 μl of wash buffer. Following the second wash, 100 μl of bDNA amplifier molecule is added to each well, and the plate is resealed and incubated for 60 min at 46°. After incubation, the amplifier is aspirated, and the wells are again washed twice with approximately 400 μl of wash buffer. Subsequently, 100 μl of label probe is added to each well, and the plate is sealed and incubated for 60 min at 46°. The hybridization of label probe is stopped by aspirating the wells, which are then washed twice with 400 μl of wash buffer per well. The proprietary chemiluminescent substrate (dioxetane) is added to each well, and the reaction is carried out for 30 min at 46°. The plate is read with a Model FL600 microplate fluorescence reader (Bio-Tech Instruments, Winooski, VT) in luminometric mode at 45°. The level of expression of CYP genes is normalized to the constitutively expressed

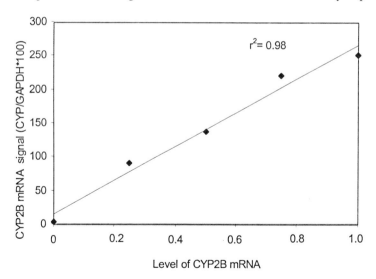

FIG. 2. Relationship between mRNA level and chemiluminescence in the bDNA signal amplification assay. Messenger RNA from rat hepatocytes treated with phenobarbital (which induces CYP2B mRNA) was mixed in various proportions with control hepatocytes (which contain little or no CYP2B mRNA). As the proportion of "phenobarbital" sample increased from 0 to 100%, the proportion of "control" sample decreased from 100 to 0%. The signal generated with a CYP2B oligonucleotide probe set increased linearly with the amount of CYP2B mRNA in the "phenobarbital" sample.

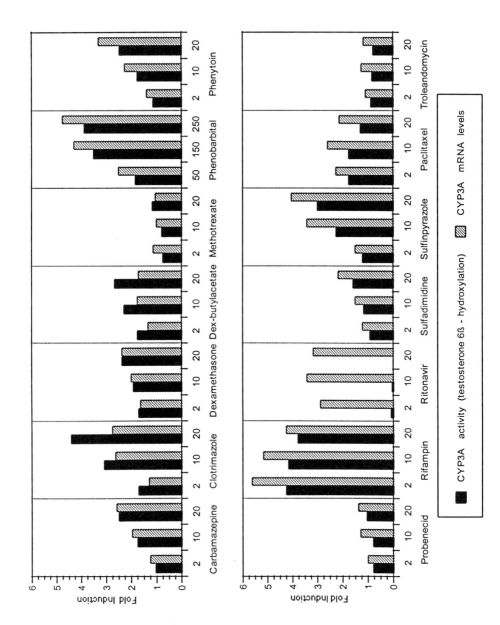

gene glyceraldehyde-3-phosphate dehydrogenase (GAPDH). Like the CYP probe sets, the GAPDH probe set is a product of XenoTech LLC (Lenexa, KS). The hybridization steps of the assay can be performed in common laboratory incubators or hybridization ovens, and luminescence can be read with a variety of inexpensive plate readers.

Linearity of Branched DNA Assay Response

To test the linearity of the response of the bDNA assay, cell lysates from control and phenobarbital-treated rat hepatocytes are mixed in known quantities (0, 25, 50, 75, and 100% of the "phenobarbital" lysate). Both lysates contain GAPDH mRNA, but only the lysate from phenobarbital-treated hepatocytes contains high levels of CYP2B1/2 mRNA. As shown in Fig. 2, the GAPDH-normalized CYP2B1/2 signal increases linearly with the amount of "phenobarbital" lysate ($r^2 = 0.98$). The rat CYP2B1/2 probe set was a generous gift of Dr. C. D. Klaassen (University of Kansas Medical Center, Kansas City, KS).[2]

The result in Fig. 3, which depicts the induction of CYP3A mRNA and CYP3A activity in human hepatocytes treated with various inducers, serves to illustrate the utility of the human probe sets. Freshly isolated human hepatocytes are cultured and treated with various CYP3A inducers as described earlier. Data are normalized to GAPDH and are presented as fold induction of CYP3A mRNA. With one notable exception, there is a good correspondence between the induction of CYP3A activity (measured as testosterone 6β-hydroxylase activity) and CYP3A mRNA levels. The exception is Ritonavir, which increases CYP3A mRNA levels but not CYP3A activity, presumably due to the fact that Ritonavir is a potent inhibitor of CYP3A activity.[7]

Conclusions

The CYP induction studies performed in our laboratory with the bDNA signal amplification assay provided a reliable alternative to traditional techniques, such as enzyme analysis and immunoblotting, to detect the induction of CYP enzymes in human hepatocytes. The protocol we developed allows human hepatocytes cultured in one 60-mm petri dish to be analyzed for changes in expression of up to 30 genes, in duplicate. In addition to probe sets for the major human CYPs, we have also designed and tested probe sets for numerous human UGT and SULT genes. Further information is available on the World Wide Web at www.xenotechllc.com.

[7] V. A. Eagling, D. J. Back, and M. G. Barry, *Br. J. Clin. Pharmacol.* **44**, 190 (1997).

FIG. 3. Application of the bDNA assay to measure the induction of CYP3A mRNA levels in primary cultures of human hepatocytes. Human hepatocytes were treated with various CYP3A inducers at three concentrations (typically 2, 10, and 20 μM). After 3 days of treatment, CYP3A mRNA levels were measured with the bDNA assay and CYP3A activity was measured based on 6β-hydroxylation of testosterone. See text for discussion.

[18] Ligand-Induced Coactivator Recruitment to Peroxisome Proliferator-Activated Receptorα Characterized by Fluorescence Resonance Energy Transfer

By Bowman Miao, Shaoxian Sun, Linda D. Santomenna, Ji Hu Zhang, Peter Young, and Ranjan Mukherjee

Nuclear receptors are ligand-activated transcription factors that modulate gene expression in response to natural or synthetic ligands. On the basis of sequence homology between them, certain domains have been identified. These are the N-terminal activating function domain (AF-1), the central DNA-binding domain (DBD), and the C-terminal ligand-binding domain (LBD), which also contains the ligand-dependent activating function (AF-2). Nuclear receptors bind to specific DNA sequences (response elements) as homodimers (e.g., estrogen receptor, glucocorticoid receptor) or as heterodimers with the retinoid X receptor (RXR).

Peroxisome proliferator activated receptors (PPARs) belong to the latter category. PPARs bind as a heterodimer with RXR to specific DNA sequences, the peroxisome proliferator response elements (PPREs), which are usually a direct repeat of GGTCA spaced by one nucleotide (DR-1). There are three PPAR subtypes: PPARα, PPARβ (PPARδ), and PPARγ.[1] Fatty acids, hypolipidemic drugs (fibrates), and various plasticizers and xenobiotics are ligands for PPARα. However, they are very weak ligands (low affinities). Traditionally, ligand affinities for receptors are determined by labeled ligand displacement assays. The scintillation proximity assay (SPA) is such an assay.

PPARs are ligand-activated transcription factors. On binding of an agonist ligand, there is a conformational change, which results in recruitment of one or more coactivators, a family of accessory proteins that make favorable contacts with the basal transcription machinery and initiate transcription of the PPAR responsive gene(s). Coactivators fall into certain families, e.g., SRC-1 or p-160 family[2] and CREB-binding protein (CBP)/p300.[3,4] These coactivators interact with the C terminus (helix 12 containing the AF-2 domain) of the nuclear receptors via a common signature LXXLL motif. The AF-2 helix forms a unique hydrophobic pocket in the presence of agonist ligands favoring binding of coactivators. This structure is not present in the unliganded state and is not induced by antagonists.

---

[1] T. M. Willson, P. J. Brown, D. D. Sternbach, and B. R. Henke, *J. Med. Chem.* **43,** 527 (2000).

[2] S. A. Onate, S. Y. Tsai, M. J. Tsai, and B. W. O'Malley, *Science* **270,** 1354 (1995).

[3] D. Chakravarti, V. J. LaMorte, M. C. Nelson, T. Nakajima, I. G. Schulman, H. Juguilon, M. Montminy, and R. M. Evans, *Nature* **383,** 99 (1996).

[4] Y. Kamei, L. Xu, T. Heinzel, J. Torchia, R. Kurokawa, B. Gloss, S. C. Lin, R. A. Heynman, D. W. Rose, C. K. Glass, and M. G. Rosenfeld, *Cell* **85,** 403 (1996).

Ligand-dependent coactivator recruitment is another assay to measure binding affinities of ligands for receptors. This was first demonstrated as a coactivator-dependent receptor ligand assay (CARLA) for PPARα[5] and was later modified to a high-throughput format using the fluorescence resonance energy transfer assay.[6]

This article describes how to measure the affinity of ligands for PPARα using the fluorescence resonance energy transfer (FRET) assay utilizing a small peptide derived from coactivators containing the LXXLL receptor interaction site. We also compare the results with more traditional methods, such as scintillation proximity assay.

Materials

c-*myc* antibody (9E10) and c-*myc* antibody affinity resin are prepared by coupling the c-myc5 antibody (generated at DuPont Pharmaceuticals) to Reacti gel (Pierce Chemical Company, Rockford, IL) following the manufacturer's protocol. The Mono S 10/10 prepacked column is from Amersham Pharmacia (Piscataway, NJ). Strepavidin-conjugated allophycocyanin (APC) and Eu-labeled anti-c-*myc* antibody are from Wallac (PerkinElmer Life Sciences, Wellesley, MA). Peptides derived from coactivators SRC-1 (CPSSHSSLTERHKILHRLLQEGSPS) and CBP (SGNLVPDAASKHKQLSELLRGGSGS) are synthesized by Synpep (Dublin, CA) and labeled by biotin via an AHX spacer or fluorescein via the free amino group at the N terminus. The peptide concentration is determined by quantitative amino acid analysis. A complete protease inhibitors set is from Boehringer Mannheim Corporation (Indianapolis, IN). Nonidet P-40 (NP-40) and bovine serum albumin (BSA) are from Calbiochem (La Jolla, CA), EDTA from Gibco-BRL (Rockville, MD), and dithiothreitol (DTT) and glycerol from Sigma (St. Louis, MO). GW9578,[7] GW2331,[8] L165041,[9] BRL49653 (rosiglitazone), and fenofibric acid are synthesized at DuPont Pharmaceuticals. Ciprofibrate, gemfibrozil, and fenofibrate are from Sigma (St. Louis, MO), and WY14643 is from Chemsyn Science Labs (Lenexa, KS). ^{3}H-labeled GW2331[8] is synthesized at DuPont Pharmaceuticals.

[5] G. Krey, O. Braissant, F. L'Horset, E. Kalkoven, M. Perroud, M. G. Parker, and W. Wahli, *Mol. Endocrinol.* **11**, 779 (1997).

[6] G. Zhou, R. Cummings, Y. Li, S. Mitra, H. A. Wilkinson, A. Elbrecht, J. D. Harmes, J. M. Schaeffer, R. G. Smith, and D. F. Moller, *Mol. Endocrinol.* **12**, 1594 (1998).

[7] P. J. Brown, D. A. Winegar, K. D. Plunket, L. B. Moore, M. C. Lewis, J. G. Wilson, S. S. Sundreth, C. S. Koble, Z. Wu, J. M. Chapman, J. M. Lehman, S. A. Kliewer, and T. M. Willson, *J. Med. Chem.* **42**, 3785 (1999).

[8] S. A. Kliewer, S. S. Sundreth, S. A. Jones, P. J. Brown, G. B. Wisely, C. S. Koble, P. Devchand, W. Wahli, T. M. Willson, J. M. Lenhard, and J. M. Lehmann, *Proc. Natl. Acad. Sci. U.S.A.* **94**, 4318 (1997).

[9] J. Berger, M. D. Leibowitz, T. W. Doebber, A. Elbrecht, B. Zhang, G. Zhou, C. Biswas, C. A. Cullinan, N. S. Hayes, Y. Li, M. Tanen, J. Ventre, M. S. Wu, G. D. Berger, R. Mosley, R. Marquis, C. Santini, S. P. Sahoo, R. L. Tolman, R. G. Smith, and D. E. Moller, *J. Biol. Chem.* **274**, 6718 (1999).

Production of Recombinant PPARα

The ligand-binding domain (LDB) of human PPARα is amplified from human small intestine cDNA using *Pfu* polymerase (Stratagene, La Jolla, CA) and gene-specific primers (GAATGTAGAATCTGCGGGGAC and GGATCCTCAGTACA-TGTCCCTGTAGATC). The resulting polymerase chain reaction (PCR) product is sequence confirmed and used as a template in a subsequent PCR to add a 5′ *Swa*I restriction site, a methionine and N-terminal *myc* tag (ATTTAAATATGGCGGAA-CAGAAACTGATTTCTGAAGAAGATCTGGAATGTAGAATCTGCGGGGAC and GGATCCTCAGTACATGTCCCTGTAGATC). The epitope-tagged LBD is cloned into a T7 expression vector as a *Swa*I/*Bam*HI fragment, sequence confirmed, and then transformed into the BL21(DE3) strain of *Escherichia coli*. This is grown in M9 medium supplemented with 2% (w/v) casamino acids and 3% (w/v) glycerol in a 10-liter BioLafitte fermenter maintained at 30° until the OD_{600} reaches 12. The temperature is then shifted to 20°, and the expression of PPARα is induced by the addition of 1 mM isopropylthiogalactoside (IPTG). Cells are harvested 5 hr after induction.

Thirty grams of cell pellet is suspended in 300 ml lysis buffer containing 50 mM Tris, pH 8.0, 100 mM NaCl, 10% glycerol, 1 mM EDTA, 1 mM DTT, and a complete set of protease inhibitors (Boehringer Mannheim Corporation, Indianapolis, IN). Cells are lysed by passing through a microfluidizer (Microfluidics Corporation), and the cell debris is removed by centrifugation. The supernatant is collected and loaded on a c-*myc* antibody affinity column (Pharmacia XK26) at a flow rate of 2 ml/min. The resin is washed with lysis buffer, and PPARα is eluted with a buffer containing 50 mM Tris, pH 8, 500 mM NaCl, 500 μM cmyc peptide, 10% glycerol, and 1 mM DTT. Fractions containing PPARα are pooled and dialyzed against dialysis buffer (50 mM Tris, pH 8.0, 2 mM DTT, 10% glycerol) overnight at 4°. The dialyzed PPARα pool is then loaded on a Mono S 10/10 column. After extensive washing with dialysis buffer, PPARα is eluted with a linear gradient of 0–1 M NaCl. PPARα elutes off at ∼300 mM NaCl. PPARα fractions are pooled, adjusted to 10 mM fresh DTT and 50% glycerol, and stored at −20°. All purification steps are carried out at 4°.

Protein concentration is determined using the Bio-Rad protein assay kit (Bio-Rad Laboratories, Hercules, CA) with horse IgG (immunoglobulin G) as the standard (Pierce Chemical Company, Rockford, IL). The purity of PPARα is analyzed by Coomassie Blue-stained SDS–PAGE gel and by Western blotting with c-*myc* antibody 9E10. The purity is estimated >95%.

Scintillation Proximity Assay

The assay is performed in a 96-well format, using white OptiPlate (Packard Bio-science Company, Gronigen, The Netherlands). The reaction mix (100 μl per well) contains assay buffer (PBS, 2 mM EDTA, 1 mM DTT, 2 mM CHAPS, and

10% glycerol, pH 7.2), 0.25 mg polylysine-coated yttrium silicate SPA beads (Amersham Pharmacia Biotech), 20 nM [^{3}H]GW2331, and 70 nM recombinant c-*myc*-hPPARα. The Optiplate is first spotted with 1.0 μl of test compound per well, prepared in dimethyl sulfoxide (DMSO). The remaining components of the reaction mix are then dispensed, and the plate is incubated for 1 hr at room temperature on a shaking platform. The beads are then allowed to settle for 30 min at room temperature prior to counting. Radioactivity is quantified in a Packard Topcount scintillation counter using parameters recommended by Packard for yttrium silicate SPA beads. Each data point is the mean of duplicate determinations. Curve fitting is performed using Prism (Graphpad Software, Inc., San Diego, CA).

Fluorescence Resonance Energy Transfer Assay

Ligand-dependent recruitment of coactivator SRC-1 and CBP peptides by PPARα is characterized by a FRET assay. The assay is performed in 96-well black plates (Packard) and analyzed with a Discovery instrument (Packard). c-*myc*-tagged PPARα (20 nM) is mixed with 1 nM Eu-anti-c-*myc* antibody, 100 nM streptavidin–allophycocyanin (APC), and 0.4–1 μM biotin–peptide in the presence of ligand in 100 μl of assay buffer (50 mM Tris, pH 8, 50 mM KCl, 1 mM DTT, 1 mM EDTA, 1 mg/ml BSA, 0.5 % NP-40, and 10 % glycerol) at room temperature. Ligands are dissolved and diluted serially in DMSO. The final DMSO concentration in the assay is 0.5%. Wells containing the same assay components except the ligand are treated as the background. The FRET signal is measured as the emission intensity ratio at 665 to 620 nm multiplied by a factor of 10^4. The FRET signal is fully developed after incubation for 4 hr at room temperature and is stable for at least 3 hr. Measurements are performed under equilibrium conditions. The background signal (usually about 250) is subtracted from the FRET reading and then normalized to the maximal FRET signal from a standard agonist set at 100%. EC$_{50}$ values are obtained using a nonlinear curve-fitting program (Prism).

Description of FRET Assay

Fluorescence resonance energy transfer measures the energy transfer between two fluorophores. For the energy transfer to occur, two conditions have to be satisfied. First, emission spectra of the donor fluorophore must overlap with absorption spectra of the acceptor fluorophore; second, the two fluorophores must be in close proximity. FRET has been used to monitor conformational changes in proteins and intermolecule interactions such as receptor–ligand binding and protein–protein interactions. For lathanide ions such as europium, excitation and emission maxima can differ by about 300 nm. They also have a long fluorescence lifetime of $\sim$1 msec. These properties allow measurements to be made in a time-resolved mode. When lathanides are used as fluorophore donors, FRET is

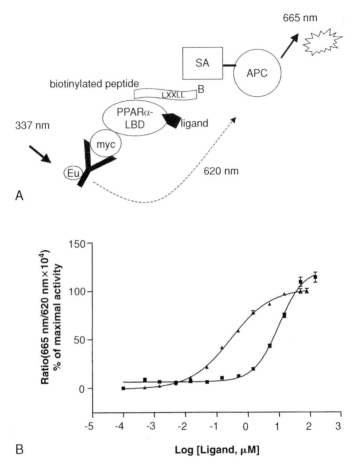

FIG. 1. Scheme for the FRET assay using coactivator peptides (A). PPARα agonist ligands induce a conformational change that results in recruitment of the coactivator peptide and the subsequent resonance energy transfer and emission. GW9578 (▲) and fenofibric acid (■) recruits SRC-1 peptide to PPARα (B). The FRET assay was performed with increasing concentrations of ligands as described in the text. EC_{50} values from dose–response curves were calculated using a nonlinear curve-fitting program (Prism) and the values indicated in Table I. GW9578 is the standard.

measured as time-resolved fluorescence at the emission wavelength of the acceptor fluorophore allophycocyanin (APC). This also reduces short-lived background fluorescence.

Figure 1A shows the strategy for ligand-induced coactivator recruitment measured by the FRET assay. The ligand-binding domain of PPARα has a *myc*-tag at the N terminus, which enables the europium cryptate-labeled anti-c-*myc* antibody to bind to PPARα. The biotin-conjugated coactivator peptide (25 amino acids long) binds noncovalently to strepavidin-conjugated APC. In the presence of agonist ligand, the coactivator peptide is recruited to PPARα. If europium is now excited by radiation at 337 nm, it emits at 620 nm. This energy is transferred to the second fluorophore APC if the two are in close proximity. APC then emits at 665 nm. The intensity of FRET is the ratio of the emission at 665 to 620 nm multiplied by 10^4. This is a measure of the distance and orientation between the two fluorophores and occurs only if coactivator peptide is induced to bind to PPARα by an agonist ligand.

Results

Two agonists of PPARα, fenofibric acid and GW9578, were tested in the FRET assay (Fig. 1B) using a peptide from the coactivator SRC-1. A dose-dependent increase in the FRET signal was observed reflecting the recruitment of the coactivator peptide. The EC_{50} for fenofibric acid was 10 μM and that of GW9578 was 0.3 μM, demonstrating that GW9578 is more potent than fenofibric acid in recruiting SRC-1.

To determine if EC_{50} values vary with different coactivator peptides, we tested seven PPAR ligands for their ability to recruit peptides derived from the coactivators SRC-1 and CBP (Fig. 2). The EC_{50} values are shown in Table I. Dose-dependent recruitment of coactivator peptides was seen with all ligands except fenofibrate. The rank order of the EC_{50} values with SRC-1 was GW9578 < GW2331 < fenofibric acid < WY14643 = ciprofibrate < L-165041 < gemfibrozil. Although BRL49653 (rosiglitazone) showed recruitment of coactivator peptides at high concentrations, a saturation dose–response curve was not obtained, hence an EC_{50} could not be calculated. Interestingly, the rank order was the same for the CBP peptide, although the EC_{50} values were lower than the corresponding values with SRC-1. The maximal FRET signals at saturation ligand concentration were different for CBP and SRC (e.g., 2600 and 1500, respectively, for GW2331), suggesting that

TABLE I
COMPARISON OF POTENCIES OBTAINED IN SPA AND FRET ASSAYS

Ligand	EC_{50} (μM) in FRET assays		K_d (μM) SPA
	SRC-1	CBP	
GW9578	0.3		0.12
Fenofibric acid	10		>100
GW2331	0.39	0.06	0.17
L165041	133	11	5.8
Gemfibrozil	184	29	>100
WY 14643	14	2	24
Ciprofibrate	14.6	2.3	58

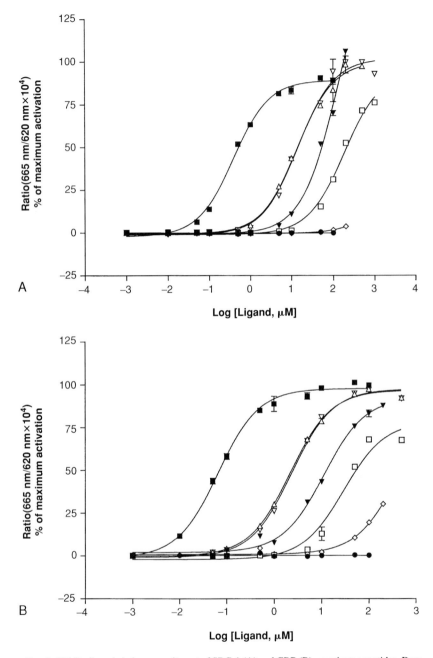

FIG. 2. PPARα ligands induce recruitment of SRC-1 (A) and CBP (B) coactivator peptides. Dose–response curves were generated with GW2331 (■), L165041 (▼), fenofibrate (●), gemfibrozil (□), WY14643 (△), ciprofibrate (▽), and BRL49653 (◇). EC_{50} values were calculated as indicated in Fig. 1 and shown in Table I. GW2331 is the standard.

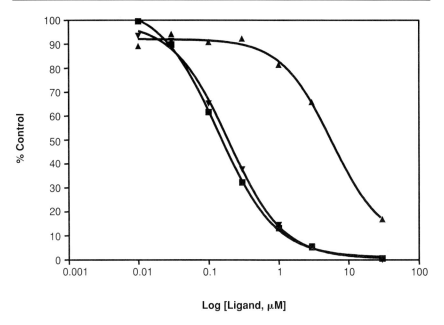

FIG. 3. Affinity of PPARα ligands using the scintillation proximity assay. The assay was performed as described in the text. Nonspecific binding was determined using a 500-fold excess of unlabeled GW2331 (▼). ■, GW9578; ▲, L165041.

the three-dimensional conformation of the SRC-1-bound receptor is different from the CBP-bound receptor.

To determine how EC_{50} values obtained by FRET assays compare to the affinities determined by traditional labeled ligand displacement assays, we determined the K_d values by the scintillation proximity assay. The K_d values for GW9578, GW2331, and L165041 are close to the EC_{50} values obtained with the FRET assay (Fig. 3 and Table I). However, no displacement of labeled ligand was obtained with fenofibric acid or gemfibrozil using doses as high as 100 μM (Table I and data not shown). The K_d values for Wy14643 and ciprofibrate were 24 and 58 μM, respectively (Table I and data not shown). Together, these data suggest that different ligands induce distinct conformations to the receptor and that the presence of the coactivator peptide may alter ligand-binding equilibrium.

Conclusion

The FRET assay has several advantages over conventional assays for determining affinities of PPARα ligands. It is a nonradioactive assay and is more sensitive than the SPA for certain ligands, such as gemfibrozil and fenofibric acid. A known, high-affinity (labeled) ligand is not required as in SPA and is therefore ideal for identifying agonist ligands, especially for orphan receptors. The FRET assay can also be performed in the presence of a known agonist to detect receptor antagonists. However, the reagents are relatively expensive. Because the peptide is only a small part of the whole coactivator and the coactivators have multiple receptor interacting domains, the affinities determined with a peptide may be different from that using the whole coactivator protein.

[19] Fluorescence-Based Ligand-Binding Assays for Peroxisome Proliferator-Activated Receptors

By DOUGLAS J. A. ADAMSON and COLIN N. A. PALMER

Introduction

CYPs are known to be regulated by a wide range of xenobiotics, and it appears this phenomenon is generally receptor mediated.[1] Elucidation of the mechanism of induction of CYPs has been hampered by the fact that most xenobiotics induce CYPs only at very high concentrations (μM to mM) that are not suitable for the biochemical isolation of specific receptors. A notable exception to this is the aryl hydrocarbon receptor (AhR) and the use of radiolabeled TCDD in its characterization.[2] Ironically, it was in the search for the AhR that a novel nuclear receptor was characterized that was activated by a very large range of xenobiotics, known as peroxisome proliferators (PPs).[3] Peroxisome proliferation is mediated by PPARα, although many PPs also activate PPARδ and/or PPARγ. PPARs appear to be fatty acid receptors that regulate whole body energy homeostasis and are therefore attractive targets for the design of drugs for the treatment of metabolic diseases such as diabetes, obesity, and atherosclerosis.[4] Compounds that bind and activate PPARs are fatty acid mimetics, and it has been observed that a wide range of compounds that have been characterized as modulators of eicosanoid metabolism or signaling can also interact with PPARs. The most noted interactions are those of LTB$_4$ with PPARα, PGI$_2$ with PPARδ, and 15-deoxy $\Delta^{12,14}$-PGJ$_2$ with PPARγ. Therefore, PPARs appear to modulate energy homeostasis and are also involved in cell signaling in inflammation and differentiation.

[1] U. Savas, K. J. Griffin, and E. F. Johnson, *Mol. Pharmacol.* **56,** 851 (1999).
[2] W. F. Greenlee and A. Poland, *J. Biol. Chem.* **254,** 9814 (1979).
[3] I. Issemann and S. Green, *Nature* **347,** 645 (1990).
[4] S. Kersten, B. Desvergne, and W. Wahli, *Nature* **405,** 421 (2000).

Ligand Binding to PPARs

PPARγ was the first PPAR to be shown to bind directly to agonists. The group at Glaxo Wellcome demonstrated that the tritiated rosiglitazone could bind recombinant PPARγ.[5] In these experiments, protein-bound radioactive rosiglitazone was separated from an unbound radioactive compound by rapid size-exclusion chromatography. Specific binding was determined by competition with an excess of unlabeled rosiglitazone. The Glaxo Wellcome group then went on to show that prostaglandins, such as PGJ, and also non steroidal anti-inflammatory drugs (NSAIDs), such as indomethacin, also bind directly to PPARγ.[6–8]

The lack of availability of high-affinity radioactive ligands for PPARs meant that many other groups had to look for alternative means to assay for ligand binding to PPARs, especially for the other isoforms PPARα and PPARδ. Several groups have used protease sensitivity as an indicator of conformation changes in ligand binding.[9–12] Other groups used the ligand dependency of PPAR binding to coactivator proteins as an indicator of ligand binding.[13] The ability of PPAR to form a heterodimer on DNA was also shown to be ligand specific and was used to demonstrate the binding of a wide range of fatty acids and PPs to PPARα and PPARδ.[14]

Binding of drugs and fatty acids to all three isoforms has been confirmed by X-ray crystallography by groups at Glaxo Wellcome and Astra-Zeneca.[15–19] It has

[5] J. M. Lehmann, L. B. Moore, O. T. Smith, W. O. Wilkison, T. M. Willson, and S. A. Kliewer, *J. Biol. Chem.* **270,** 12953 (1995).

[6] S. A. Kliewer, S. S. Sundseth, S. A. Jones, P. J. Brown, G. B. Wisely, C. S. Koble, P. Devchand, W. Wahli, T. M. Willson, J. M. Lenhard, and J. M. Lehmann, *Proc. Natl. Acad. Sci. U.S.A.* **94,** 4318 (1997).

[7] J. M. Lehmann, J. M. Lenhard, B. B. Oliver, G. M. Ringold, and S. A. Kliewer, *J. Biol. Chem.* **272,** 3406 (1997).

[8] S. A. Kliewer, J. M. Lenhard, T. M. Willson, I. Patel, D. C. Morris, and J. M. Lehmann, *Cell* **83,** 813 (1995).

[9] J. Berger, M. D. Leibowitz, T. W. Doebber, A. Elbrecht, B. Zhang, G. Zhou, C. Biswas, C. A. Cullinan, N. S. Hayes, Y. Li, M. Tanen, J. Ventre, M. S. Wu, G. D. Berger, R. Mosley, R. Marquis, C. Santini, S. P. Sahoo, R. L. Tolman, R. G. Smith, and D. E. Moller, *J. Biol. Chem.* **274,** 6718 (1999).

[10] H. S. Camp, O. Li, S. C. Wise, Y. H. Hong, C. L. Frankowski, X. Shen, R. Vanbogelen, and T. Leff, *Diabetes* **49,** 539 (2000).

[11] A. Elbrecht, Y. Chen, A. Adams, J. Berger, P. Griffin, T. Klatt, B. Zhang, J. Menke, G. Zhou, R. G. Smith, and D. E. Moller, *J. Biol. Chem.* **274,** 7913 (1999).

[12] P. Dowell, V. J. Peterson, T. M. Zabriskie, and M. Leid, *J. Biol. Chem.* **272,** 2013 (1997).

[13] G. Krey, O. Braissant, F. L'Horset, E. Kalkhoven, M. Perroud, M. G. Parker, and W. Wahli, *Mol. Endocrinol.* **11,** 779 (1997).

[14] B. M. Forman, J. Chen, and R. M. Evans, *Proc. Nat. Acad. Sci. U.S.A.* **94,** 4312 (1997).

[15] P. Cronet, J. F. Petersen, R. Folmer, N. Blomberg, K. Sjoblom, U. Karlsson, E. Lindstedt, and K. Bamberg, *Structure (Camb)* **9,** 699 (2001).

[16] H. E. Xu, M. H. Lambert, V. G. Montana, K. D. Plunket, L. B. Moore, J. L. Collins, J. A. Oplinger, S. A. Kliewer, R. T. Gampe, Jr., D. D. McKee, J. T. Moore, and T. M. Willson, *Proc. Natl. Acad. Sci. U.S.A.* **98,** 13919 (2001).

been shown that all these compounds bind to an unusually large internal ligand-binding cavity, and the efficacy of the ligand in activating the PPAR appears to be determined by the effect of the ligand on the conformation of the carboxy-terminal helix, helix 12. This conformation change determines the ability of the PPAR to bind to coactivator proteins and thus determines the transcriptional activity of the receptor.

Use of Fluorescent Probes in Study of Fatty Acid-Binding Proteins

Fatty acid-binding proteins are ubiquitous throughout nature, as free fatty acids are very insoluble and toxic and therefore have to be shuttled throughout aqueous cellular environments to fulfill roles in energy generation, membrane structure, and signaling. Mammalian intracellular fatty acid-binding proteins were originally known by the tissue from which they were first isolated, such as adipose (aFABP), liver (LFABP), and intestinal (iFABP). This nomenclature has been changed to a systematic numbering system, FABP1–9, which is more appropriate, as multiple FABPs are now known to be expressed in each tissue. Another important fatty acid-binding protein is serum albumin. Albumin acts as a systemic carrier of hydrophobic molecules, including fatty acids and steroids. Many primitive organisms also express fatty acid-binding proteins of unknown function. The fatty acid-binding activity of many of these proteins has been probed using fluorescent fatty acids, and it is from these studies that relevant techniques for the analysis of PPAR ligand binding have been developed.[20–23]

Preparation of Recombinant PPAR

The utility of fluorescent-binding assays requires the availability of pure protein for analysis. This determines that the majority of the signal produced is specific to the interaction with the protein under study and also allows for precise control of the amount of receptor present, which is essential for the design and interpretation of the studies. We have produced recombinant PPAR ligand-binding domains

---

[17] H. E. Xu, M. H. Lambert, V. G. Montana, D. J. Parks, S. G. Blanchard, P. J. Brown, D. D. Sternbach, J. M. Lehmann, G. B. Wisely, T. M. Willson, S. A. Kliewer, and M. V. Milburn, *Mol. Cell* **3,** 397 (1999).

[18] R. T. Nolte, G. B. Wisely, S. Westin, J. E. Cobb, M. H. Lambert, R. Kurokawa, M. G. Rosenfeld, T. M. Willson, C. K. Glass, and M. V. Milburn, *Nature* **395,** 137 (1998).

[19] J. Uppenberg, C. Svensson, M. Jaki, G. Bertilsson, L. Jendeberg, and A. Berkenstam, *J. Biol. Chem.* **273,** 31108 (1998).

[20] F. Schroeder, P. S. Myers, J. T. Billheimer, and W. G. Wood, *Biochemistry* **34,** 11919 (1995).

[21] G. Nemecz, T. Hubbell, J. R. Jefferson, J. B. Lowe, and F. Schroeder, *Arch. Biochem. Biophys.* **286,** 300 (1991).

[22] R. S. Sha, C. D. Kane, Z. Xu, L. J. Banaszak, and D. A. Bernlohr, *J. Biol. Chem.* **268,** 7885 (1993).

[23] H. J. Keuper, R. A. Klein, and F. Spener, *Chem. Phys. Lipids* **38,** 159 (1985).

(LBD) using the histidine tag approach.[24,25] This is the most commonly used expression and purification system in the study of nuclear receptor ligand binding and crystallization. In this system the cDNA encoding the receptor protein is cloned in frame with a linker encoding six histidines. With the pET15b vector, this results in a fusion protein with the receptor having the six histidines fused at the amino terminus. The expression of this fusion protein is driven by the bacteriophage T7 RNA polymerase in bacterial strains that carry T7 RNA polymerase on an episomal plasmid under control of the lactose repressor (LacR). This allows for the regulatable expression of the fusion protein on the addition of the lactose analog, isopropylthiogalactoside (IPTG). A standard strain for this purpose is the BL21(DE3) series.

A suitable protocol for the purification of His-tagged human PPARγ is as follows.

Reagents

Phosphate-buffered saline (PBS): Prepare as described[26] and autoclave
Ampicillin: Stocks of 100 mg/ml prepared in sterile water and aliquots frozen at $-20°$
IPTG: Added fresh as solid to cultures
Phenylmethylsulfonyl fluoride (PMSF): 100 mM in dimethylformamide; store at $4°$
2-Mercaptoethanol: Added fresh to solutions from 13 M stock (analytical reagent grade) Chelating Sepharose: Hitrap chelating columns from Amersham/Pharmacia (Piscataway) are a convenient reagent for nickel affinity chromatography. These may be loaded by syringe, low-pressure pump, or fast protein liquid chromatography (FPLC) 1 M imidazole in PBS, filtered

Procedure

pET15b-hPPARγ LBD plasmid is transformed into *Escherichia coli* strain BL21(DE3) pLYSs. A single colony of the transformed bacteria is inoculated into 25 ml of LB broth containing ampicillin (100 μg/ml). Cultures are incubated overnight at $37°$ with shaking at around 200–300 rpm. One liter of LB broth is inoculated with 10 ml of the overnight culture and incubated as just described until the culture reaches an OD$_{600}$ of $\sim$1.0. The culture is cooled to room temperature ($20–30°$), and IPTG is added to 0.5 mM. The culture is then incubated at $30°$ within shaking (200 rpm maximum) for a further 3 hr. Cells are harvested by centrifugation

[24] M. Causevic, C. R. Wolf, and C. N. Palmer, *FEBS Lett* **463,** 205 (1999).
[25] C. N. A. Palmer and C. R. Wolf, *FEBS Lett* **431,** 476 (1998).
[26] J. Sambrook, E. F. Fritsch, and T. Maniatis, "Molecular Cloning: A Laboratory Manual," Cold Spring Harbor Laboratory Press, Cold Spring Harbor, NY, 1989.

at 3000g for 20 min at 4°, and the cell pellet is resuspended in 20 ml ice-cold PBS containing 0.2 mg/ml lysozyme. Suspended cells are left on ice for 15 min to allow digestion of the cell wall. Spheroplasts are harvested by centrifugation at 3000g for 20 min at 4° and the pellet is resuspended in 20 ml PBS. The suspension is frozen at −20° overnight. Frozen spheroplasts are thawed on cold water, and protease inhibitors and 2-mercaptoethanol are added. They are then sonicated three times for 30 sec at full power, and the lysates are cleared by centrifugation at 100,000g for 45 min at 4°. The supernatant is then made up to 10% glycerol and applied to a nickel-loaded chelating Sepharose column (1 ml resin per liter of culture) at a flow rate of 1–2 ml/min. The column is then washed with 30 column volumes of loading buffer containing 10 mM imidazole (pH 8.8) followed by 10 column volumes of loading buffer containing 25 mM imidazole (pH 8.8). The hPPARγLBD protein is eluted with loading buffer containing 250 mM imidazole (pH 8.8). The imidazole is removed, and the protein is concentrated using a Centricon concentrator (10 K cutoff). Any aggregated protein is then removed by ultracentrifugation at 100,000g for 45 min. The glycerol concentration in the final protein preparation is increased to 50% (v/v), and aliquots were stored at −20°. Protein preparations stored in this manner are active for at least 1 year.

Spectrophotometric Analysis of cis-Parinaric Acid Binding to PPARγ

cis-Parinaric acid (CPA) has a distinct spectrum in aqueous solution with a peak of 319 nm. This peak is shifted to approximately 324 nm on binding to proteins such as albumin. This property was first described as an assay for the quantitation of free fatty acids in serum.[27] We found that PPARγ LBD also produced a spectral shift on binding CPA (Fig. 1). This shift was consistent with one molecule of CPA binding one molecule of PPAR. This was evident, as at concentrations above 2 μM receptor and fatty acid, equimolar concentrations produced complete binding. It is then possible to monitor competition by test compounds by following the spectrum shift back toward 319 nm. This is calculated most easily by monitoring the absorbance readings at 319 and 329 nm. This has a ratio of 3 with the CPA in solution and a ratio of 1 with the CPA completely bound by protein. The percentage occupancy can then be plotted against a dose range of competing ligand to determine a K_i for the test compound. This assay is highly suitable for the analysis of low-affinity compounds and has been used to explore the interactions of very low-affinity agonists such as phenyl acetate.[28]

---

[27] C. B. Berde, J. A. Kerner, and J. D. Johnson, *Clin. Chem.* **26**, 1173 (1980).
[28] D. Samid, M. Wells, M. E. Greene, W. Shen, C. N. A. Palmer, and A. Thibault, *Clin. Cancer Res.* **6**, 933 (2000).

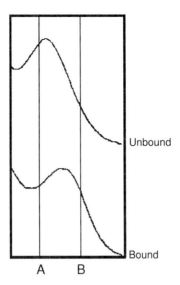

FIG. 1. Spectral properties of *cis*-parinaric acid. The spectrum between 312 and 340 nm of CPA in solution (unbound) and CPA completely bound to hPPARγ LBD (bound) is shown. The ratio of 319 nm (A) and 329 nm (B) is indicative of the degree of occupancy. Bound, 1.0; unbound, 3.0.

Reagents

cis-Parinaric acid (CPA) can be obtained from Molecular Probes (Eugene, OR) and Cayman Chemical Company. Due to the susceptibility of CPA to oxidation, CPA is purchased in special 10-mg aliquots and is stored at $-20°$. Dissolve aliquots, as required, in DMSO and confirm the concentration of CPA in stock solutions before each procedure using the extinction coefficient at 304.2 nm in ethanol of 78,000.[27] Solutions of CPA can be stored under nitrogen at $-20°$ for up to 1 month.

Procedure

cis-Parinaric acid (1.5 μM) is added to 25 mM Tris–HCl, pH 7.5, at room temperature and scanned against buffer between 312 and 340 nm using a Shimadzu UV-3000 scanning spectrophotometer. The 319/329 ratio should be at least three and a distinctive CPA spectrum should be obtained (Fig. 1). Purified hPPARγ LBD is then added to 2 μM to both sample and reference cuvettes. They are then rescanned, and the 319/329 ratio should be approximately 1.2. Increasing amounts of test ligands are then added to both the sample and the reference cuvettes. The 319/329 ratio is determined with each sequential addition of test ligand. Incorporation of up to 2% DMSO does not produce any significant changes in observed spectra and is the maximum used in the addition of competitor compounds. The concentration of competitor compound is increased until a stable plateau of competition is reached (usually between 2.0 and 2.5). A control plot with increasing amounts of DMSO should be performed to confirm the stability of the protein/CPA complex under individual laboratory conditions.

Fluorescence Displacement Assays for PPAR Binding

Probably the simplest and most sensitive method for determining PPAR binding is by using fluorescence displacement. In these assays, a fluorescent signal is obtained on the addition of protein and on the addition of a competing ligand, the fluorescence is diminished. The sequential addition of increasing amounts of test compound will give a dose response of fluorescence diminution. This system has the advantage over the spectrophotometric assay that it does not require a substantial degree of occupancy of CPA and PPARγ. Rather, it only requires sufficient fluorescence signal to plot the effect of the competitor in a robust fashion. In practice, we find that 100 nM CPA and PPARγ protein is the lower limit for this assay. We have used this method to describe a K_i of 650 nM for diclofenac, which was not possible using the spectrophotometric assay.[29]

Procedure

Purified hPPAR LBD protein (100 nM) is added to *cis*-parinaric acid (100 nM) in 25 mM Tris–HCl (pH 7.5) at room temperature, and the resulting fluorescence is measured immediately using a PerkinElmer LS-3 fluorescence spectrophotometer (excitation wavelength of 318 nm and emission wavelength of 410 nm). The fluorescence resulting from protein alone and *cis*-parinaric acid alone were added and deducted from the measured value.

Data Analysis

The simplest assay for screening for new ligands is the addition of a single concentration of a competitor compound and the calculation of the amount of CPA that is still bound to the protein. An example of several NSAIDs examined in such a fashion is shown in Fig. 2. The degree of occupancy may be determined by the following formula: %CPA bound = $(R_{observed} - R_{min})/(R_{max} - R_{min}) \times 100$, where R is the 319/329 ratio, R_{max} is the fully displaced ratio, and R_{min} is the ratio of CPA/PPAR alone. The degree of displacement can therefore also be plotted as 100-(%CPA bound).

Dose responses may also be plotted to determine an IC$_{50}$, and it is vitally important to cover a large range of concentrations in order to include several points

[29] D. J. A. Adamson, R. Tatoud, D. Frew, C. R. Wolf, and C. N. A. Palmer, *Mol. Pharmacol.* **61,** 7 (2002).

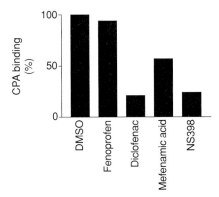

FIG. 2. The spectrophotometric assay can be used for a single concentration "screen" for ligands. Values obtained for CPA binding are shown in the presence of 20 μM of each NSAID, fenoprofen, diclofenac, mefenamic acid, and NS398. Also shown is the value obtained with an equal volume of solvent alone (DMSO).

to define the minimum and maximum values. A good example of this is shown in Fig. 3A, in which sufficient data are present for the determination of threshold and plateau values, which are essential for the definition of the IC_{50}. In Fig. 3B a good dose response is observed, but as only 1 or 2 points at plateau have been measured, this results in an overestimation of B_{max} and IC_{50}. K_i may be calculated by the formula $K_i = IC_{50}/(1+[ligand]/K_d)$. The K_i obtained is only valid when the IC_{50} is sufficiently above the concentration of the receptor and fluorescent ligand and when the concentrations of ligand and receptor are below K_d. This is the only way in a solution assay such as those described here to ensure that the fraction of the protein-bound reporter ligand is not confounding the assay. If the value obtained for IC_{50} is similar to the concentrations of ligand or receptor in the assay, then the value represents stoichiometric binding. Under these conditions, the IC_{50} is marginally above [ligand] and may be well above K_d. This can result in a very small value for K_i. This has inappropriately led to certain reports that conclude that fatty acids have a binding affinity for PPARs in the very low nanomolar range.[30]

Potential Physical Problems with Assay

The first and most important aspect of the assay is the quality of the protein preparation. We have found that if the protein preparation is not undertaken in a prompt fashion, then the resulting protein may not be of sufficient quality. In practice, this means the time taken from loading the column to storing the purified

[30] Q. Lin, S. E. Ruuska, N. S. Shaw, D. Dong, and N. Noy, *Biochemistry* **38**, 185 (1999).

A

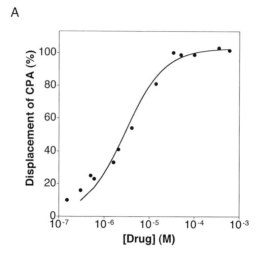

B

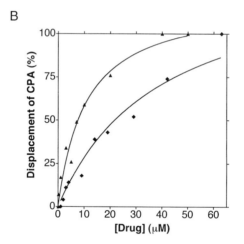

FIG. 3. Dose response of displacement of CPA by NSAIDs. Values obtained with the addition of increasing concentrations of (A) diclofenac (●) and (B) NS398 (▲) and mefenamic acid (◆). Plots shown for NS398 and mefenamic acid do not describe the maximal value accurately due to lack of sampling at high concentrations. This inevitably leads to an overestimation of IC_{50}.

protein in 50% glycerol at $-20°$ should be less than 48 hr. The quality of the protein preparation can be determined by the stoichiometry of binding to CPA. Another variable is the temperature of the spectrophotometer. We find that at temperatures over 25° the protein/CPA complex is not stable enough to perform the assays. It may be necessary to use a cooled block to perform the assays at 20°. Another common problem with the assay is that many chemicals, especially NSAIDs, are very yellow in color. This means that they absorb light in the region of the assay. This is generally only a problem at very high concentrations of most compounds, but it does limit the upper end of the assay for test chemicals and may not allow the generation of a complete binding curve. This phenomenon is generally observed as a decrease in the 319/329 ratio on addition of the test compound. In extreme cases, the spectrophotometer cannot read an accurate difference spectrum and will not "zero." A good example of an anti-inflammatory compound that interferes with the assay is indomethacin, in which only a partial plot can be obtained. Curcumin (turmeric) is understandably one of the worst compounds for such interference and this completely blocks the assay at 1 μM. In these cases the fluorescence displacement assay may or may not be useful. In each case the intrinsic fluorescence of the competing ligand has to be evaluated.

Other Fluorescent Fatty Acid-Binding Assays

A variety of other fluorescent fatty acids are suitable for the analysis of PPAR ligand binding. We have used 12-anthraceneoleic acid as a fluorescent probe for PPARs and it has a relatively similar binding activity to all three PPARs.[24] *trans*-Parinaric acid has been shown to be a good probe for PPARα binding[30]; however, this fatty acid is no longer available commercially. The synthesis details are available on request from Molecular Probes (Oregon). Another fatty acid that we have found useful is DAUDA[31] (Molecular Probes), and although this fatty acid has not been published as a ligand for PPAR, we have found that it is a suitable probe for PPARα and PPARγ binding.

Acknowledgments

This work was funded by the Biotechnology and Biological Sciences Research Council of Great Britain (Award No. MOLO4650). D.J.A.A. is an Imperial Cancer Research Fund/Bobby Moore Cancer Research Fellow.

[31] D. C. Wilton, *Biochem. J.* **270,** 163 (1990).

[20] Developing Toxicologically Predictive Gene Sets Using cDNA Microarrays and Bayesian Classification

By RUSSELL S. THOMAS, DAVID R. RANK, SHARRON G. PENN,
MARK W. CRAVEN, NORMAN R. DRINKWATER,
and CHRISTOPHER A. BRADFIELD

Introduction

The application of DNA microarray technology to the field of toxicology has increased significantly. Recent applications include studies identifying gene expression changes related to dioxin exposure[1] and correlating carbon tetrachloride-induced hepatotoxicity with interleukin-8 (IL-8) expression.[2] In most cases, the emphasis of research has been to monitor global changes in gene expression in order to provide insight into the cellular mechanisms of toxicity. Although monitoring global gene expression changes may prove to be important when characterizing the action of a particular chemical, it is not necessarily predictive of how that chemical or any other chemical will behave toxicologically without constructing appropriate statistical models.

The potential applications of predictive statistical models in toxicology based on gene expression measurements are numerous. For example, short-term studies measuring gene expression could be used to predict long-term toxicity studies like those still performed by the National Toxicology Program (NTP). Given that chronic exposure studies can cost between 2 and 4 million dollars, the cost benefits of rapid predictive approaches are obvious. Other short-term gene expression studies could be used to predict which chemicals would be teratogenic or cause more subtle developmental changes after human prenatal exposure. In either case, the application of microarray analysis and predictive statistical models has the potential to be extremely useful from both economic and human health perspectives.

Background and Considerations in Deriving Predictive Gene Sets

The development of a toxicologically predictive statistical model based on gene expression profiles is similar to the development of other multivariate classification models. In this case, two or more predictor variables (i.e., expression measurements of multiple genes) are used along with a single criterion variable that is categorical

---

[1] F. W. Frueh, K. C. Hayashibara, P. O. Brown, and J. P. Whitlock, Jr., *Toxicol. Lett.* **122,** 189 (2001).
[2] P. R. Holden, N. H. James, A. N. Brooks, R. A. Roberts, I. Kimber, and W. D. Pennie, *J. Biochem. Mol. Toxicol.* **14,** 283 (2001).

with either binomial or multinomial classes (i.e., toxicological class or end point). The goal is to use the data vectors in the test set of chemicals, which are composed of the predictor variables, to build a model that correctly classifies the criterion variable and can eventually be used to predict other untested chemicals. As a result, the capability of the model to correctly classify the criterion variable is directly related to the amount of information the predictor variables contain about the criterion variable.

Not only is the ultimate success of a model dependent on the information contained within the predictor variables, the removal of uninformative predictor variables and the selection of a diagnostic subset of genes are also important for both biological and statistical reasons. In the biological sense, the classification of a set of chemicals into a toxicological class or end point based on gene expression is difficult due to the variety of potential mechanisms that underlie the toxicity of these chemicals. For example, both 2,3,7,8-tetrachlorodibenzo-p-dioxin and N,N-dimethyl-4-aminoazobenzene (DAB) are carcinogenic in laboratory animals. However, they produce the same toxic end point, cancer, through different mechanisms. Dioxin is a nongenotoxic carcinogen that acts through a ligand-activated transcription factor known as the Ah receptor, whereas DAB is metabolized to an active form that is mutagenic. Even though the toxicological end point may be the same, these two chemicals will undoubtedly activate and repress different biochemical and molecular pathways leading to different global patterns in gene expression. Therefore, in order to construct a predictive model for chemically induced cancer, the global gene expression data set must be reduced to a subset of genes that are diagnostic across all chemicals and mechanisms.

From a statistical perspective, selection of a subset of diagnostic genes is also important. Statistical models with few predictors relative to the overall sample size typically yield more accurate (i.e., less biased) and more precise estimators.[3,4] For Bayesian models, underlying assumptions are made about the dependency of the variables. If the assumptions are violated in the full data set, better classification results can be obtained by removing uninformative predictor variables, making the dependency assumptions less severe.

Although the method presented in this article can be used on any microarray data set related to toxicology and virtually any toxicological end point, the original application and development were performed on a microarray study to classify 24 prototype chemical treatments into five well-characterized toxicological classes. A more detailed description of the treatments and toxicological classes is provided in the original study.[5]

[3] S. C. Hora and J. B. Wilcox, *J. Market. Res.* **19,** 57 (1982).
[4] C. J. Huberty, "Applied Discriminant Analysis." Wiley, New York, 1994.
[5] R. S. Thomas, D. R. Rank, S. G. Penn, G. M. Zastrow, K. R. Hayes, K. Pande, E. Glover, T. Silander, M. W. Craven, J. K. Reddy, S. B. Jovanovich, and C. A. Bradfield, *Mol. Pharmacol.* **60,** 1189 (2001).

Microarray Construction and Image Analysis

Gene expression for the original set of toxicological treatments is measured using a microarray that contains approximately 1200 minimally redundant cDNAs from both an internal expressed sequence tag (EST) project and the public EST effort. In the internal EST project, mice are treated with various hepatotoxicants, and EST are sequenced from the livers of control and treated mice. As a result, these microarrays contain a high percentage of genes that are both expressed in the liver and induced following chemical treatment. ESTs supplemented from the public effort are genes believed to be toxicologically important, but missing from our EST collection.

To construct cDNA microarrays, an aliquot from each glycerol stock is transferred to a 96-well polymerase chain reaction (PCR) plate and lysed at 95° in 1× PCR buffer. The resulting lysate is amplified using PCR. PCR reactions are purified using Millipore glass filter plates (MAFB N0B, Bedford, MA). Briefly, 200 μl of binding buffer (150 mM potassium acetate, pH 4.9; 5.3 M guanidine hydrochloride) is added to the PCR reactions and passed over the glass filters using vacuum filtration. The filters are washed four times with 80% ethanol and eluted with 65 μl distilled water. The purified PCR products are dissolved in 5 M NaSCN to create the final spotting solution. Prior to spotting, cleaned microscope slides (VWR, West Chester, PA) are exposed for 2 hr to the vapors of 3-aminopropyltrimethoxysilane (United Chemical Technologies, Bristol, PA) in a stream of nitrogen. The slides are washed with water, cured overnight at 100°, and spotted in replicate with the purified PCR products in NaSCN. After spotting, the slides are baked at 80° for 2 hr. Prior to hybridization, slides are agitated gently for 10 min in 2-propanol and boiled in water for 5 min.

The labeled cDNA probe is produced from 1 μg poly(A) RNA by incorporation of Cy-dCTP (Amersham Pharmacia, Piscataway, NJ) during a standard reverse transcriptase reaction primed with oligo(dT)$_{12–18}$ and random nonomers (Life Technologies, Gaithersburg, MD). The poly(A) RNA and oligonucleotides are heated at 70° for 10 min and transferred to ice. Reverse transcription is performed in a solution containing 1× Superscript II buffer (Life Technologies), 10 mM dithiothreitol (DTT), 100 mM dATP, 100 mM dTTP, 100 mM dGTP, 50 mM dCTP, 50 mM Cy-dCTP, and SuperScript II enzyme. The reaction is incubated at 42° for 2 hr. Following incubation, 1 μl RNase H (Life Technologies) is added to the reaction, and the reaction is incubated at 37° for 30 min. Purification of labeled cDNA is carried out using Qiaquick PCR purification columns (Qiagen, Valencia, CA). Fifty picomoles of each dye is dried down in a Speed-Vac and resuspended with 40 μl of hybridization solution containing a mixture of 50% formamide, 4× SSC (0.15 M sodium citrate, 15 mM sodium chloride), 0.1% Sodium dodecyl sulfate (SDS), 50 μg/ml human COT-1 DNA (Life Technologies), and 50 μg/ml poly(A)$_{80}$ DNA (Amersham Pharmacia). The hybridization solution is added to the slide, a coverslip is added, and hybridization is performed for 18 hr at 42°. Following hybridization, the slides are washed for 5 min in 2× SSC/0.1%

SDS at 42°, 5 min in 1× SSC/0.1% SDS at 25°, 5 min in 0.1× SSC at 25°, and a brief rinse in deionized water. The slides are scanned using a microarray scanner at an excitation wavelength of 532 nm (Cy3) or 632 nm (Cy5).

For each hybridization, fluorescence data from the replicate spots corresponding to each cDNA are averaged and normalized. To eliminate any dye bias, all samples are typically analyzed at least twice, with one experiment using Cy3 to label the control mRNA and Cy5 to label the treated mRNA and in the replicate experiment Cy5 to label the control and Cy3 to label the treated. Results from replicate hybridizations are averaged.

Data Reduction

Prior to statistical analysis, the genes (i.e., predictor variables) are first collapsed into a fully nonredundant set by averaging the results from multiple copies of the same gene. After collapsing, the data set is screened for genes that do not respond to any of the treatments used in the study and do not contribute significantly to the classification. A threshold of twofold change in gene expression is typically used as the cutoff value and is similar to a standard threshold level used in other studies (e.g., Schena et al.[6]).

In data sets with time course or dose–response experiments, additional screening is typically performed to identify a subset of genes with a stable expression over all time points and doses. This provides computational savings by eliminating variables that would have little predictive value and is achieved by collapsing all time points or doses into a single average value representing the average change in that gene. After collapsing the data, only genes that showed greater than a twofold change in more than one treatment are selected for further analysis. After this data reduction step, the individual time points or doses are broken out and analyzed separately in the classification analysis. Collapsing the data is only used as a data screening tool and not for statistical classification analysis. Finally, the gene expression values are discretized such that genes upregulated greater than twofold are given a value of one, genes downregulated greater than twofold are given a value of minus one, and genes with less than a twofold change are given a value of zero. A flow chart describing data reduction is provided in Fig. 1.

Statistical Classification Analysis

The type of classification model applied to this problem can be one of a number of alternatives. We have chosen to use the Naïve Bayes model structure based on previous work by Kontkanen and colleagues,[7] which has been shown to

[6] M. Schena, D. Shalon, R. Heller, A. Chai, P. O. Brown, and R. W. Davis, *Proc. Natl. Acad. Sci. U.S.A.* **93,** 10614 (1996).

[7] P. Kontkanen, P. Myllymaki, T. Silander, and H. Tirri, "Proceedings of the Fourth International Conference on Knowledge Discovery and Data Mining," p. 254. AAAI Press, Menlo Park, CA, 1998.

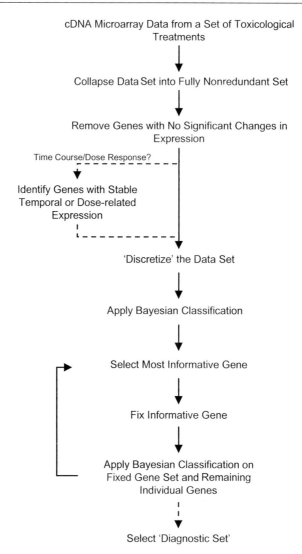

FIG. 1. A flow chart outlining the method for data reduction and classification analysis leading to the diagnostic set of genes.

perform well in comparison with other approaches.[8,9] In the Naïve Bayes method, the predictor variables $X_1, \ldots, X_k$ are assumed to be independent of each other when

[8] P. Langley, W. Iba, and K. Thompson, "Proceedings of the Tenth National Conference on Artificial Intelligence," p. 223. AAAI, San Jose, CA, 1992.
[9] N. Friedman, D. Geiger, and M. Goldszmidt, *Machine Learn.* **29**, 131 (1997).

conditioned on the class variable C. Our model M is constructed by the joint probability distribution for a data vector $(x, c) = (X_1 = x_1, \ldots, X_k = x_k, C = c)$ and can be written as follows:

$$P(x, c) = P(C = c) \prod_{i=1}^{k} P(X_i = x_i \mid C = c) \tag{1}$$

Prior to incorporating any data, we also assume that all classes are equally probable (i.e., the probability that a toxicological treatment will belong to a certain class or end point is the same for each class) and that within each class the gene expression values of each gene are also equally probable. Given these assumptions, we can use Bayesian probability theory to calculate the conditional predictive distribution for the class c given x and the data set D by

$$P(c \mid x, D) = \frac{P(c, x \mid D)}{P(x \mid D)} \tag{2}$$

where the numerator is calculated as

$$P(c, x \mid D) = \frac{t_c + 1}{N_t + NC} \prod_{i=1}^{k} \frac{f_{cxi} + 1}{F_c + V_{xi}} \tag{3}$$

where t_c is the number of toxicological treatments in class c, N_t is the total number of toxicological treatments, NC is the total number of classes, f_{cxi} is the number of cases in class c having a value equal to x_i, F_c is the number of toxicological treatments in class c, and V_{xi} is the number of possible values of x_i. The denominator is the same for all c and is calculated as

$$P(x \mid D) = \sum_{c'} P(c', x \mid D) \tag{4}$$

The result of this conditional predictive distribution is then used to classify the data vector (i.e., the combination of all the gene expression measurements for that toxicological treatment).

Parameter Selection

For the parameter or gene selection, a modification of the forward parameter selection process outlined in Huberty[4] is employed. Specifically, an iterative process is used where genes are run individually using the Naïve Bayes model and the gene with the best internal classification rate and highest confidence (represented by the sum of all probabilities for correctly classified treatments) is selected. The internal classification rate is defined as the number of data vectors that are classified correctly with the model without performing cross-validation divided by the total number of data vectors. In the subsequent round, the selected gene is fixed and the remaining parameters are added individually to find which gene, along with the first selected gene, produces the highest internal classification rate and confidence. This process is repeated until all genes that pass the data reduction process are added to the model in the order of their internal classification rate. This type of selection ranks the genes in the order of their estimated predictive value and adds them sequentially to the model. It should be noted that the forward selection approach does not necessarily yield the best set or even the smallest set. In addition, genes with similar expression profiles may also be added to the set if they increase the internal classification rate or confidence significantly. The approach is simply a heuristic to look for a diagnostic set of predictors with a high classification rate.

To estimate the predictive accuracy of this approach, the process of parameter selection is integrated with leave-one-out cross-validation where one of the treatments is removed from the analysis and the model is constructed and then used to predict the left-out treatment. The predictive accuracy is then assessed after each parameter is added, and the number of genes in the "diagnostic set" is chosen based on the peak predictive accuracy and confidence of the model (Fig. 2). The final "diagnostic set" is derived by following the same procedure on the complete data set (i.e., no treatment left out). A flow chart describing the classification analysis and parameter selection is outlined with the data reduction steps in Fig. 1.

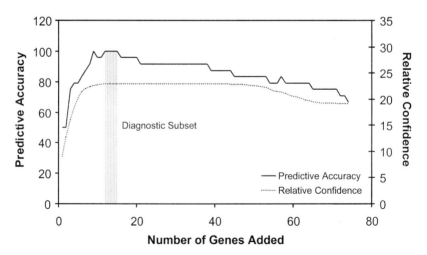

FIG. 2. An example of the estimated accuracy and relative confidence of the classification model for predicting a toxicological end point as a function of genes added to the model. Genes are added to the model using a forward selection scheme, and predictive accuracy is estimated using leave-one-out cross-validation. The diagnostic set of genes is highlighted by the hatched area.

All statistical analyses are scripted in Perl, and the general code is available by emailing the corresponding author (bradfield@oncology.wisc.edu).

Conclusions

The application of predictive statistical models to chemically induced gene expression is the next logical step in the developing field of toxicogenomics. The development of these models will eventually open the door to a new era of toxicological testing where relatively short and inexpensive microarray studies will allow the assessment of the human health risks associated with a previously untested chemical. This would mean significant savings in both animal usage and financial resources while also reducing the disparity between the number of tested and untested chemicals in commerce today. However, the accuracy and applicability of these models are highly dependent of the quality of the training sets used in their development. As the public gene expression database grows, more chemicals can be added to training the models and those models will become more predictive.

Acknowledgments

This work was supported by The Burroughs Wellcome Foundation, the National Institutes of Health (Grants ES05703, T32CA09681, CA07175, and GM23750), and a postdoctoral fellowship cosponsored by the Society of Toxicology and the Colgate-Palmolive Corporation.

[21] Direct Expression of Fluorescent Protein-Tagged Nuclear Receptor CAR in Mouse Liver

By TATSUYA SUEYOSHI, RICK MOORE, JEAN-MARC PASCUSSI, and MASAHIKO NEGISHI

The nuclear orphan receptor constitutive active receptor (CAR) mediates induction of *CYP2B* genes by phenobarbital (PB) and various PB-type compounds in liver.[1–6] CAR is a nuclear receptor acting as a heterodimer with retinoid X receptor (RXR).[7] In response to PB exposure, a CAR:RXR complex activates a 51-bp enhancer sequence called the PB responsive enhancer module (PBREM) that is

---

[1] P. Honkakoski, I. Zelko, T. Sueyoshi, and M. Negishi, *Mol. Cell. Biol.* **18**, 5652 (1998).

[2] T. Kawamoto, T. Sueyoshi, I. Zelko, R. Moore, K. Washburn, and M. Negishi, *Mol. Cell. Biol.* **19**, 6318 (1999).

conserved in mouse, rat, and human *CYP2B* genes. CAR undergoes multiple steps to be activated in liver cell following PB exposure, the first of which is nuclear translocation.

Immunohistochemical analysis of liver sections of control and PB-induced mice showed that the anti-CAR antibody stains nuclei of PB-induced liver sections. Western blotting of liver nuclear extracts prepared from control and PB-induced mice revealed the absence of CAR in control nuclear extracts and increased accumulation in PB-treated counterparts.[1] Evidently CAR is retained in the cytoplasm in control liver and translocates into the nucleus following PB induction[2]. Thus, defining the molecular and cellular mechanisms that regulate the PB-elicited nuclear translocation of CAR is an emerging issue for understanding the receptor activation leading to CYP induction. To investigate nuclear translocation mechanisms, it is essential to develop a method by which genetically engineered CAR can be expressed in liver cells and its intracellular localization can be followed after PB treatment.

This article describes a simple experimental procedure to directly express fluorescent protein-tagged CAR in mouse liver *in vivo*. The procedure includes construction of a fluorescent protein-tagged CAR expression plasmid, delivering the plasmid to the liver through tail vein injection, preparation of liver frozen sections, and detection of the expressed receptor under conventional and/or confocal fluorescent microscopes.

Experiments using this method have shown that the fluorescent protein-tagged CAR and glucocorticoid receptor (GR) are coexpressed in the cytoplasm and nucleus of mouse liver cell, respectively. Subsequently, we have delineated PB-inducible translocation activity to a short peptide near the C terminus of the CAR protein.[8]

General Consideration

Green Fluorescent Protein

Originally isolated from the jellyfish *Aequoria victoria,* green fluorescent protein (GFP) is a small protein with 238 amino acid residues that emits a bright green florescence under UV light. GFP and GFP-tagged fusion protein can be expressed in cells and observed under a fluorescence microscope, providing us

[3] T. Sueyoshi, T. Kawamoto, I. Zelko, P. Honkakoski, and M. Negishi, *J. Biol. Chem.* **274,** 6043 (1999).

[4] P. Honkakoski and M. Negishi, *Biochem. J.* **347,** 321 (2000).

[5] I. Zelko and M. Negishi, *Biochem. Biophys. Res. Commun.* **277,** 1 (2000).

[6] T. Sueyoshi and M. Negishi, *Annu. Rev. Pharmacol. Toxicol.* **41,** 123 (2001).

[7] I. Tzameli and D. D. Moore, *Trends Endocrinol. Metab.* **12,** 7 (2001).

[8] I. Zelko, T. Sueyoshi, T. Kawamoto, R. Moore, and M. Negishi, *Mol. Cell. Biol.* **21,** 2838 (2001).

TABLE I
FILTER SETS FOR FLUORESCENT PROTEINS

Fluorophore	Filters
GFP[a]	XF100, XF104 (FITC)
YFP[a,b]	XF104 (FITC)
CFP[a,b]	XF114
Hoechst S-33258 DAPI	XF03

[a] GFP, green fluorescent protein; YFP, yellow fluorescent protein; CFP, cyan fluorescent protein.
[b] Ideal combination for dual expression of two different proteins.

with a powerful tool to study the fate of a given protein in cells.[9,10] Expression of genetically altered GFP–CAR fusion proteins may help us understand processes of PB-inducible CAR nuclear translocation. Site-directed mutagenesis generates a group of GFP variants with different excitation and emission spectra: yellow fluorescent protein (YFP), cyan fluorescent protein (CYP), and blue fluorescent protein (BFP). With these fluorescent proteins, expression of two different proteins can be visualized simultaneously, making it possible to investigate intracellular localization of two different proteins in single or multiple cells (see Table I). The experimental versatility of fluorescent proteins is further increased by the simultaneous visualization of three different proteins that now becomes possible by using a new set of red fluorescent proteins in addition to GFP variants.[9]

Delivery of Expression Plasmid to Liver

To study nuclear translocation using fluorescent protein-tagged CAR, we need to establish a system in which the receptor can be retained in cytoplasm and translocated to the nucleus in response to PB. Transformed cells such as HepG2 and HEK293 are not a suitable system for this because GFP–CAR fusion protein are not retained in the cytoplasm, but spontaneously accumulate in the nucleus.[2] Using mouse primary hepatocytes is also problematic at the present time, as discussed later. Therefore, we have chosen to express the GFP–CAR fusion protein in mouse liver *in vivo* using gene delivery through tail vein injection of plasmid. The total volume of plasmid solution should be about one-tenth of the body weight of the mouse used. Mice with around 30 g of body weight are most suitable for these experiments. When a 30-g mouse is used, injection of 3 ml of plasmid solution

[9] www.clontech.com/clontech/GFPRefs.html
[10] Clontech, "Living Colors User Manual" (2000) and references therein.

should be completed within 7 to 10 sec. Injections of either shorter or longer duration may decrease the efficiency of DNA delivery. Employing this method, a maximum of 5% transfection of liver cells can be achieved based on the number of cells that express GFP fusion protein.

Slide Preparation

One of the key considerations is to avoid denaturing the GFP fusion protein during slide preparation because denatured GFP loses its fluorescence. Nail polish should not be used to affix the cover glass because it bleaches the GFP fluorescence.[10] Paraffin-embedded tissue should not be used because heat can also denature the GFP protein. The following procedure provides the best results.

1. A small block of liver is submerged in OCT compound and frozen.
2. A liver section is prepared, placed on a glass slide, and fixed with paraformaldehyde.
3. A glycerol solution (80%) is dropped on the section, and a cover glass is applied.

The glycerol solution does not require the addition of antifade, as GFP fluorescence is stable against photo-dependent bleaching. The thickness of the liver section is another important consideration. When a conventional fluorescent microscope is used, a thinner section is better to obtain a well-focused image. An 8-μm section, which is about one-third the size of the hepatocyte, is practical. To avoid counting the same hepatocytes twice, we use one out of every three or four consecutive sections. When a confocal microscope is used, sections 20–30 μm or thicker are better suited for experimental purposes, as sections of this thickness may include the whole hepatocyte, and a three-dimensional distribution of GFP fusion protein can be obtained.

Detection of GFP–CAR Fusion Protein

A conventional fluorescent microscope equipped with filters can be used to analyze GFP fusion protein in cells. Filters specific for GFP, YFP, or CFP can be purchased from several suppliers.[9,10] For detection of YFP or GFP, a filter for fluorescein isothiocyanate (FITC) also works well. Two different proteins tagged with different GFP derivatives can be expressed simultaneously and detected separately in the same cell. YFP and CFP are the best combination for this type of experiment, as their fluorescence can be separated completely. A laser scanning confocal microscope should be used for precise details of intracellular localization of a given protein. For nuclear fluorescent staining, Hoechst S-33258 or DAPI is added to 80% (v/v) glycerol and has no effect on GFP fluorescence. Fluorescence from S-33258 or DAPI can be separated easily from that of GFP derivatives.

Experimental Protocol: Simultaneous Expression of CFP–hCAR and YFP–hCAR$_{181-348}$ in Mouse Liver

Construction of Plasmid

Human CAR (hCAR) cDNA is amplified using primers and creates *Xho*I and *Eco*RI sites to the 5' and 3' ends, respectively. The amplified hCAR cDNA is cloned into pECFP-c1 (Clontech, Palo Alto, CA) to construct hCAR/pECFP-c1. Similarly, a truncated hCAR cDNA (from residues 181 to 348) is amplified and cloned into pEYFP-c1 to construct hCAR$_{181-348}$/pEYFP-c1. These plasmids are transformed into *Escherichia coli* Top 10 (Invitrogen, Carlsbad, CA), purified with the Qiagen plasmid Maxi kit (Qiagen, Valencia, CA), and dissolved in water.

Tail Vein Injection

A 190 μl water solution containing 5 μg each of hCAR/pECFP-c1 and hCAR181-348/pEYFP-c1 is mixed with 10 μl TransIt In Vivo polymer solution (Mirus, Madison, WI) and incubated for 5 min at room temperature. Prior to injection, the mouse (30 g of body weight) is anesthetized using Metofane (2,2-dichloro-1,1-difluoroethyl methyl ether), and the tail vein is dilated with warm water by pressing on the proximal side of the vein. The mixture is than diluted with 2.8 ml of 1× Mirus Delivery solution and is injected using a $26\frac{1}{2}$-gauge needle through the tail vein of the mouse (30 g body weight).

Treatment with Inducer

A potent PB-type inducer chlorpromazine (CPZ) is used to elicit the nuclear translocation of CAR. Fifty and 25 mg/kg body weight of CPZ are administered intraperitoneally at 2 and 5 hr, respectively, after plasmid injection. The mouse is sacrificed 3 hr after the second CPZ administration, the liver is excised, and a block of the liver is submerged in Tissue-Tek OCT compound (Sakura Finetek, Torrence, CA) and frozen. The duration of CPZ treatment is chosen based on our previous study that CAR is maximally accumulated about 3 hr after inducer treatment. It is recommended that all animal treatments be completed within 8 to 10 hr after initial injection of plasmid. Further treatment can be destructive to liver cells. For example, some of the liver cells that express GFP begin to die 24 hr after plasmid injection, which may be caused by toxicity from overexpression of GFP.

Slide Preparation

Blocks of the liver are submerged in OCT compound in a disposable vinyl specimen mold (25 × 20 × 5 mm, Tissue-Tek Cryomold, Miles, Inc., Elkhard, IN) and are frozen on an aluminum plate on dry ice. A section of frozen liver in the OCT block is prepared using a cryostat (Microm-Heidelberg). Frozen tissue can be stored at −70° for 6 months without GFP degradation. One of every four consecutive sections is placed on a slide glass, fixed with 4% paraformaldehyde at room temperature, washed once with phosphate-buffered saline, dehydrated in methanol, and the nucleus stained in 80% glycerol containing 0.5 μg/ml Hoechst S-33258.

Detection of CAR in Cells

Liver sections are analyzed under an Axiovert 35 inverted microscope (Zeiss) equipped with a filter from Omega Optics (Brattleboro, VT). YFP, CFP, and Hoechst S-33258 fluorescence is visualized through XF104, XF114, and XF03 filters, respectively. The fluorescent image is captured with a Spot II cooled charge-coupled device camera (Diagnostic Instruments, Sterling Heights, MI). The blue fluorescence of Hoechst S-33258 is changed to red using the accompanying software. Fluorescent images of the hCAR fusion proteins are shown in Fig. 1.

The intracellular fluorescence of YFP- or CFP-tagged proteins is also analyzed by confocal laser scanning microscopy (CLSM) using a Zeiss LSM 510 microscopy system (Carl Zeiss, Thornwood, NY) based on an Axiovert 100M inverted microscope equipped with an argon ion laser as the excitation source. Samples on the slides are excited at 514 nm (YFP) or 458 nm (CFP). YFP emission is detected using a 530-nm long-pass filter, and CFP fluorescence is detected using a 475-nm long-pass filter. A 40× water immersion objective (numerical aperture 1.2) is used for scanning with a pixel size of 0.1 μm in the *xy* plane and a distance of 1.0 μm between serial sections along the *z* axis. The pinhole setting is 50 to 60 μm in diameter, which yields a theoretical confocal slice thickness of approximately 1 μm. An example of the confocal microscopy analysis is shown in Fig. 2.

Results and Discussion

Both wild-type CAR and truncated $CAR_{181-348}$ were retained in the cytoplasm and were accumulated in the nucleus after CPZ treatment. hCAR contains 348 amino acid residues and consists of DNA and ligand-binding domains (DBD and LBD, respectively) and a hinge region that connects the two domains. The truncated $CAR_{181-348}$ lacks DBD, hinge, and helix 1 through helix 3 of LBD based on the multiple amino acid sequence alignments of nuclear receptors.[11] Amino acid sequence analysis shows a putative bipartite nuclear translocation signal (NLS) overlapping the hinge and LBD regions of CAR. The intracellular localization and nuclear accumulation of $CAR_{181-348}$ indicate that this NLS is not required for CAR nuclear translocation in mouse liver *in vivo*. Consistent with this finding, studies

[11] V. Giguere, *Endocr. Rev.* **20,** 689 (1999).

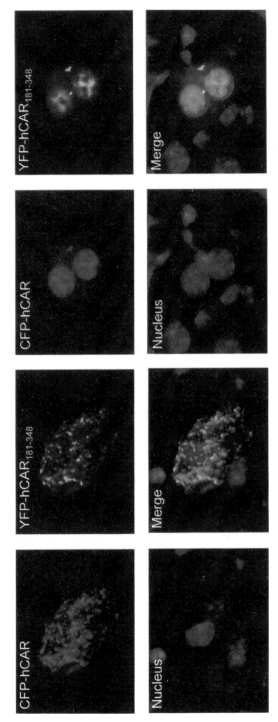

FIG. 1. Simultaneous expression of CFP–hCAR and YFP–hCAR$_{181-348}$ in mouse liver. Expression plasmids for CFP–hCAR and YFP–hCAR$_{181-348}$ were injected through the mouse tail vein as described in the experimental protocol. These images, representing fusion proteins and the nucleus, were captured with a fluorescent microscope (Axiovert 35) equipped with Spot II and a cooled CCD camera using the filters XF114, XF104, and XF03, respectively. The three images were merged using Spot II software package.

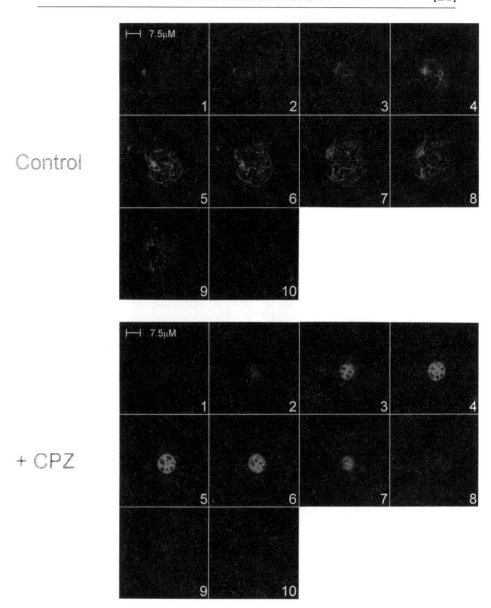

FIG. 2. Confocal microscopy images of CFP–hCAR expressed in mouse liver. These tomograms represent intracellular localization of the fusion protein in sequential z-axis stages of a 20-μm liver section with a 1-μm increment and show every other stage.

have shown that the LXXLXXL peptide motif near the C terminus, but not within the AF2 domain, regulates the PB-inducible nuclear translocation of the receptor.[8]

Prospective

Using the tail vein injection method, fluorescent protein-tagged CAR can be effectively expressed in mouse liver cells. The expressed receptor is properly retained in the cytoplasm of the control livers, and PB treatment translocates the receptor into the liver nucleus. The method is experimentally simple and is a practical tool to investigate intracellular localization of a given protein and its molecular/cellular mechanisms.

The method also has limitations. Because the whole animal is used, it is difficult to trace a time-dependent movement of protein in the cells. It is difficult to examine the effects of various chemicals (inhibitors and activators) on intracellular protein localization. For example, okadaic acid is known to repress the nuclear accumulation of CAR in PB-treated mouse primary hepatocytes. However, treatment of whole mouse with okadaic acid is impractical. These problems would be solved if primary hepatocytes can be used for all the cell system to express GFP–CAR fusion protein. However, using primary hepatocytes also presents problems. Due to a relatively poor detection sensitivity of GFP, compared with luciferase activity, for example, hepatocytes must be cultured for longer durations, Under our culture conditions, hepatocytes begin to lose their PB responsiveness. Another difficulty is the homogeneity of hepatocyte preparation and selective DNA transfection into the well-attached and extended cells. These cells are less responsive to PB treatment, and GFP–CAR fusion protein tends to localize in both the cytoplasm and the nucleus. The round-shaped hepatocytes are less effective in DNA transfection, and the localization of GFP–CAR expressed in these cells is difficult. Despite these difficulties, primary hepatocytes have advantages in the experimental system. Development of primary hepatocytes suitable for the experiments is an urgent issue in order to examine further details of the regulatory mechanism of PB-elicited nuclear translocation of CAR.

[22] Application of Fluorescent Differential Display and Peroxisome Proliferator-Activated Receptor (PPAR)α-Null Mice to Analyze PPAR Target Genes

By Susanna Sau-Tuen Lee, Li Tian, Wing-Sum Lee, and Wing-Tai Cheung

Introduction

The peroxisome proliferator-activated receptor (PPAR) is a nuclear receptor that was originally found to play an important toxicological role in mediating peroxisome proliferator (PP)-induced hepatotoxicity. For example, when rodents were treated with PPs including Wy-14,643 and clofibrate for a short period of time (1–2 weeks), hepatic peroxisome proliferation and hepatomegaly, hypolipidemia, and gene expression alteration occurred.[1,2] More importantly, prolonged treatment (>11 months) of rodents with PPs resulted in hepatocarcinoma formation.[3] Direct evidence that PPARα, one of the isoforms of PPARs, is the mediator involved in both short-term and long-term pleiotropic responses was provided by studies using the *in vivo* PPARα-null mouse model, which was created in 1995.[4] These mice, which lack PPARα receptor expression, are resistant to the actions of PP-induced short-term[4] and long-term[5] pathogenic effects.

In addition to its toxicological role, PPARα was implicated in playing a pivotal physiological role in regulating fatty acid metabolism and homeostasis during energy deprivation, a condition in which a high rate of fatty acids metabolism is demanded. Under short-term (12–48 hr) cellular fasting conditions, PPARα-null mice develop hepatomegaly, fatty liver, and hypoketonuria.[6–8] These studies clearly demonstrated that PPARα plays an important role in regulating the trafficking of lipid catabolism during fasting.

What remains unclear is the spectrum of genes that are regulated by PPARα during PP-induced pathogenic conditions and during cellular fasting. Although it

[1] A. J. Cohen and P. Grasson, *Food Cosmet. Toxicol.* **19,** 585 (1981).

[2] D. E. Moody, J. K. Reddy, B. G. Lake, J. A. Popp, and D. H. Reese, *Fundam. Appl. Toxicol.* **6,** 233 (1991).

[3] J. K. Reddy and N. D. Lalwani, *CRC Crit. Rev. Toxicol.* **12,** 1 (1983).

[4] S. S. T. Lee, T. Pineau, J. Drago, E. J. Lee, J. W. Owens, D. L. Kroetz, P. M. Fernandez-Salguero, H. Westphal, and F. J. Gonzalez, *Mol. Cell. Biol.* **15,** 3012 (1995).

[5] J. M. Peters, T. Aoyama, R. C. Cattley, U. Nobumitsu, T. Hashimoto, and F. J. Gonzalez, *Carcinogenesis* **19,** 1989 (1998).

[6] D. L. Kroetz, P. Yook, P. Costet, P. Bianchi, and T. Pineau, *J. Biol. Chem.* **273,** 31581 (1998).

[7] T. C. Leone, C. J. Weinheimer, and D. P. Kelly, *Proc. Natl. Acad. Sci. U.S.A.* **96,** 7473 (1999).

[8] S. Kersten, J. Seydoux, J. M. Peters, F. J. Gonzalez, B. Desvergne, and W. Wahli, *J. Clin. Nutr.* **103,** 1489 (1999).

has been demonstrated that PPARα regulates some of the genes located in peroxisomes, mitochondria, endoplasmic reticulum, and cytoplasm, it is not known how these PPARα target genes are regulated by each other and what other novel PPAR target genes are involved in these processes.

As an initial step in gaining insight into the molecular mechanism of how PPARα mediates PP-induced hepatotoxicity and regulates lipid homeostasis during fasting, we have applied the fluorescent differential display (fluoroDD) and PPARα-null mouse model to look for PPARα target genes involved in Wy-14,643-induced short-term hepatotoxicity and fatty acid trafficking during 72 hr of cellular fasting. Identification and characterization of differentially expressed genes are major first steps toward the elucidation of molecular mechanisms underlying a variety of biological processes. In addition, the identification of PPARα target genes has progressed rapidly since the creation of PPARα-null mice in 1995. With the availability of this animal model, a systematic screening for PPARα target genes involved in the peroxisome proliferator-induced liver injury and cellular fasting response has been carried out successfully in our laboratory.

We have employed the fluorescent DD system that was originally developed by Liang and Pardee[9] and was modified by the Genomyx Corp. (Foster City, CA). Differential display (DD) has been widely used to understand changes in gene expression in response to a number of normal physiological processes, as well as abnormal pathologic conditions, such as diseases and cancer. The advantages of this method are that it is fast, simple, and sensitive. Furthermore, DD is able to compare multiple groups of RNA samples simultaneously and it is capable of identifying both up- and downregulated differentially expressed genes in one experiment. FluoroDD represents the second generation of conventional DD systems in which radioisotopes are not used. The fluoroDD system has the advantages of being nonradioactive and identifying large DNA fragments that are not necessarily at the 3' end of the gene.

The fluorescent analysis of differentially expressed cDNA band patterns described here requires the use of the fluoroDD kit (Beckman Coulter Corp., Fullerton, CA) in conjunction with the HIEROGLYPH mRNA profile kit (Genomyx Corp.) for the synthesis of cDNA fragments from total RNA. In addition, the GenomyxLR DNA sequencer electrophoresis system (Beckman Coulter Corp.) and the GenomyxSC fluorescent imaging scanner (Beckman Coulter Corp.) are required for separation of cDNA mixtures and for production of fluorescent imaging of the cDNA patterns, respectively. In brief, fluoroDD involves isolation of high-quality total RNA from samples. Polyadenylated [poly(A)$^+$] mRNAs are selectively reverse transcribed into single-strand DNA with the use of specific anchored oligopolydeoxythymidine [oligo(dT)] primers. Subsequently, cDNAs are amplified by polymerase chain reaction (PCR) with the use of same anchored

[9] P. Liang and A. B. Pardee, *Science* **257,** 967 (1992).

oligo(dT) primer and an upstream arbitrary primer. For fluoroDD, the anchored oligo(dT) primer used in PCR amplification is labeled with a fluorescent tag, tetramethylrhodamine (TMR). PCR products are then separated on a long-range polyacrylamide sequencing gel in the GenomyxLR electrophoresis system. After gel separation, fluorescent bands are visualized by scanning the fluoroDD gel with the GenomyxSC fluorescent imaging scanner and the fluorescent images are captured with a computer. Because the fluorescent tag is present only on the 3'-anchored oligo(dT) primer, fluoroDD PCR bands represent products generated from the poly(A)$^+$ mRNA, and the signal intensity of TMR-labeled bands is indicative of the actual number of cDNA bands.

This chapter provides a step-by-step description of the experimental strategy and fluorescent differential display protocols that have been used successfully in our laboratory to identify the PPARα target genes involved in the actions of Wy-14,643-induced hepatotoxicity and lipid homeostasis during cellular fasting. The fluoroDD manual[10] provides an excellent in-depth description of the protocol for fluoroDD analysis and the reader is referred to it for details.

Experimental Strategies and Protocols

The overall procedures used for analyzing PPARα target genes by fluorescent differential display and PPARα (−/−) mice in our laboratory are summarized in Fig. 1.

Step 1: Tail Genotyping

For all experiments, 3-month-old male purebred PPARα (+/+) and PPARα (−/−) mice with a 129/Sv genetic background are used. It is essential to perform the experiments using purebred mice with homogeneous and identical genetic background between PPARα (+/+) and PPARα (−/−) mouse groups because it is difficult to replicate the results obtained from a mixed, heterogeneous genetic background. Both PPARα (+/+) and PPARα (−/−) male mice are the offspring of breeder mice obtained from the National Cancer Institute (Bethesda, MD) in 1995.[4] They have been bred and reared at the Chinese University of Hong Kong Laboratory Animal Service Facility since then. Two to three mice are housed per cage and provided with ozonated water *ad libitum*. Lighting is on a 12-hr light–12-hr dark cycle.

Due to the lack of obvious external phenotypes between PPARα (+/+) and PPARα (−/−) mice, we designed a PCR method to confirm the genotypes of these mice before using them for experiments. Using OLIGO 6 primer analysis software (Molecular Biology Insights, Inc., Cascade, CO), a 18-mer forward primer (5'-GCCTGGCCTTCTAAACAT-3') before the *Pst*I site (1330–1347) and a 18-mer reverse primer (5'-ACTCGGTCTTCTTGATGA-3') after the *Sph*I site

---

[10] FluoroDD Manual, Beckman Coulter, Inc., Fullerton, CA (1999).

(1503–1520) in exon 8 of the ligand-binding domain of the mouse PPARα gene[11,12] were chosen for the PCR reaction (Fig. 2A). In PPARα (+/+) mice, the expected PCR fragment is 191 bp, whereas a longer PCR fragment of 1248 bp is expected for PPARα (−/−) mice. This is due to the deletion of the 83-bp gene sequences between the PstI and the SphI sites in exon 8 of the ligand-binding domain and the replacement of a 1140-bp neomycin fragment in PPARα (−/−) mice.[4]

Preparation of Genomic Tail DNA. We routinely use the following method adapted from Laird *et al.*[13] for the extraction of genomic DNA from mouse tail in our laboratory.

1. Prepare 50 ml of incomplete lysis buffer [0.1 M Tris–HCl, pH 8.5, 5 mM EDTA, 0.2% (w/v) sodium dodecyl sulfate (SDS), and 0.2 M NaCl] in a sterile 50-ml Falcon tube as follows (this solution can be kept at room temperature for several months).

Component and stock concentration	Volume/50 ml	Final concentration
Tris–HCl (1 M, pH 8.5)	5 ml	0.1 M
EDTA (500 mM)	500 μl	5 mM
SDS (20%)	500 μl	0.2%
NaCl (1 M)	10 ml	0.2 M
Autoclaved, distilled H$_2$O	34 ml	—

2. Shortly before excision of the mouse tails (at least 3 mice per treatment group), prepare an appropriate amount of complete lysis buffer (0.5 ml per mouse tail) by freshly dissolving 100 μg proteinase K (Boehringer Mannheim, Germany) per milliliter of incomplete lysis buffer.

3. Excise about 1 cm of mouse tail tip from each mouse and immediately drop the tip into a 15-ml Falcon tube containing 0.5 ml of complete lysis buffer. Incubate at 55° with vigorous shaking overnight.

4. The following morning vortex the tubes vigorously for 2 min and incubate at 55° for an additional 2–4 hr to ensure complete digestion of the mouse tails.

5. Centrifuge tubes at 3000 rpm for 10 min at 4° to pellet large cellular debris. Decant the supernatant into 1.5 ml-microcentrifuge tubes and discard the pellets. Centrifuge tubes at full speed (12,000g) in a microcentrifuge for 10 min at 4° (very small blackish cellular debris will be seen in bottom of tubes). Decant the supernatants into 1.5-ml microcentrifuge tubes each containing 0.5 ml of 2-propanol.

[11] I. Issemann and S. Green, *Nature* **347**, 645 (1990).
[12] K. L. Gearing, A. Crickmore, and J.-A. Gustafsson, *Biochem. Biophys. Res. Commun.* **199**, 255 (1994).
[13] P. W. Laird, A. Zijderveld, K. Linders, M. A. Rudnicki, R. Jaenisch, and A. Berns, *Nucleic Acids Res.* **19**, 4293 (1991).

Step 1: Tail Genotyping
↓
Step 2: Animal Treatment
↓
Step 3: RNA Isolation and DNase I Treatment
↓
Step 4: First-Strand cDNA Synthesis

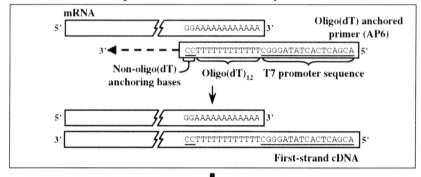

↓
Step 5: Synthesis of Double-Stranded cDNA Fragments

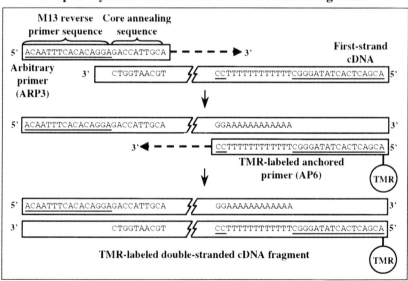

↓
Step 6: FluoroDD Gel Electrophoresis
↓

FIG. 1. Schematic diagram illustrating the overall procedures used for the analysis of PPARα target genes by fluorescent differential display. The details of each step are described in the text.

Step 7: Excision of Differentially Expressed cDNA Fragments

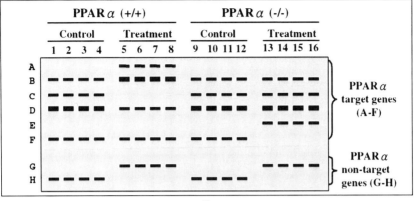

↓

Step 8: Reamplification of Differentially Expressed cDNA Fragments

↓

Step 9: Subcloning of Reamplified cDNA Fragments

↓

Step 10: Sequencing of Reamplified cDNA Fragments

↓

Step 11: Confirmation of Differentially Expressed cDNA Fragments by Northern Blot Analysis

↓

Characterization of Differentially Expressed cDNA Fragments

- Library screening to obtain full-length cDNA
- Tissue and temporal expression
- Functional analysis
  - *In Vitro* (Antisense and overexpression)
  - *In Vivo* (Transgenic and knockout mice)

Fig. 1. (*continued*)

A

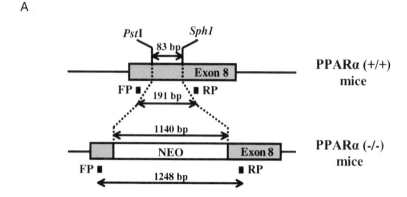

B

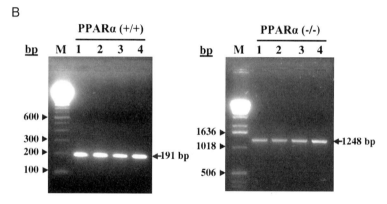

FIG. 2. Confirmation of PPARα (+/+) and PPARα (−/−) genotypes by PCR. (A) A schematic representation of PPARα (+/+) and PPARα (−/−) mouse genomic organizations and the strategy of PCR primer design. Exon 8 encodes the PPARα ligand-binding domain. In PPARα (−/−) mice, the 83 bp between *Pst*I and *Sph*I restriction enzyme sites in exon 8 was deleted and replaced with an 1140-bp neomycin gene fragment. The forward primer (FP) was located before the *Pst*I site, whereas the reverse primer (RP) was located after the *Sph*I site. The expected PCR fragments were 191 and 1248 bp for PPARα (+/+) and PPARα (−/−) mice, respectively. (B) Tail genotyping of PPARα (+/+) and PPARα (−/−) mice. The PCR products of PPARα (+/+) and PPARα (−/−) mice were electrophoresed on 2 and 1% agarose 0.5× TBE gels, respectively, with ethidium bromide staining. The sizes of PCR products for PPARα (+/+) and PPARα (−/−) mice were as expected. M, DNA marker.

Immediately shake the tube laterally until the solution viscosity is completely gone (DNA precipitation is complete).

6. Use a sterile, disposable pipette tip connected to a P20 microliter pipette to swirl and transfer DNA strands to a new 1.5-ml microcentrifuge tube (leave the lid open). Air-dry the brownish DNA pellet for 20 min and redissolve the DNA pellet in 50–150 μl of TE buffer (10 mM Tris–HCl and 0.1 mM EDTA, pH 7.5)

depending on the size of the DNA pellet. Incubate the tubes overnight at 55° to enhance complete dissolution of DNA pellets. The next morning, mix the DNA by gently shaking the tubes and spin the tubes quickly. Resuspend the DNA pellet by pipetting up and down until the viscosity of DNA has been eliminated. If the DNA solution remains viscous, add more TE in 20-μl aliquots until the viscosity has been eliminated. Measure the absorbances of the DNA solutions at OD_{260} using a spectrophotometer. Dilute the DNA solutions to a final concentration of 0.1 μg/μl and use 100 ng DNA for subsequent PCR reaction.

Polymerase Chain Reaction

1. Set up the PCR reaction in a 0.2-ml thin-walled PCR tube as follows.

Component and stock concentration[a]	Volume (μl)/50 μl reaction	Final concentration
Autoclaved, distilled H_2O	37.5	—
PCR buffer II (10×)	5	1×
$MgCl_2$ (25 mM)	3	1.5 mM
dNTP mix (1 : 1 : 1 : 1) (10 mM each)	1	200 μM each
Forward primer (10 μM)	1	0.2 μM
Reverse primer (10 μM)	1	0.2 μM
AmpliTaq (5 U/μl) (PerkinElmer, Norwalk, CT)	0.5	0.05 U/μl
Tail DNA (0.1 μg/μl)	1	—

[a] Prepare the core mixture for a large number of samples to avoid pipetting errors.

2. Mix gently by tipping the tube and quick-spinning in a microcentrifuge (2000g, 15 sec, room temperature).

3. Perform PCRs in a PerkinElmer 9700 thermal cycler (Norwalk, CT) with a heated lid at 95° for 2 min followed by 34 cycles at 92° for 15 sec, 55° for 30 sec, and 72° for 2 min; 72° for 7 min; and hold at 4°.

4. Electrophorese 5 to 10 μl of PCR product on 2 and 1% agarose gels for PPARα (+/+) and PPARα (−/−) tail DNA, respectively.

5. A typical genotyping result of the PPARα (+/+) and PPARα (−/−) tail DNA is shown in Fig. 2B.

Step 2: Animal Treatment

For the peroxisome proliferator study, mice (six per treatment group) from both genotypes are fed either 0.0% (w/w) control or 0.1% (w/w) Wy-14,643 ([4-chloro-6-(2,3-xylidino)-2-pyrimidinylthio]acetate) (Chemsyn Science Laboratories, Lenexa, KS) rodent diet (Bioserv Company, Frenchtown, NJ) for 2 weeks *ad libitum*.[4] For the fasting study, mice are deprived of food for 72 hr; during that time they have free access to water. Controls for each group are fed laboratory

Diet Brand Mouse Diet #5015 (PMI Nutrition Inc., St. Louis, MO) and water *ad libitum*. At the end of the experiments, the animals are sacrificed by decapitation, and livers are excised and weighed, wrapped in aluminum foil, snapped frozen, and stored in liquid nitrogen until further analysis.

Step 3: RNA Isolation and DNase I Treatment

Isolation of total RNA from mouse livers is the first step for fluoroDD analysis. The most critical factor contributing to the success of fluoroDD analysis is the quality of RNA. Poor RNA quality can lead to poor fluoroDD results and severe problems in subsequent downstream analyses. If any trace of RNA degradation is observed, do not use the RNA for fluoroDD. The use of purified poly(A)$^+$ in place of total RNA is unnecessary for fluoroDD and may be undesirable because of possible contamination and interference by oligo(dT) and because of the potential loss of low copy number mRNA species.[10] Two steps are involved in preparation of total RNA. The first step is to isolate high-quality RNA from tissues using TRIzol reagent (Life Technologies, Carlsbad, CA) according to the manufacturer's instructions. The second step involves treating the total RNA sample with DNase I (Amersham Pharmacia Biotech, Piscataway, NJ), followed by phenol–chloroform (pH 5.2) extraction and ethanol precipitation to remove any traces of contaminating DNA. This DNase I treatment is often crucial to the success of the fluoroDD screening because trace amounts of contaminating DNA will be amplified during fluoroDD RT-PCR reaction and contribute false positives to the fluoroDD banding pattern. The quality of the DNase I-treated total RNA must be checked by electrophoresis before proceeding with the subsequent cDNA synthesis. The 28S rRNA band at 4.7 kbp should be about twice the intensity of the 18S rRNA band at 1.9 kbp. Isolating RNA from at least two different mice for each treatment group is absolutely necessary to detect intraanimal variability.

RNA Isolation

1. For all manipulations of RNA, wear clean gloves and use aerosol-barrier, sterile, nuclease-free pipette tips, and sterile disposable plasticwares. All reagents should be RNase free.

2. Aliquot 10 ml of TRIzol reagent (for whole liver of ~0.8–1 g per mouse) into a 50-ml Falcon tube (one tube per mouse liver) at room temperature. Remove the aluminum foil-wrapped liver sample from the liquid nitrogen tank as quickly as possible. Without unwrapping the aluminum foil, immediately break up the frozen liver into several small chunks using a pair of long forceps. Unwrap the liver tissue quickly and transfer it into the 50-ml Falcon tube containing 10 ml of TRIzol reagent.

3. Homogenize the tissue while it is still frozen using a tissue tearer (Dremel, BioSpec Products, Inc., Bartlesville, OK). The tissue tearer should be cleaned

thoroughly with distilled water and then rinsed with TRIzol reagent a few minutes before use. Transfer 1-ml aliquots of the tissue homogenate into 1.5-ml microcentrifuge tubes each containing 0.2 ml of chloroform (a total of ~10 tubes per 0.8–1 g of mouse liver tissue).

4. Centrifuge the mixture at 12,000g for 15 min at 4° and transfer the supernatant to a new microcentrifuge tube containing 0.5 ml of 2-propanol. Mix the contents vigorously and precipitate the RNA at −80° overnight.

5. In the morning, centrifuge the tubes at 12,000g for 15 min at 4°. Discard the supernatant and wash the RNA pellet with 1 ml of 70% (v/v) ethanol three times. Air-dry and resuspend the RNA pellet from one tube in nuclease-free water (Promega, Madison, WI) for further procedures. Store the remaining RNA pellets in 1.5 ml of absolute ethanol (for long-term storage) at −80° until use.

6. Determine the quantity of isolated RNA by measuring the absorbances at OD_{260} and OD_{280} using a spectrophotometer. Before measurement, do not heat the diluted RNA as suggested in the fluorodd manual to avoid overestimation of the RNA concentrations in the samples.

7. Electrophorese 10–20 μg RNA on a 1% formaldehyde-agarose gel (step 11, below) to check the integrity of the sample. Do not use RNA samples with noticeable degradation for DNase I treatment.

DNase I Treatment

1. Set up a DNase I digestion as follows.

Component and stock concentration	Volume (μl)/50 μl reaction	Final concentration
Total RNA in nuclease-free water	26.5 (1 μg/μl)	0.53 μg/μl
Tris–HCl (1 M, pH 7.5)	2.5	50 mM
MgCl$_2$ (25 mM)	20	10 mM
RNase-free DNase I FPLC purified (10 U/μl) (Amersham Pharmacia Biotech.)	1	0.2 U/μl

2. Mix carefully by pipetting up and down (do not vortex and create bubbles) and incubate at 37° for 30 min.

3. Add 50 μl of nuclease-free water to increase the tube volume before extraction and add 200 μl of phenol/chloroform/isoamyl alcohol (v : v : v, 25 : 24 : 1; Life Technologies) for extraction of any traces of contaminating DNA. Vortex for 10 sec and spin in a microcentrifuge at 12,000g for 15 min at 4°.

4. Transfer the upper aqueous phase (90 μl) to a new autoclaved microcentrifuge tube. Avoid taking the interphase and do not transfer any organic phase to the new tube. Add 1/10 volume (9 μl) of autoclaved 3 M sodium acetate, pH 5.2, and 2.2 volume (198 μl) of absolute ethanol to precipitate the RNA. Mix by inverting the tubes several times and leave at −80° overnight.

5. The next day, centrifuge at 12,000g for 15 min at 4°. Pipette off and discard the supernatant. Wash the RNA pellet (it may be difficult to see) twice with ice-cold 70% ethanol. Add 1.5 ml of 70% ethanol and store the pellet at −80° until needed.

6. When ready to use DNase I-treated RNA, centrifuge the tube at 12,000g for 15 min at 4° to pellet the RNA at the bottom of the tube. Remove the 70% ethanol and let the sample air-dry for 10 min. Be sure that no residual ethanol remains before redissolving the RNA pellet. Dissolve the RNA thoroughly by pipetting the mixture up and down in 20 μl of nuclease-free water until no gelatinous-like pellet can be seen in the bottom of the tube.

7. Determine the quantity and purity of DNase I-treated RNA by measuring the absorbances at OD_{260} and OD_{280} using a spectrophotometer. Before measurement, do not heat the diluted RNA as suggested in the fluoroDD manual to avoid overestimation of the RNA concentrations in the samples. It is important to obtain accurate OD measurement for each RNA sample at this step in order to eliminate false differential display gene expression patterns due to loading of unequal RNA concentrations for fluoroDD analysis. For good RNA quality, the ratio of OD_{260} to OD_{280} should be greater than 1.8.

8. Electrophorese 1 μg of the RNA on a 0.8% agarose, 1× TAE gel at 80–100 V for 20–30 min to check the integrity of the RNA samples. The 28S rRNA band at 4.7 kbp should be about twice the intensity of the 18S rRNA band at 1.9 kbp. Do not use any RNA with noticeable degradation for first-strand cDNA synthesis.

9. Store the good quality RNA samples at −80° in 5-μg aliquots (at a concentration of at least 1 μg/μl). For long-term storage, store the RNA as ethanol precipitates at −80°.

Step 4: First-Strand cDNA Synthesis

We follow the fluoroDD protocol (Beckman Coulter Corp.) for performing first-strand DNA synthesis. Reverse transcription reactions (RT) are carried out with the 12 3′-nonfluorescent oligo(dT)-anchored primers (AP) provided by the HIEROGLYPH mRNA profile kit (Genomyx Corp.) (Table I). There are different two-base combinations on the upstream of the poly(A)$^+$ region, which represents a considerably smaller subset of the mRNA pool rather that single-base permutations. Each respective two-base AP would encompass approximately one-twelfth of the mRNA pool constituency and would render approximately 800–1200 different first-strand cDNAs.[10] Twelve first-strand cDNA pools can be obtained from each RNA sample. Because there is no way to predict which of the 12 APs in combination with a particular arbitrary primer (ARP) will produce the greatest number of differentially expressed cDNA fragments with a particular total RNA

TABLE I
ARBITRARY AND ANCHORED PRIMERS USED FOR
FluoroDD-PCR REACTION

Primer	Sequence
AP6[a]	5′-<u>ACGACTCACTATAGGGC</u>**TT**TTTTTTTTTTTTCC-3′
AP9[a]	5′-<u>ACGACTCACTATAGGGC</u>**TT**TTTTTTTTTTTTAC-3′
ARP3[b]	5′-<u>ACAATTTCACACAGGA</u>GACCATTGCA-3′

[a] The 17 nucleotides of the T7 promoter sequence are underlined and the two bases located upstream are in bold.
[b] The 16 nucleotides of the M13 reverse (−48) sequence are underlined.

sample, we randomly select the AP for the first-strand cDNA synthesis. Also, it is more efficient to work systematically with a subset of AP at one time, in combination with different ARPs, rather than with all subsets of the APs at once. To identify intraanimal variability, RNAs from at least two controls and two treated animals of both genotypes should be used for the RT reaction. In order to test for genomic DNA contamination that can contribute false-positive fluoroDD-PCR bands, negative RT-PCR (without reverse transcriptase) should be performed.

1. Dilute the DNase I-treated total RNA to 0.1 μg/μl (requires great accuracy) with nuclease-free water just before first-strand cDNA synthesis. Dilute only the amount of RNA needed for RT reactions. Excess diluted RNAs should not be refrozen for later use because the synthesized first-strand cDNA can tolerate only one freeze–thaw cycle.

2. Combine the following in a 0.2-ml thin-walled PCR tube (tube A) as follows.

Component and stock concentration	Volume (μl)/20 μl reaction	Final concentration
Total RNA (DNA free, freshly diluted)	2 (0.1 μg/μl)	10 ng/μl
HIEROGLYPH-anchored primer (2 μM)	2	0.2 μM

3. Mix the aliquots carefully with a pipette (do not vortex) and quick-spin the tubes briefly. Incubate at 70° for 5 min in a thermal cycler with a heated lid (PerkinElmer 9700). Quick-chill the tubes on ice. Spin briefly to collect contents at the bottom of the tubes.

4. Set up the core mix described below on ice in a sufficient volume for the number of RT reactions to be carried out.

Component and stock concentration	Volume (µl)/ RT reaction	Volume (µl)/-RT reaction	Final concentration
Sterile nuclease-free H_2O	7.8	8.0	—
SuperScript II RT buffer (5×)	4.0	4.0	1×
dNTP mix (1 : 1 : 1 : 1) (250 µM each)	2.0	2.0	25 µM each
DTT (100 mM)	2.0	2.0	10 mM
SuperScript II RNase H⁻ RT (200 U/µl) (Gibco-BRL, Gaithersburg, MD)	0.2	—	2 U/µl

5. Add 16 µl of RT core mix into tube A (as described earlier) for a final volume of 20 µl per tube. Mix well with gentle pipetting and quick-spin.

6. Perform RT reactions in a thermal cycler (PerkinElmer 9700) with a heated lid at 42° for 5 min; 50° for 50 min; 70° for 15 min; and hold at 4°. Save the RT reactions in 5-µl aliquots (for use with different ARP in fluoroDD-PCR) and store in a −20° constant temperature, nonfrost-free freezer for use in fluoroDD. The first-strand cDNA should be used in fluoroDD-PCR within 1 week.

Step 5: Synthesis of Double-Stranded cDNA Fragments

FluoroDD-PCR reactions are performed using the HIEROGLYPH mRNA profile (Genomyx Corp.) and fluoroDD (Beckman Coulter Corp.) kits. The fluoroDD kit contains 12 APs identical in DNA sequence to the HIEROGLYPH APs and 3′ end labeled with the fluorescent tag, TMR. The HIEROGLYPH mRNA profile kit contains 20 different arbitrary primers (ARP), which can be used for fluoroDD-PCR reaction (Table I). Both APs (31 nucleotides) and ARPs (26 nucleotides) are particularly long to allow for stringent annealing conditions during the fluoroDD-PCR step, and thus prevent mispriming, and formation of nonspecific products and false positives during the reamplification process.[14,15] Because there is no way to predict which of the ARP prime in combination with a particular AP will generate more bands with a particular total RNA sample, we first perform second-stranded cDNA synthesis with a 3′ nonfluorescent AP and then analyze the nonfluorescent DD-PCR patterns on a 1% agarose minigel. If obvious differential cDNA patterns are detected with one particular AP-ARP primer set, then we proceed to perform the fluorescent DD-PCR with the corresponding 3′ fluorescent AP from the fluoroDD kit. To verify PCR reproducibility and for the accurate identification of unique cDNA fragments, all fluoroDD-PCRs should be carried in duplicate per RNA sample.

[14] L. Mou, H. Miller, J. Li, E. Wang, and L. Chalifour, *Biochem. Biophys. Res. Comm.* **199,** 564 (1994).
[15] S. Zhao, S. L. Ooi, and A. B. Pardee, *Biotechniques* **18,** 842 (1995).

1. Set up the fluoroDD-PCR reactions in 0.2-ml thin-walled PCR tubes as follows.

Component and stock concentration	Volume (μl)/10 μl reaction	Final concentration
Sterile nuclease-free H$_2$O	1.95	—
PCR buffer II (10×)	1.0	1×
MgCl$_2$ (25 mM)	1.5	3.75 mM
dNTP mix (1 : 1 : 1 : 1) 250 μM each	2.0	50 μM each
5'-ARP (2 μM) (exclude from core mix)	1.75	0.35 μM
3'-TMR-AP (5 μM)	0.70	0.35 μM
(fluorescent version of same 3'-AP in RT mix)		
RT mix (derived with the unlabeled version of the same 3'-AP)	1.0	—
AmpliTaq enzyme (5 U/μl) (PerkinElmer)	0.10	0.05 U/μl

2. Perform PCR reactions in a thermal cycler (PerkinElmer 9700) with a heated lid at 95° for 2 min; 4 cycles at 92° for 15 sec, 50° for 30 sec, and 72° for 2 min; 30 cycles at 92° for 15 sec, 60° for 30 sec, and 72° for 2 min; 72° for 7 min; and hold at 4°. To avoid loss of fluorescence signal due to dilution of samples from condensation within the cycler, do not leave the samples in the thermal cycler at 4° after the completion of the PCR reaction. Instead, wrap the samples with foil to protect from light and store the samples in a −20° freezer overnight.

Step 6: FluoroDD Gel Electrophoresis

Following fluoroDD-PCR, TMR-labeled cDNA fragments are separated electrophoretically on a 5.6% clear denaturing HR-1000 gel matrix (Beckman Coulter Corp.) under denaturing conditions. The GenomyxLR DNA sequencer (Beckman Coulter Corp.) is required for full resolution of fluoroDD-PCR cDNA fragments.

1. Gently mix 70 ml of 5.6% clear denaturing HR-1000 gel with 320 μl of freshly prepared 10% ammonium persulfate and 32 μl of TEMED. Cast the gel, wrap it with plastic wrap, and allow the gel to polymerize overnight at room temperature.

2. The next day, transfer 4.0 μl of each fluoroDD-PCR sample into new 0.2-ml thin-walled PCR tubes and add 1.5 μl of fluoroDD loading dye (Beckman Coulter Corp.) to each PCR tube. Save the remaining portion of the fluoroDD-PCR samples, wrap completely in foil to keep light out, and store at −20°. For the TMR-MW DNA standard marker for each gel, mix 1.5 μl of fluoroDD loading dye with 1.5 μl of TMR-standard DNA marker (Beckman Coulter Corp.) in a 0.2-ml thin-walled PCR tube.

3. Place all tubes uncapped in the thermal cycler, and leave the lid of the cycler completely open. Heat the tubes in a thermal cycler at 95° for 5 min to denature and concentrate the samples so that they can be loaded entirely on the gel in a small volume (~2.5 to 3 μl per lane). Quick-spin the tubes after heating and keep on ice.

4. Right before loading the samples onto the 5.6% polyacrylamide gel, use a 50-ml syringe and a 16-gauge needle to flush urea out of the wells in the gel. Carefully load the entire denatured sample into the well on a 48-well comb with a sterile long capillary flat gel-loading tip. Arrange duplicate PCRs of the same mouse RNA in adjacent lanes and samples of the same treatment group in consecutive lanes (Fig. 1, step 7).

5. Electrophorese the fluoroDD-PCR products through a GenomyxLR DNA sequencer (Genomyx Corp.) at 3000 V and 1000 W for 4.5 hr at 50° until the 300-bp marker migrates near the bottom of gel.

Step 7: Excision of Differentially Expressed cDNA Fragments

cDNA bands that consistently show differential expression changes (either up or down) between control and treated samples in PPARα (+/+) but not in PPARα (−/−), or vice versa, are considered putative PPARα target genes (Fig. 1, step 7). Excise only cDNA bands that are reproducible in duplicate fluoroDD-PCR for each mouse and for a total of two mice from the same treatment group.

1. After gel electrophoresis, remove urea from the gel thoroughly with distilled water as suggested by the fluoroDD manual. Residual urea can interfere with cDNA fragment reamplification. Dry the gel for ~10 min after each of the three rinse steps. Be sure the gel is completely dry (~25 min) after the final rinse step.

2. Scan the fluoroDD gel in the GenomyxSC fluorescent scanner (Beckman Coulter Corp.) according to the protocol provided in the GenomyxSC operating manual. A physical grid (Beckman Coulter Corp.) is used to locate the position of interested bands, and a Excision workstation (Beckman Coulter Corp.) is used to hold the fluoroDD gel. Keep the surface of the dried gel as clean as possible prior to band excision to avoid surface contamination with environmental nucleic acids.

3. Carefully excise the cDNA bands that are of interest from the fluoroDD gels as described by the fluoroDD manual. Use a new, sterile razor blade for each excised band and use only a sterile, nuclease-free tube for band collection. After gel bands are excised, rescan the fluoroDD gel in order to check for the accuracy of band excision. Incubate the gel slice in 30 μl of sterile, nuclease-free TE (10 mM Tris–HCl, pH 7.4, and 0.1 mM EDTA) at 37° for 1 hr in order to allow diffusion of DNA out of the gel slice. Keep the EDTA concentration in the

gel eluate to 0.1 mM to prevent it from interfering with the subsequent reamplification efficiency. Store the gel band eluate at $-20°$ until use for reamplification.

Step 8: Reamplification of Differentially Expressed cDNA Fragments

Following elution of cDNA from the HR-1000 gel slice, cDNA fragments are reamplified using full-length M13 reverse (-48) 24-mer (5'-AGCGGATAACAA TTTCACACAGGA-3') and full-length T7 promoter 22-mer (5'-GTAATACGACT CACTATAGGGC-3') primers (Genomyx Re-Amp kit). Each of the ARP used in fluoroDD RT-PCR incorporates a 16 nucleotide segment of the M13 reverse (-48) 24-mer priming sequence, which allows directional sequencing from the 5' end of the original transcript. Furthermore, each of the AP used in fluoroDD RT-PCR incorporates a 17 nucleotide segment of the T7 promoter 22-mer priming sequence, which allows directional sequencing from the 3' end of the original transcript. The expected size of the reamplified PCR product is 13 nucleotides longer than the size of the fragments appeared on fluoroDD gels because the M13 reverse and T7 promoter primers were extended by 8 and 5 nucleotides, respectively.

1. Set up the PCR reaction in a 0.2-ml, thin-walled PCR tube as follows.

Component and stock concentration	Volume (μl)/20 μl reaction	Final concentration
Sterile nuclease-free H$_2$O	10.45–7.45	—
PCR buffer II (10×)	2.0	1×
MgCl$_2$ (25 mM)	0.75	0.94 mM
dNTP mix (1 : 1 : 1 : 1) (250 μM each)	1.6	20 μM each
M13 reverse 24-mer primer (2 μM) (Genomyx Corp.)	2.0	0.2 μM
T7 promoter 22-mer primer (2 μM) (Genomyx Corp.)	2.0	0.2 μM
AmpliTaq (5 U/μl) (PerkinElmer)[a]	0.2	0.05 U/μl
Gel band eluate in TE	1–4[b]	—

[a] Exclude from the core mix when performing hot-start PCR.
[b] Depend on the intensity of the cDNA band appeared on the fluroDD gel.

2. Perform hot-start PCR in a thermal cycler (PerkinElmer 9700) with a heated lid at $95°$ for 2 min and then at $80°$ for 10 min during which time the AmpliTaq is added, followed by 4 cycles at $92°$ for 15 sec, $50°$ for 30 sec, and $72°$ for 2 min; 25 cycles at $92°$ for 15 sec, $60°$ for 30 sec, and $72°$ for 2 min; $72°$ for 30 min; and hold at $4°$.

3. Run 10 μl of reamplified product on a 1–2% agarose gel depending on the size of PCR products. Although the expected PCR products will be 13 bp longer than the original fluoroDD-PCR band, this size change may not be always

discernible on a low percentage agarose gel. If no bands or an insufficient amount of reamplified band is generated, try to increase the gel eluate to 4 μl for reamplification. It is not preferable to perform secondary reamplification from the primary reamplifications in order to generate more reamplified products. If there are extra bands present, it will be necessary to isolate and purify the appropriately sized reamplified PCR product before use in the subsequent subcloning step. Store the reamplified PCR product at 4° and subclone in a TA vector within 24 hr.

Step 9: Subcloning of Reamplified cDNA Fragments

The AdvanTAge PCR cloning kit (Clontech, Palo Alto, CA) is used for subcloning the reamplified cDNA fragments according to the manufacturer's instructions. About 0.5–2 μl of reamplified PCR products is used for ligation with 50 ng of the pT-Adv vector. For each reamplified cDNA fragment, a total of five recombinant clones are picked and subjected to EcoRI restriction enzyme digestions. The restriction enzyme digest is then electrophoresed on a 1–2% agarose gel to check for the size of the insert. For accurate identification of the correct insert, it is important to include 2–4 μl of the reamplified PCR product that is used for subcloning on the agarose gel as an expected size standard. Mini-prep DNA (Qiagen) is then prepared from at least two recombinant clones with the correct insert size. DNA should be resuspended in nuclease-free water or sterile 18 Ω water (Sigma, St. Louis, MO) for subsequent DNA sequencing. EDTA should not be present, as it can inhibit the sequencing reaction.

Step 10: Sequencing of Reamplified cDNA Fragments

Although many investigators perform Northern blot analysis or reverse Northern blots before sequencing putative differentially expressed bands, we find that it is more efficient to do DNA sequencing prior to Northern blot analysis. Sequencing is a faster method in comparison to Northern blotting and it can tell us whether the gene is known or novel before we decide whether the gene is of sufficient interest to do further confirmation and study. Often times, we found that different fluoroDD fragments excised from either the same or different fluoroDD gels are homology to a same gene. In this case, we can focus only on one cDNA fragment, preferably the longest one, of the same gene for downstream confirmation.

The CEQ 2000 dye terminator cycle sequencing kit (Beckman Coulter Corp.) and CEQ 2000 automated DNA sequencer (Beckman Coulter Corp.) are used to perform the sequencing reactions. The Qiagen purified minipreparation DNA (100 fmol) from each recombinant clone is used as template for cycle sequencing using the M13 reverse (5′-AAACAGCTATGACCATG-3′) or M13 sequencing primer (5′-GTAAAACGACGGCCAGT-3′) located in both orientations of the TA cloning vector (pT-Adv). The PCR sequencing reactions are then purified according to the manufacturer's directions and are loaded onto the CEQ 2000 capillary electrophoresis-automated DNA sequencer. DNA sequences obtained are edited and then submitted to the Basic Local Alignment Search Tool (BLAST) for searching the homology of sequences against the known genes or ESTs in the GenBank databases (www.ncbi.nlm.nih.gov/BLAST/).

Step 11: Confirmation of Differentially Expressed cDNA Fragments by Northern Blot Analysis

Confirmation of differential gene expression should be carried out with RNA samples generated from an independent but not the original fluoroDD experiment. Total RNA from each group is subjected to nonradioactive Northern blot analysis using the random-primed DIG-labeled cDNA fragment as a probe. GAPDH is used as a loading control.

DIG-Labeled cDNA Probe Preparation

1. Digest 20 μg of Qiagen minipreparation purified subcloned cDNA from the pT-Adv vector with 5 μl of *Eco*RI (20 U/μl) at 37° overnight. The reaction volume for the restriction enzyme digest is 150 μl. Electrophorese the restriction enzyme digest on a 1–1.5% preparative agarose gel, depending on the size of the cDNA insert, at 35 V overnight (~17 hr).

2. Excise the cDNA insert from the agarose gel with a sterile razor blade under a hand-held UV lamp. Trim off as much as agarose as possible and weigh the excised gel slice in a 50-ml Falcon tube. Extract the DNA from the agarose gel slice using a Qiagen QIAquick gel extraction kit. Precipitate the eluted DNA in the presence of 3 M sodium acetate, ethanol, and glycogen and resuspend the DNA in 16 μl of nuclease-free water. Determine the quantity of DNA by diluting 1 μl of resuspended DNA to 100 μl with nuclease-free water at OD_{260} using a spectrophotomter.

3. Denature the remaining 15 μl DNA (10–3000 ng) at 100° for 15 min and quick-chill on ice. Mix denatured DNA with 2 μl of hexanucleotide mix, 2 μl of DIG-labeled dNTP mix, and 1 μl of Klenow enzyme at 37° for 20 hr according to the random-primed DIG DNA-labeling kit (Boehringer Mannheim). Add 2 μl of 0.2 M EDTA, pH 8.0, to stop the reaction, and use the random-primed DIG-labeled DNA probe immediately or store at $-20°$ until use.

Formaldehyde-Agarose Gel Electrophoresis. Resuspend the RNA pellet in 100% formamide (Boehringer Mannheim) in order to reduce the degradation of the RNA samples.[16] Determine the quantity of RNA by measuring the absorbances at OD_{260} and OD_{280} using a spectrophotometer. Dilute the total RNA samples to 2 μg/μl in formamide. Three different mouse samples are used for each treatment

---

[16] P. Chomczynski, *Nucleic Acids Res.* **20,** 3791 (1992).

group, and a total of 12 RNA samples are used for each Northern blot analysis for confirmation.

1. Prepare the total RNA samples in 1.5-ml Eppendorf tubes as follows.

Component and stock concentration	Volume (μl)/20 μl reaction	Final concentration
Sterile, nuclease-free H$_2$O	5	—
Total RNA sample (2 μg/μl)	10	20 μg
Formaldehyde (37%)	3	2.2 M
MOPS (10×)	2	1×

2. Prepare 100 ng (5 μl) DIG-labeled RNA molecular weight marker I (Boehringer Mannheim) the same way as the total RNA samples except that no nuclease-free water is added and 10 μl of formamide is used. Denature the total RNA and RNA marker samples at 65° for 15 min and quick-chill on ice. Add 2 μl of loading dye and 1.5 μl of 0.2 mg/ml ethidium bromide to each of the aforementioned tubes. Mix and quick-spin.

3. Load the RNA samples onto 13 wells on a 1% formaldehyde-agarose gel (1.5 g agarose, 15 ml of 10× MOPS buffer, 2.7 ml of 37% formaldehyde and add autoclave water to a final volume of 150 ml). Electrophorese the gel in 1× MOPS buffer without formaldehyde at 35 V at room temperature overnight until the bromphenol blue dye reaches the bottom of the gel (~15 hr). After electrophoresis, remove gel, take photo, and then wash gel in 10× SSC for 45 min with agitation at room temperature. Flip the gel upside down and then transfer RNA onto a Hybond N+ nylon membrane (Amersham Pharmacia Biotech) in 10× SSC overnight by the upward capillary transfer method. The next day, mark the wells on the nylon membrane with black pen and then wash the membrane in 5× SSC for 15 min at room temperature with agitation. Bake membrane at 80° for 2 hr.

Prehybridization, Hybridization, and Detection

1. Wash membrane in 2× SSC at room temperature with agitation for 15 min. Transfer membrane to an 8 inch × 12 inch hybridization bag (Kapak Sealpak) and add 35 ml DIG Easy Hyb (Boehringer Mannheim). Remove bubbles, seal the hybridization bag, and prehybridize membrane at 42° overnight with gentle agitation in a hybridization oven.

2. Denature the DIG-labeled cDNA probe at 100° for 15 min and quick-chill on ice. Pour out the DIG Easy Hyb from the prehybridized membrane in a 50-ml Falcon tube and add the denatured DIG-labeled probe into the DIG Easy Hyb solution (the same as the one used for prehybridization). Mix and then add the probe solution back into the prehybridized membrane in the hybridization bag.

Hybridize membrane at 42° overnight with gentle agitation in a hybridization oven. After hybridization, transfer the probe solution to a 50-ml Falcon tube and store at −20°. This probe solution is stable for >1 year when store frozen at this temperature and can be reused several times for additional membranes.

3. For washing, blocking, incubation with antibodies, and color development, it is important to eliminate all alkaline phosphatase contamination from your hands and wash trays. Wear gloves at all times, and clean and bake all glass wash trays at 180° before use. After hybridization, wash the membrane in 2× SSC/0.1% SDS twice for 15 min each at room temperature with gentle agitation and then in 0.5× SSC/0.1% SDS twice for 15 min each at 60° with gentle agitation. Wash the membrane in washing buffer [0.1 M maleic acid, 0.15 M NaCl, pH 7.5, and 0.3% (v/v) Tween 20] at room temperature for 15 min.

4. Block the membrane in 100 ml of blocking buffer (Boehringer Mannheim) for 1 hr with agitation at room temperature. After blocking, incubate the membrane in 100 ml of anti-DIG antibody alkaline phosphatase conjugate at a dilution of 1 : 10,000 (Boehringer Mannheim) for 1 hr with agitation at room temperature. Wash the membrane in washing buffer twice for 15 min each at room temperature with agitation. Equilibrate the membrane in 100 ml of detection buffer (0.1 M Tris–HCl and 0.1 M NaCl, pH 9.5) for 5 min and transfer the membrane to a new wash tray. Develop the membrane in 40 ml of detection buffer containing 200 μl of nitroblue tetrazolium (100 mg/ml) and 150 μl of 5-bromo-4-chloro-3-indolyl-phosphate (50 mg/ml) from 2 hr to overnight of exposure in the dark. Scan the purple color signal on the membrane and take a black-and-white photo for records.

Normalization with GAPDH. In order to assess RNA loading and transfer efficiency, membranes are stripped and reprobed with the mouse glyceraldehyde-3-phosphate dehydrogenase (GAPDH) probe for normalization. A partial mouse GAPDH cDNA (191–1017) is synthesized by RT-PCR based on the published mouse GAPDH cDNA sequence.[17] A 22-mer forward (5′-TCCACTCACGGCAA ATTCAACG-3′) (191–212) primer and a 20-mer reverse (5′-TCCACCACCCTGT TGCTGTA-3′) (992–1017) primer are chosen for synthesis of mouse GAPDH cDNA using the RT-PCR reaction. The size of the expected PCR product is 827 bp. After color development, remove the purple color precipitates on the membrane by incubating in 25 ml of prewarmed N,N-dimethylformamide (Fluka) at 60° with gentle agitation. After the purple color has been removed, rinse the membrane thoroughly in distilled water. To remove the DIG-labeled probe from the membrane, wash it in preboiled 0.1% SDS for 15 min at room temperature with agitation, followed by washing in washing buffer for 5 min at

[17] D. E. Sabath, H. E. Broome, and M. B. Prystowsky, *Gene* **91,** 185 (1990).

room temperature. The membrane is then ready for prehybridization and hybridization with the random-primed DIG-labeled GAPDH probe as described earlier.

Experimental Results and Discussion

To identify the spectrum of PPARα target genes that are involved in peroxisome proliferator-induced hepatotoxicity, and cellular response to fasting, fluorescent differential display was carried to compare liver RNA samples from control and treated PPARα (+/+) and PPARα (−/−) mice with different anchored and arbitrary primer combinations. Each AP and ARP combination generated a unique FDD banding pattern. Figures 3A and 4A show representative results of Wy-14,643 and fasting experiments, respectively. The fluoroDD cDNA fragments ranged in size from 350 to 1500 bp, and the FDD bands, showing different intensities among the RNA samples, were presumably derived from differentially expressed cDNA transcripts. We had great success (>98%) in reamplifying PCR fragments of 350–900 bp, while only occasionally for 1000–1500 bp. In most cases, only one amplified product corresponding to the size of the excised fluoroDD-PCR band was obtained. However, in some cases, in addition to the major expected PCR product, there were a few other faint bands present. This might be from traces of contaminating nucleic acid species on the surface of the dried gel or might be due to the comigration of different cDNA species during electrophoresis. All of the reamplified cDNA fragments were subcloned and sequenced successfully. Some of the subclones generated from the same fluoroDD fragment yielded different sequences, which could be due to the presence of comigrating bands on the excised fluoroDD fragment. It was not uncommon to find that several different fluoroDD fragments, which were generated from either the same AP and ARP or different AP and ARP combinations, were homology to a same gene. In this case, we found that their differential expression patterns on Northern blot analyses always agreed with their fluoroDD banding patterns. We did not get a 100% success rate on the confirmation of fluoroDD fragments by Northern blot analyses. These negative results included the lack of differential expression signals on Northern blot confirmations with the DIG-labeled fluoroDD fragments, and no expression signals were detected on the Northern blots. However, these false positives were also reported in DD and other methods used for the identification of differential expression gene patterns.[9] In order to improve and speed up the most time-consuming Northern confirmation step, we are in the process of adopting recently developed DNA microarray technology to perform the reverse Northern for large-scale preliminary confirmation of differential expressed subclones simultaneously from the reamplified cDNA fragment.

Most of the differentially expressed genes involved in the action of Wy-14,643 were either up- or downregulated in PPARα (+/+) but not in PPARα (−/−) mice after Wy-14,643 treatment (Fig. 3A). Because the up- or downregulations of these genes was abolished in PPARα (−/−) receiving the same treatment, indicating that PPARα is required in the transcriptional regulation of these genes, they were considered as putative PPARα target genes. Whether these PPARα target genes are regulated directly or indirectly by the PPARα could not be ascertained just based on fluoroDD results. Some of these putative PPARα target genes are known to contain the peroxisome proliferator response element and therefore are regulated directly by the PPARα. These genes included the peroxisomal enoyl-CoA: hydrotase-3-hydroxyacyl-CoA bifunctional enzyme (H1)[18,19] and *Cyp4a14* (H3)[20,21] (Fig. 3). It was not known if the α_1 protease inhibitor 1/serine protease inhibitor (H5), another downregulated putative PPARα target gene that was identified in our study (Fig. 3) and other previous studies,[22,23] was regulated directly by the PPARα or not just based on our study. The differential expression patterns of these PPARα target genes were confirmed by Northern blot analysis (Fig. 3B).

Similar to the Wy-14,643 experiment, some of the putative PPARα target genes involved in the 72-hr cellular fasting response were either upregulated (fragment B1 in Fig. 4A) or downregulated (data not shown) in PPARα (+/+) but not in PPARα (−/−) mice after food deprivation. In contrast to the Wy-14,643 experiment, many of the putative PPARα target genes were either upregulated (fragment B2, P8 protein, in Fig. 4A) or downregulated (fragment B3, novel gene, in Fig. 4A) in PPARα (−/−) but not in PPARα (+/+) mice during fasting. These genes were in response to the cellular fasting, and their transcriptional regulations were dependent on the presence of PPARα. The differential expression patterns of these PPARα target genes were confirmed by Northern blot analysis (Fig. 4B). These results illustrated the usefulness of the PPARα (−/−) knockout mouse model and the advantage of the DD method, which allows simultaneous comparisons of multiple RNA samples in identifying batches of rarely discovered PPARα target genes in response to starvation.

[18] O. Bardot, T. C. Aldridge, N. Latruffe, and S. Green, *Biochem. Biophys. Res. Commun.* **192,** 37 (1993).

[19] S. L. Marcus, K. S. Miyata, B. Zhang, S. Subramani, R. A. Rachubinski, and J. P. Capone, *Proc. Natl. Acad. Sci. U.S.A.* **90,** 5723 (1993).

[20] E. F. Johnson, C. N. A. Palmer, K. J. Griffin, and M. H. Hsu, *FASEB J.* **10,** 1 (1996).

[21] Y. M. Heng, C. W. S. Kuo, P. S. Jones, R. Savory, R. M. Schulz, S. R. Tomlinson, T. J. B. Gray, and D. R. Bell, *Biochem. J.* **325,** 741 (1997).

[22] P. R. Devchand, H. Keller, J. M. Peters, M. Vazquez, F. J. Gonzalez, and W. Wahli, *Nature* **384,** 39 (1996).

[23] S. P. Anderson, R. C. Cattley, and J. C. Corton, *Mol. Carcinogen.* **26,** 226 (1999).

A

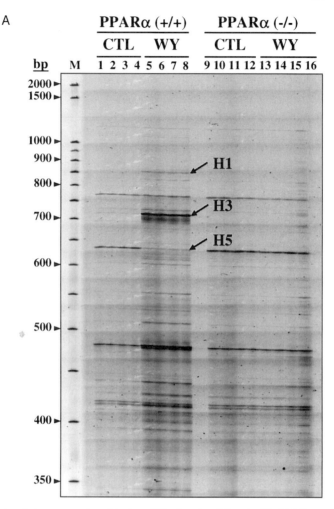

FIG. 3. A typical representation of the fluoroDD gel produced from the Wy-14,643 experiment. (A) Fluorescent differential display of liver RNA from PPARα (+/+) and PPARα (−/−) mice treated with either a control (CTL) or a 0.1% Wy-14,643 (WY) diet for 2 weeks. Total RNA was reverse transcribed using the AP9 primer. PCR reactions were performed in duplicate for each mouse in the presence of 3′-TMR-labeled AP9 and 5′-ARP3 primers. TMR-labeled fluorescent PCR products were run on a 5.6% denaturing polyacrylamide gel. Each lane represents one PCR reaction. A total of two mice (i.e., four PCR reactions) were used for each treatment group. Arrows indicate fragments that were excised from the gel for characterization. M, TMR-labeled molecular weight DNA marker. [Reprinted with permission from L. W. Sum, MPhil Thesis, The Chinese University of Hong Kong (2000).] (B) Confirmation of fluorescent differential display expression patterns of cDNA fragments H1 (peroxisomal enoyl-CoA: hydrotase-3-hydroxyacyl-CoA bifunctional enzyme), H3 (cytochrome P450 4a14), and H5 (α1 protease inhibitor 1/serine protease inhibitor) by Northern blot analysis. Twenty micrograms of total liver RNA from PPARα (+/+) and PPARα (−/−) mice fed either a control (CTL) or a 0.1%

B

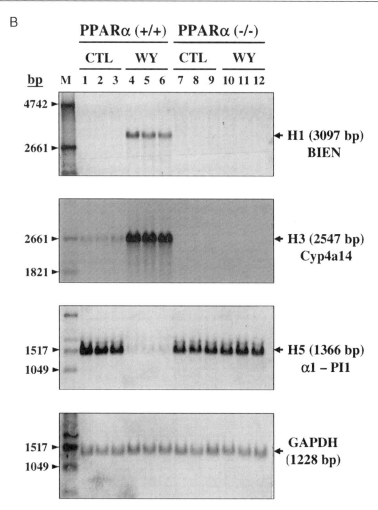

Wy-14,643 (WY) diet for 2 weeks was separated on a 1.0% formaldehyde-agarose gel with ethidium bromide staining and then transferred to a nylon membrane. The membrane was hybridized with random-primed DIG-labeled cDNA fragments (H1, 850 bp; H3, 715 bp; and H5, 640 bp), and the signal was detected by a colorimetric method with reagents nitroblue tetrazolium chloride and 5-bromo-4-chloro-3-indolyl-phosphate. For normalization, the membrane was decolorized, stripped, and reprobed with the random-primed DIG-labeled partial GAPDH cDNA fragment (827 bp). The expression levels of H1 and H3 were upregulated in PPARα (+/+) mice after treatment with 0.1% Wy-14,643 for 2 weeks, but no such induction was observed in PPARα (−/−) mice fed the same diet. In contrast, the expression level of H5 was downregulated in PPARα (+/+) mice after treatment with 0.1% Wy-14,643 for 2 weeks, whereas no such suppression was observed in PPARα (−/−) mice receiving the same treatment. M, DIG-labeled RNA molecular weight marker I.

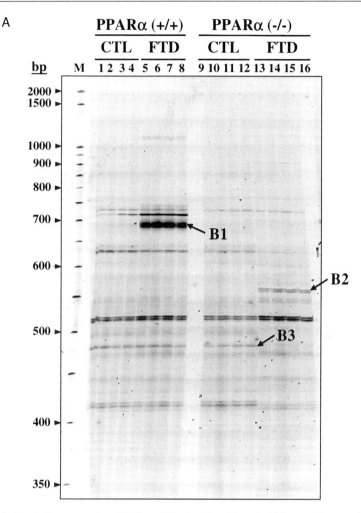

FIG. 4. A typical representation of the fluoroDD gel produced from the 72-hr starvation experiment. (A) Fluorescent differential display of liver RNA from PPARα (+/+) and PPARα (−/−) mice fed with either control (CTL) rodent diet or fasted for 72 hr (FTD). Total RNA was reverse transcribed using the AP6 primer. PCR reactions were performed in duplicate for each mouse in the presence of 3′-TMR-labeled AP6 and 5′-ARP3 primers. TMR-labeled fluorescent PCR products were run on a 5.6% denaturing polyacrylamide gel. Each lane represents one PCR reaction. A total of two mice (i.e., four PCR reactions) were used for each treatment group. Arrows indicate fragments that were excised from the gel for characterization. M, TMR-labeled molecular weight DNA marker. (B) Confirmation of fluorescent differential display expression patterns of cDNA fragments B1 (cytochrome P450 4a14), B2 (P8 protein), and B3 (novel gene) by Northern blot analysis. Twenty micrograms of total liver RNA from PPARα (+/+) and PPARα (−/−) mice fed either control (CTL) rodent diet or fasted for 72 hr (FTD) was separated on a 1.0% formaldehyde-agarose gel with ethidium bromide staining and then

B

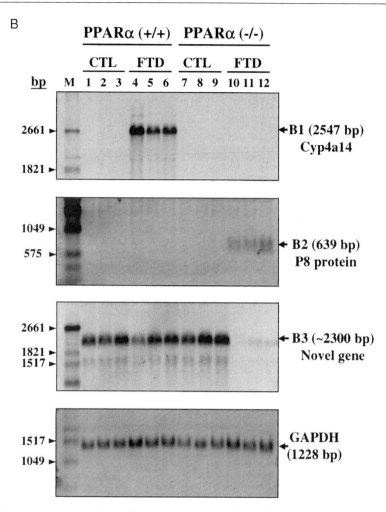

transferred to a nylon membrane. The membrane was hybridized with random-primed DIG-labeled cDNA fragments (B1, 680 bp; B2, 560 bp; and B3, 475 bp), and the signal was detected by a colorimetric method with the substrates nitroblue tetrazolium chloride and 5-bromo-4-chloro-3-indolyl-phosphate. For normalization, the membrane was decolorized, stripped, and reprobed with the random-primed DIG-labeled partial GAPDH cDNA fragment (827 bp). The expression level of B1 was upregulated in PPARα (+/+) after fasting for 72 hr, but no such induction was observed in PPARα (−/−) mice receiving the same treatment. No constitutive expressions of B2 were observed in both PPARα (+/+) and PPARα (−/−) mice fed a control diet, whereas upregulation of B2 was observed only in PPARα (−/−) mice after 72 hr of fasting. In contrast, B3 was expressed in both PPARα (+/+) and PPARα (−/−) mice fed a control diet, whereas 72 hr of fasting resulted in the suppression of B3 expression in PPARα (−/−) but not PPARα (+/+) mice. M, DIG-labeled RNA molecular weight marker I.

It is of interest to note that some of the PPARα target genes that were transcriptionally upregulated in response to the action of Wy-14,643 were similarly transcriptionally activated under the 72-hr cellular fasting condition. One example was *Cyp4a14,* which was upregulated by both Wy-14,643 (fragment H3 in Fig. 3A) and under the 72-hr fasting condition (fragment B1 in Fig. 4A) in the PPARα (+/+) in comparison to their corresponding controls, whereas no such induction was observed in PPARα (−/−) mice after Wy-14,643 or fasting treatments. These data were in agreement with other previously published reports.[4,6] Thus our result clearly demonstrated that the fluoroDD technique is sensitive and can be used to identify putative PPARα target genes that are involved in the action of Wy-14,643 and under the physiological fasting condition.

Summary

In conclusion, we have applied the fluorescent differential method and the PPARα-null mouse model for the rapid isolation of expression tags of PPARα target genes that are involved in the action of peroxisome proliferators and in the regulation of lipid homeostasis under energy deprivation. Identification of a wide spectrum of PPARα target genes will provide new insights into the diverse cellular pathways regulated by these receptor, and this information will be critical for understanding the complicated biological interactions among members of the PPARα target genes. With the recent technological advancement, a newer method, such as DNA microarray, has emerged in the identification of differential gene expressions. This new DNA microarray method, in conjunction with the differential display method, is the first important step toward understanding the molecular mechanisms of gene interactions in any biological systems and can speed up the search for differential gene expressions.

Acknowledgments

The work described in this article was partially supported by grants from the Research Grants Council of the Hong Kong Special Administrative Region (Projects No. CUHK4069/97M, CUHK4157/99M, and CUHK4241/00M). The authors acknowledge the excellent graphical and technical assistance of Ka Pik Chen, Wan Chi Leung, Kam Chun Lo, and Wai Hung Chan.

[23] Histological and Metabolism Analysis of P450 Expression in the Brain

By Suguru Kawato, Yasushi Hojo, and Tetsuya Kimoto

Introduction

Neurosteroids, synthesized *de novo* in the brain, are promising neuromodulators that influence learning and memory.[1,2] Neurosteroidogenesis has, however, not been well elucidated due to the extremely low levels of steroidogenic cytochrome P450s in the brain.[3] This article demonstrates the sensitive methods for the immunostaining of steroidogenic P450s, as well as the measurement of their steroidogenic activity, particularly in the hippocampus.[4] The hippocampus, which is involved essentially in the learning and memory processes, is a target for the neuromodulatory actions of neurosteroids, which are synthesized locally in hippocampal neurons, in addition to those of steroid hormones produced in the adrenal glands and gonads.[1,4,5] Neurosteroids include pregnenolone (PREG), dehydroepiandrosterone (DHEA), testosterone, and 17β-estradiol and their sulfated esters (PREGS, DHEAS, etc.). Neurosteroids may modulate neurotransmission acutely in an either excitatory or inhibitory manner.[6] The acute actions of neurosteroids are thought to be mediated through ion-gated channel receptors, such as N-methyl-D-aspartate (NMDA)-type glutamate receptors, rather than through the nuclear steroid receptors that promote the classic genomic actions of steroid hormones.[1,7–10]

Procedures

Preparation of Rat Hippocampal Slices

Male Wistar rats (1- to 12-week-old) are deeply anesthetized with pentobarbital and perfused transcardially with phosphate-buffered saline (PBS) [0.1 *M*

[1] S. Kawato, M. Yamada, and T. Kimoto, *Adv. Biophys.* **37**, 1 (2001).

[2] E.-E. Baulieu, *Recent Prog. Horm. Res.* **52**, 1 (1997).

[3] A. G. Mensah-Nyagan, J. L. Do-Rego, D. Beaujean, V. Luu-The, G. Pelletier, and H. Vaudry, *Pharmacol. Rev.* **51**, 63 (1999).

[4] T. Kimoto, T. Tsurugizawa, Y. Ohta, J. Makino, H. Tamura, Y. Hojo, N. Takata, and S. Kawato, *Endocrinology* **142**, 3578 (2001).

[5] M. Schumacher, R. Guennoun, P. Robel, and E.-E. Baulieu, *Stress* **2**, 65 (1997).

[6] F. S. Wu, T. T. Gibbs, and D. H. Farb, *Mol. Pharmacol.* **40**, 333 (1991).

[7] M. Joels, *Front. Neuroendocrinol.* **18**, 2 (1997).

[8] F. L. Moore and S. J. Evans, *Brain Behav. Evol.* **54**, 41 (1999).

[9] E. Falkenstein, H. C. Tillmann, M. Christ, M. Feuring, and M. Wehling, *Pharmacol. Rev.* **52**, 513 (2000).

[10] H. Mukai, S. Uchino, and S. Kawato, *Neurosci. Lett.* **282**, 93 (2000).

phosphate buffer and 0.14 M NaCl (pH 7.3)], followed by a fixative solution (4% paraformaldehyde in PBS) at 4°. After dissection from the skulls, hippocampi are postfixed for 24–48 hr in a fixative solution at 4° and cryoprotected in PBS with 30% sucrose. Hippocampi are frozen-sliced coronally at a thickness of 20 μm, with a cryostat at $-17°$.

Immunohistochemical Staining

Staining of cytochrome P450s (P450scc, P45017α, and P450arom), hydroxysteroid sulfotransferase, and StAR is performed using the avidin–biotin–peroxidase complex (ABC) technique according to the free-floating method. Endogenous peroxidase activity is blocked with 0.3% H_2O_2 (v/v) in methanol. The slices are preincubated in PBS containing 5% (v/v) normal goat serum, 3% (w/v) fat-free skim milk, and 0.5% (v/v) Triton X-100 for 2 hr at room temperature with gentle shaking in order to eliminate nonspecific binding sites for primary antibodies. Then the slices are treated with primary antibodies for 18–24 hr (P45017α and StAR) or 36–48 hr (P450scc, P450arom, and sulfotransferase) at 4° in the presence of 3% skim milk and 0.5% Triton X. Dilution of antibodies is 1 : 200 for anti-rat P450scc IgG (peptide antibody against amino acids 421–441, Chemicon, Temecula, CA),[11] 1 : 1000 for anti-guinea pig P45017α IgG, 1 : 1000 for anti-human P450arom IgG, 1 : 1000 for anti-rat sulfotransferase IgG, and 1 : 500 for anti-mouse StAR IgG (peptide antibody against amino acids 88–98), respectively. Primary antibodies are pretreated with 0.5% liver acetone powder and 3% skim milk for 18 hr at 4°. The slices are then washed three times with a PBS solution containing 0.05% Tween 20. Biotinylated anti-rabbit IgG (Vector Lab, Burlingame, CA), diluted in a PBS solution containing 0.5% skim milk, is then applied for 30 min, followed by incubation for 30 min with the avidin–horseradish peroxidase complex (Vector Lab). Immunoreactivity is detected by immersing the slices for 2 min in a detection solution [0.1 M Tris–HCl (pH 7.2) containing 0.05% diaminobenzidine, 0.1% H_2O_2, and 0.3% ammonium nickel sulfate]. After dehydration and embedding, immunoreactive cells are examined.

Fluorescence double immunostaining of steroidogenic proteins is carried out in the same manner as ABC staining except that the streptavidin–Oregon Green 488 complex (Molecular Probe, Eugene, OR) is substituted for the avidin–horseradish peroxidase complex. A low dilution of primary antibodies of 1 : 200 is used for P45017α IgG and P450arom IgG because the signal intensity of fluorescence immunostaining is weaker than that of diaminobenzidine-nickel staining. The fluorescence of Oregon Green 488 is measured using a MRC-1024 confocal microscope equipped with an Ar–Kr laser (Bio-Rad, Hercules, CA). The distributions of neurons, astroglial cells, and oligodendroglial cells are visualized by immunostaining with monoclonal antibodies to NeuN (1 : 100), GFAP (1 : 3,000),

[11] K. F. Roby, D. Larsen, S. Deb, and M. J. Soares, *Mol. Cell. Endocrinol.* **79,** 13 (1991).

and MBP (1 : 10), respectively (all of these antibodies are from Chemicon). The detection of neuronal/glial marker proteins is achieved using Cy3-labeled anti-mouse IgG (Jackson Immunoresearch Lab, West Grove, PA) without avidin–biotin amplification.

Preparation of Mitochondria, Microsomes, and Cytosolic Fractions

The hippocampus, testis, ovary, and lung from 1- to 12-week-old male Wistar rats are excised, minced, and homogenized in a glass–Teflon homogenizer at 4° in homogenization buffer [50 mM potassium phosphate buffer (pH 7.4), 250 mM sucrose, 5 mM EDTA, 0.5 mM phenylmethylsulfonyl fluoride, 0.1 mM leupeptin, and 3 mM 2-mercaptoethanol]. After the removal of nuclei by centrifugation at 3000g for 10 min, mitochondrial fractions are pelleted by centrifugation at 10,000g for 10 min at 4°. Purification is repeated to obtain the final mitochondrial fractions. After the removal of nucleic fractions and mitochondria by centrifugation at 9500g for 20 min, microsomal fractions are pelleted by centrifugation at 105,000g for 60 min. Purification is repeated to obtain the final microsomal fractions. The supernatant is used as the cytosolic fractions.

Western Immunoblot Analysis

Mitochondria, microsomes, and cytosolic fractions are diluted to 10 mg protein/ml with a Tris buffer composed of 62.5 mM Tris–HCl (pH 6.8), 6% sodium dodecylsulfate, 5% sucrose, 5% 2-mercaptoethanol, and 0.01% bromphenol blue. The samples are denatured for 5 min at 90° and subjected to electrophoresis. Ten percent polyacrylamide gels are employed for P450scc, P45017α, and P450arom. For StAR and the sulfotransferase, 12.5 and 12% polyacrylamide gels are used, respectively. After electrophoresis, proteins are transferred to polyvinylidene fluoride membranes (Immobilon-P; Millipore, Bedford, MA) with the TE70 semidry blotting apparatus (Amersham Pharmacia Biotech, Piscataway, NJ) for 90 min at 2.2 mA/cm^2. Blotted membranes are washed three times with PBS containing 0.05% Tween 20 and are blocked for 15 min in PBS containing 0.05% Tween 20 and 10% fat-free skim milk. Blots are then probed with antibodies against P450scc (1 : 5000), P45017α (1 : 5000), P450arom (1 : 3000), sulfotransferase (1 : 5000), and StAR (1 : 1000) for 12–18 hr at 4° in PBS containing 0.05% Tween 20 and 2% skim milk. After the primary antibody treatment, blots are washed and treated with biotinylated goat anti-rabbit IgG (1 : 3000) for 1 hr. Finally, the membranes are treated with streptavidin–horseradish peroxidase complex (1 : 3000) for 1 hr. Protein bands are detected with ECL (enhanced chemiluminescence) plus Western blotting detection reagents (Amersham Pharmacia Biotech).

Radioimmunoassay of Neurosteroid Synthesis

Male Wistar rats (aged 1 to 12 weeks) are decapitated and trunk blood is collected in heparinized tubes. Blood is centrifuged at $1800g$ for 20 min at $4°$ to obtain plasma. Just after blood collection, hippocampi are excised, sliced ($1 \times 1 \times 1$-mm pieces) using a razor, and transferred into a low Mg^{2+} physiological saline [low Mg^{2+} PSS, composed of 137 mM NaCl, 2.5 mM CaCl$_2$, 1 mM NaHCO$_3$, 0.34 mM Na$_2$HPO$_4$, 0.44 mM KH$_2$PO$_4$, 5.7 mM KCl, 0.1 mM MgSO$_4$, 22 mM glucose, and 5 mM HEPES (pH 7.2)] into which O$_2$ gas is bubbled at $4°$. To measure the NMDA-inducible synthesis, the hippocampal cubic slices are then incubated at $37°$ for 30 min in low Mg^{2+} PSS in the presence and absence of NMDA (stimulator) and inhibitors such as MK-801. During a 30-min incubation period, the incubation medium is gassed constantly with 95% O$_2$ and 5% CO$_2$. After the steroidogenic reactions are terminated by the addition of 1 N NaOH, the slices are homogenized with a glass–Teflon homogenizer. For the measurement of PREG(S), 2 μM trilostane (Mochida, Japan) and 20 μM SU-10603 (Novartis Pharma, Switzerland) are used to inhibit 3β-hydroxysteroid dehydrogenase (3β-HSD) and P45017α, respectively. For the measurement of estradiol, 500 μM metyrapon is used to inhibit P45011β activity.

Extraction and purification of PREG(S) are performed by mixing the homogenates with methanol solution containing 1% acetic acid (1 : 10, v/v), followed by sonication for 10 min.[12] The extraction is performed over 1–2 days, after which the mixture is centrifuged at 10,000g for 30 min. The extraction is repeated, the collected supernatant is combined, and the solvents are evaporated under a N$_2$ stream. Dried residues are reconstituted in methanol–water (40 : 60, v/v) and subjected to solid-phase extraction on C$_{18}$ Amprep minicolumns (500 mg, Amersham Pharmacia Biotech) in order to separate nonconjugated steroids and sulfated steroids. For PREG measurements, unconjugated steroid fractions are used. PREG is purified by System I Celite column chromatography (ICN Biomedicals, Costa Mesa, CA), according to the manufacturer's instructions. PREG is then reconstituted in a radioimmunoassay (RIA) buffer, which consists of 0.15 M NaCl, 0.1% gelatin, 0.02% NaN$_3$, and 0.1 M sodium phosphate (pH 7.0). The mass of the PREG is measured by RIA, using a PREG RIA kit of ICN Biomedicals. The average PREG recovery should be 60–70%. For PREGS, the sulfated steroid fraction is used. The solvent is evaporated, and the steroids are reconstituted in ethyl acetate. Sulfuric acid is added (final concentration 2 mM), and PREGS is converted to PREG by overnight solvolyzation. After washing once with 1 N NaOH and twice with water, the PREG is purified using Celite columns. Finally, the mass of the PREG is measured by RIA. The recovery of PREGS should be about 50%.

---

[12] P. Liere, Y. Akwa, S. Weill-Engerer, B. Eychenne, A. Pianos, P. Robel, J. Sjovall, M. Schumacher, and E.-E. Baulieu, *J. Chromatogr. B* **739,** 301 (2000).

Extraction of 17β-estradiol is performed by mixing the homogenates with ethyl acetate : hexane (3 : 2) and shaking for 10 min. Extraction is repeated, and the collected organic supernatant is evaporated under a N_2 stream. The dried residue is reconstituted in a HPLC solvent (hexane : 2-propanol : acetic acid = 98 : 2 : 1) and applied to HPLC (JASCO, Japan) with a silica gel column (0.46 × 15 cm, Cosmosil 5SL, Nacalai Tesque, Japan) for purification after filtration with a membrane filter (0.45 μm pore, Ultrafree-MC, Millipore). The amount of estradiol is then measured by RIA using the estradiol RIA kit of ICN. The average estradiol recovery should be about 50%.

HPLC Analysis of Metabolism from PREG to DHEA and Estradiol

The activity of P45017α in the hippocampus is observed by measuring the conversion of [7-^{3}H]pregnenolone (specific activity, 22.5 Ci/mmol, New England Nuclear, Boston, MA) to [^{3}H]DHEA. Hippocampal cubic slices from two rats (the same as those used for RIA) are incubated at 20° for 0, 1, 3, or 5 hr in 2 ml of low Mg^{2+} PSS containing 10^7 cpm of [^{3}H]pregnenolone and 10 μM of trilostane. The incubation medium should be gassed constantly with 95% O_2 and 5% CO_2. After the reaction is terminated by the addition of 2 ml of chloroform, the slices are homogenized. After the water phase is removed, an equal volume of methanol is added in order to extract steroids. The homogenates are stirred for 15 min and centrifuged at 3000g for 5 min at 4°. The organic phase is then collected, and the extraction step is repeated twice. The combined organic extracts are dried, dissolved in 200 μl of an elution solvent, and filtrated through a membrane filter. Steroid metabolites are separated by a HPLC system with a silica gel column using an elution solvent consisting of hexane : 2-propanol : acetic acid (97 : 3 : 1, v/v). The steroids are eluted at a flow rate of 1.0 ml/min for 30 min, and the eluate is fractionated every 0.5 min. Fractions with a retention time of 9–10 min are assigned as DHEA fractions because the retention time of control [^{3}H]DHEA and [^{14}C]DHEA is 9.5 min. The radioactivity of the fractions is measured with a liquid scintillation spectrometer (LSC-701, Aloka, Japan).

The activity of P450arom is evaluated by measuring the conversion of [7-^{3}H]DHEA (specific activity, 60.0 Ci/mmol, New England Nuclear) to [^{3}H]estradiol. Slices from two rats are incubated at 20° for 5 hr in 2 ml of a low Mg^{2+} PSS containing 10^7 cpm of [7-^{3}H]DHEA. Extraction procedures of steroid metabolites are the same as those used for the RIA of estradiol. The dried residue is reconstituted in a HPLC solvent [hexane : 2-propanol : acetic acid = 98 : 2 : 1 (solvent A) or 99 : 1 : 1 (solvent B)] and subjected to HPLC after filtration with a membrane filter. The eluate is fractionated every 1 min from 0 to 60 min for solvent A and every 20 sec from 45 to 70 min for solvent B. The eluate with retention times of 21 and 22 min (for solvent A) or 49–53 min (for solvent B) is assigned to be 17β-estradiol, as judged from the retention time of control [^{14}C]estradiol

(21 min for solvent A and 51 min for solvent B). The radioactivity of the fractions is measured with a liquid scintillation spectrometer.

Results and Discussion

Distributions of Neurosteroidogenic Proteins in the Hippocampus

An intense immunoreaction with antibodies against P450scc, P45017α, P450arom, hydroxysteroid sulfotransferase, and StAR is restricted to pyramidal neurons in the CA1–CA3 regions and granule cells in the dentate gyrus, as indicated by immunostaining with NeuN antibodies (Fig. 1).[1,4] Fluorescence double immunostaining indicates that glial cells (astroglial cells and oligodendroglial cells) do not contain a significant amount of steroidogenic proteins. A good signal-to-noise ratio in immunohistochemical investigations is achieved by considering the following points in order to get rid of nonspecific staining: (a) Rat peptide antibodies, whose epitopes are exposed to water, may be most suitable for specific staining of the rat brain. We, however, do not always have such good rat antibodies.

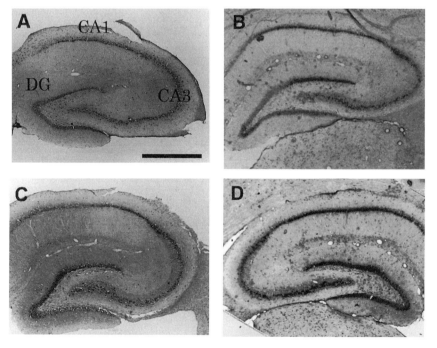

FIG. 1. Immunohistochemical staining of steroidogenic proteins in hippocampal slices from 12-week-old male rats. (A) P450scc, (B) P45017α, (C) P450arom, and (D) StAR. Staining is restricted to pyramidal neurons in the CA1–CA3 regions and granule cells in the dentate gyrus (DG). Immunoreactive cells are visualized by diaminobenzidine-nickel staining. Bar: 800 μm.

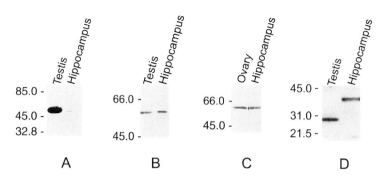

FIG. 2. Western immunoblot analysis of steroidogenic proteins in hippocampal tissues from 12-week-old male rats. (A) P450scc in mitochondria, (B) P45017α in microsomes, (C) P450arom in microsomes, and (D) StAR in mitochondria. For each panel, the left lane indicates a positive control protein band in rat testis (1 μg protein for A and D, 0.5 μg protein for B) or ovary (0.5 μg protein) and the right lane indicates a protein band in the hippocampus (50 μg protein). Numbers along the vertical axis indicate molecular weights. Only the full-length 37-kDa StAR is observed in the hippocampus, while only the truncated 30-kDa StAR is observed in testis. None of these protein bands is observed in the lung used as the negative control.

Trial-and-error processes are then necessary to find second-suitable antibodies. Anti-guinea pig P450 IgG, anti-human P450 IgG, and anti-mouse StAR IgG demonstrated good staining. Antibodies against beef P450scc and beef StAR showed some nonspecific staining in addition to specific staining. (b) When paraffin-embedded sections are used, there may be a considerable loss of antigen reactivity and/or mRNA hybridization reactivity due to the heating used to remove paraffin from slices. Fresh frozen slices should be used.

Western Immunoblot Analysis

Single bands corresponding to steroidogenic proteins are observed when mitochondria (P450scc and StAR), microsomes (P45017α and P450arom), or cytosolic fractions (the sulfotransferase) are subjected to Western blotting (see Fig. 2).[1,4] The electrophoretic mobility of the steroidogenic protein bands for the hippocampus is almost identical to that of purified steroidogenic proteins from testis and ovary. The concentration of these steroidogenic proteins in the hippocampus was around 10^{-2} of those in peripheral organs, whereas extremely low-level expression was reported for mRNAs. For example, mRNAs of P450scc and P45017α in whole brain homogenates were only 10^{-4}–10^{-5} of that in the adrenal gland.[13–15]

[13] S. H. Mellon and C. F. Deschepper, *Brain Res.* **629**, 283 (1993).
[14] J. L. Sanne and K. E. Krueger, *Gene* **165**, 327 (1995).
[15] A. Furukawa, A. Miyatake, T. Ohnishi, and Y. Ichikawa, *J. Neurochem.* **71**, 2231 (1998).

TABLE I
CONCENTRATIONS OF NEUROSTEROIDS BEFORE AND AFTER NMDA STIMULATION[a]

Source	Incubation time (min)	NMDA	MK801	PREG (fmol/mg wet weight)[b]	PREGS (fmol/mg wet weight)	17β-Estradiol (fmol/mg wet weight or μl)
Hippocampus	0	−	−	15.8 ± 1.2	28.2 ± 1.5	0.67 ± 0.05
	30	−	−	16.9 ± 0.5	29.7 ± 4.5	0.68 ± 0.09
	30	+	−	33.3 ± 3.4	57.2 ± 3.8	1.35 ± 0.18
	30	+	+	16.0 ± 1.4	30.3 ± 3.5	
Plasma	0	−	−	2.0 ± 0.1	4.0 ± 0.3	0.098 ± 0.039

[a] In the hippocampus and plasma from 12-week-old rats.
[b] Ten milligrams wet weight of the hippocampal tissue contained 0.96 mg of protein.

Neurosteroidogenic Activity Measured with RIA

The basal concentrations of PREG, PREGS, and estradiol are six- to eightfold higher in the hippocampus than those in the plasma (Table I). On stimulation with 100 μM NMDA for 30 min, the levels of PREG(S) and estradiol *increase* roughly twice. Stimulation of PREG(S) and estradiol production with NMDA is completely suppressed by the application of blockers of NMDA receptors or by the depletion of extracellular Ca^{2+}, indicating that NMDA-induced steroid production is mediated by the influx of Ca^{2+} through NMDA receptors. Aminoglutethimide (inhibitor of P450scc; 1 mM) is observed to completely block the PREG production induced by NMDA stimulation.

Neurosteroidogenic Activity Measured with HPLC

The catalytic activity of P45017α of the hippocampus from a 12-week-old rat is demonstrated as the conversion of [³H]PREG to [³H]DHEA using a HPLC system. The eluted radioactive peak of [³H]DHEA increases in a time-dependent manner from 1 to 5 hr. The presence of bifonazole or SU10603 (Novartis Pharmaceutical), specific inhibitors of P45017α, abolishes the appearance of the [³H]DHEA peak, even after 5 hr of incubation. The conversion of [³H]DHEA to [³H]estradiol by 3β-HSD and P450arom is demonstrated as the appearance of significant amounts of [³H]estradiol and [³H]testosterone in the hippocampus from a 12-week-old male rat.

In conclusion, hippocampal neurons are equipped with the machinery that synthesizes PREG(S), DHEA, and estradiol. Neurosteroidogenesis may be acutely performed on stimulation of neurons with glutamate via a NMDA receptor-mediated Ca^{2+} influx. Because neurosteroids such as PREGS and estradiol modulate acutely the Ca^{2+} conductivity of NMDA receptors, neurosteroids should be paracrine modulators of neuronal activity. The methods of investigation used for the hippocampus are also applicable to other brain regions, such as the cerebellum, cortex, hypothalamus, and olfactory bulb.[3,16,17]

Acknowledgments

Anti-guinea pig P45017α IgG is donated by Dr. S. Kominami, Hiroshima University, Japan. Anti-human P450arom IgG is donated by Dr. N. Harada, Fujita Helth University, Japan. Anti-rat sulfotransferase IgG is donated by Dr. H. Tamura, Kyoritsu College of Pharmacy, Japan. Anti-mouse StAR IgG is donated by Dr. D. M. Stocco, Texas Tech University, Texas.

[16] K. Tsutsui, K. Ukena, M. Usui, H. Sakamoto, and M. Takase, *Neurosci. Res.* **36,** 261 (2000).
[17] P. Guarneri, R. Guarneri, C. Cascio, P. Pavasant, F. Piccoli, and V. Papadopoulos, *J. Neurochem.* **63,** 86 (1994).

[24] Proteomic Analysis of Rodent Hepatic Responses to Peroxisome Proliferators

By Neil Macdonald and Ruth Roberts

Introduction

Sequence data from the human genome project will provide the opportunity to approach the most fundamental question of how genes and their protein products interact to generate a normal phenotype and how disease or xenobiotics affect these interactions. Protein profiling has the potential to reveal much information about the function of the estimated 38,000 genes in the human genome.[1] Due to differential processing of mRNA, alternative transcriptional start sites, and posttranslational modifications, it is estimated that each gene in humans may code for six or more distinct protein products.[2] Additional complexity arises from tissue-specific patterns of expression and posttranslational modification.[2] For example, we identified four distinct protein spots in two-dimensional (2D) gels of cultured rat hepatocyte proteins representing isoforms of mitochondrial aldehyde dehydrogenase.[3]

We examined the molecular mechanism of the response to peroxisome proliferators (PPs) and used protein profiling due to the additional functional information

[1] J. C. Venter, M. D. Adams, E. W. Myers, P. W. Li, R. J. Mural, G. G. Sutton, H. O. Smith, M. Yandell, C. A. Evans, R. A. Holt *et al., Science* **291,** 1304 (2001).
[2] Hochstrasser, *Clin. Chem. Lab. Med.* **36,** 825 (1998).
[3] S. Chevalier, N. Macdonald, R. Tonge, S. Rayner, R. Rowlinson, J. Shaw, J. Young, M. Davison, and R. A. Roberts, *Eur. J. Biochem.* **267,** 4624 (2000).

that may be revealed.[4] PPs constitute a class of nongenotoxic rodent hepatocarcinogens that include commercially important plasticizers, hypolipidemic drugs, and industrial solvents. In rodents, PPs induce the expression of hepatic genes such as β-oxidative enzymes and cytochrome P4504A members, cause increased hepatic DNA replication, and suppress hepatic apoptosis, leading to the formation of tumors. These effects are mediated by the peroxisome proliferator-activated receptor α (PPARα), as PPARα null transgenic mice are resistant to the effects of PPs.[5,6] PPARα is a ligand-dependent transcription factor that binds to PPARα response elements (PPREs) upstream of PP-regulated genes.[7-10] Understanding the PPARα- mediated changes in the rodent hepatic proteome in response to PPs should enable us to identify biomarkers for predevelopment toxicology and will enhance our ability to predict human risk from rodent data.

Proteomics Studies Using Two-Dimensional Gels

The word proteome was first used by Wilkins *et al.*[11] to describe the total protein content expressed by the genome of any cell, organ, or organism, and the comparison of proteomes has come to be known as proteomics. There is an increasing number of sophisticated technologies for this analysis, but the best characterized and most widely used is a combination of two-dimensional gel electrophoresis and mass spectrometry (MS)[3,12] (Fig. 1). Proteins are sorted by isoelectric point and mass using 2D gels and are then identified by specialist image analysis software coupled with MS. Although two-dimensional protein gels are an old technology,[13] huge improvements in separation, visualization, and quantitation have revived this tool for analyzing changes in protein expression and posttranslational modification.

Separation

Reliable premade isoelectric focusing (IEF) gels with predetermined optimal focusing conditions, such as Immobiline Drystrips (Amersham Pharmacia,

[4] N. L. Anderson and J. Seilhamer, *Electrophoresis* **18**, 533 (1997).
[5] S. S. Lee, T. Pineau, J. Drago, E. J. Lee, J. W. Owens, D. L. Kroetz, P. M. Fernandez-Salguero, H. Westphal, and F. J. Gonzalez, *Mol. Cell. Biol.* **15**, 3012 (1995).
[6] J. M. Peters, R. C. Cattley, and F. J. Gonzalez, *Carcinogenesis* **18**, 2029 (1997).
[7] C. N. A. Palmer, M. Hsu, K. J. Griffin, J. L. Raucy, and E. F. Johnson, *Mol. Pharmacol.* **53**, 14 (1998).
[8] N. Vu-Dac, K. Schoonjans, V. Kosyth, J. Dallongeville, J. C. Fruchart, B. Staels, and J. Auwerx, *J. Clin. Invest.* **96**, 741 (1995).
[9] S. Hasmall, N. H. James, N. Macdonald, A. Soames, and R. A. Roberts, *Arch. Toxicol.* **74**, 84 (2000).
[10] N. Woodyatt, K. Lambe, K. Myers, J. Tugwood, and R. Roberts, *Carcinogenesis* **20**, 369 (1999).
[11] S. J. Wilkins, *Electrophoresis* **17**, 830 (1996).
[12] N. Macdonald, K. Barrow, R. Tonge, M. Davison, R. A. Roberts, and S. Chevalier, *Biochem. Biophys. Res. Commun.* **277**, 699 (2000).
[13] P. H. O'Farrell, *J. Biol. Chem.* **250**, 4007 (1975).

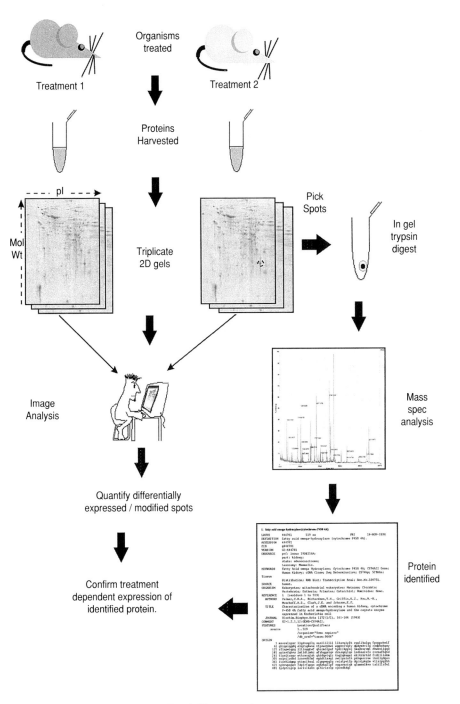

FIG. 1. The proteomics process.

Piscataway, NJ), have ensured consistency in the IEF process. In addition, the shorter pH gradients spread over a long strip have increased resolution and reduced the likelihood of multiple proteins comigrating. Isoelectric focusing must be carried out under temperature-controlled conditions, as the pH gradients in the strip are temperature sensitive. In addition, it is vital to have as low a salt concentration as possible in the starting protein sample. High salt concentrations alter the pH gradient and can cause Immobiline strips to short out. Following overnight IEF, the strips are equilibrated in a denaturing buffer containing sodium dodecyl sulfate (SDS) and dithiothreitol (DTT) to improve migration of the focused proteins out of the IEF gel and are then laid on top of a standard 12.5% SDS–PAGE gel. Again it is important that all second-dimension gels are run at a consistent temperature to obtain a consistent pattern of spots between runs. We also run our silver and Coomassie-stained SDS–PAGE gels with a PAG support (Flowgen), which prevents distortion or fragmentation of the gel during subsequent fixing and staining steps.

Visualization

The most frequently used method of obtaining images from 2D gels for analysis is silver staining followed by digitization with a desktop computer scanner. In our experience, standard silver staining protocols allow highly sensitive detection of protein spots.[3] However, silver staining cannot be regarded as truly quantitative due to its short linear range and problems with saturation of highly expressed protein spots. All these make the computer analysis of the images, which is a major bottleneck at the best of times, that much more laborious. In addition, intercalation of the silver into the proteins in the gel makes mass spectrometry of these proteins much more difficult. A solution is to run extra gels and stain them with Neuhoff colloidal Coomassie[14]—the spots can be picked from these gels for MS analysis.

Several alternatives to silver staining exist. Fluorescent stains with long linear ranges are now available, allowing more accurate quantification of spots and hence more statistically valid comparisons of treatment-dependent expression. Postelectrophoresis staining with fluorescent dyes, e.g., SyproRuby, has the advantage of being relatively easy and is available as an "off the shelf" product. Sypro dyes are claimed to have a linear quantitative range of three orders of magnitude (Bio-Rad web site: www.bio-rad.com). Alternatively, proteins can be prelabeled prior to 2D electrophoresis using dyes such as Cy5, Cy3, and Cy2. These cyanine dyes may offer advantages over the postlabeling technology, as up to three protein samples can be labeled with different dyes and run on the same gel. Due to the low efficiency of the labeling process, these dyes do not alter the electrophoretic mobility of the proteins, and the resulting "within gel" comparisons require less *in silico* manipulation to align spots. A linear range of four to five orders of magnitude is

[14] V. Neuhoff, N. Arold, D. Taube, and W. Ehrhardt, *Electrophoresis* **9**, 255 (1988).

claimed.[15] An example of quantitative analysis of spot intensity using a combination of cyanine dyes and PDQuest software is shown in Fig. 2. In our experience, cyanine dye technology permits the identification of many more regulated protein spots. The drawback at present is that this technology is only available from Amersham Pharmacia Biotech as part of an access agreement.

Quantification

A variety of software packages now exist that allow the researcher to quantify spots on an individual gel, repeat this quantification on multiple gels using protein samples from the same treatment groups, and then compare this average with the average from other treatment groups. These include packages such as PDQuest (Bio-Rad, Hercules, CA), Z3 (Compugen), and Melanie3 (Geneva Bioinformatics). Where quantitative staining has been used, these packages allow the researcher to match spots across multiple gels to provide statistically verifiable data about intertreatment variations in the intensity of individual spots.

Identifying Proteins

Observing changes in intensity of a protein spot on a 2D gel is the first step in elucidating the identity of the protein. The development of matrix-assisted laser desorption ionization time-of-flight mass spectrometry (MALDI-TOF) means that accurate mass profiles of trypsin digests of proteins can now be carried out using the relatively small amounts of protein in a spot excised from a 2D gel. The spot is excised from the surrounding gel, subjected to an in-gel trypsin digest, and the resulting peptides are mixed with a laser absorbent matrix and deposited onto a MS laser target. Laser excitement of the matrix ionizes the peptides, which fly down the MS tube; their molecular masses are determined based on time of flight from target to detector. The fragment mass profile for each spot is compared with a databases of theoretical digest patterns for known proteins based on DNA sequence data. Where peptide mass profiles do not produce a positive ID, peptides can be sequenced, using quadropole time-of-flight (Q-TOF) MS.

Despite advances in MS technology, only proteins represented by the strongest spots can be identified. In our experience, up to 90% of the spots from a total hepatic protein extract quantified using cyanine dyes remain unidentified. To identify proteins that are relatively rare as a proportion of the total proteome of a tissue, but which may control important processes, such as the cell cycle, some degree of fractionation or enrichment will be required, either based on the chemical properties of the proteins, or by organelles.

---

[15] R. Tonge, J. Shaw, B. Middleton, R. Rowlinson, S. Rayner, J. Young, F. Pognan, E. Hawkins, I. Currie, and M. Davison, *Proteomics* **1,** 377 (2001).

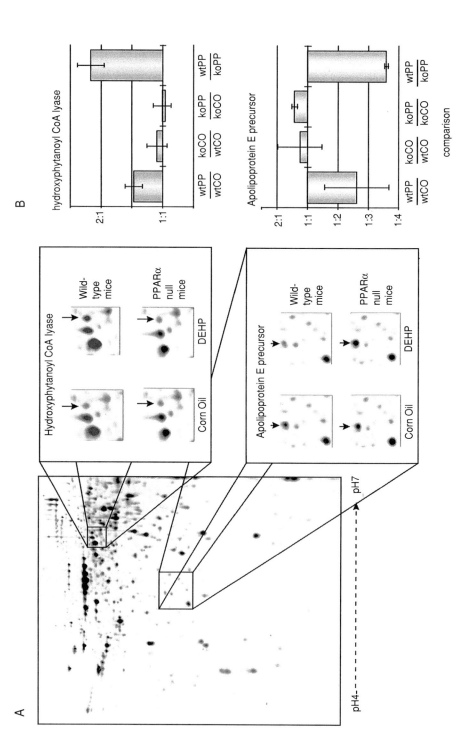

A

B

hydroxyphytanoyl CoA lyase

Apolipoprotein E precursor

comparison

Hydroxyphytanoyl CoA lyase

Wild-type mice

PPARα null mice

Corn Oil

DEHP

Apolipoprotein E precursor

Wild-type mice

PPARα null mice

Corn Oil

DEHP

pH7

pH4

Proteomics, Peroxisome Proliferators, and P450s

Of the genes identified as PP regulated based on 2D gel studies, only a small number from each study corresponds to known "classic" PP-regulated genes, such as β-oxidation enzymes, cytochrome p4504A family members, or apolipoproteins.[16,17] We observed statistically significant PPARα-dependent *in vivo* changes in spots corresponding to known PPARα-regulated proteins ApoE and fatty acid binding protein by DEHP (Fig. 2). None of these studies reported MALDI-TOF or Q-TOF identifications of any spots corresponding to other PPARα transcriptionally regulated genes such as lipoprotein lipase[18] or cytochrome P450.[5] In our quantitative *in vivo* study, the β-oxidation gene acyl-CoA oxidase was found to be regulated, but this regulation was inconsistent between gels. However, we also observed PP- and PPARα-dependent differential expression of hydroxyphytanoyl-CoA lyase (Fig. 2), a key enzyme in α oxidation.[19]

In the three proteomics studies we have carried out, only one protein spot has produced a MS pattern corresponding to a cytochrome P450, Cyp7B1 (N. Macdonald, unpublished observation, 2000), which catalyzes the first step in the conversion of cholesterol to bile acids.[20,21] This is surprising considering that 37 mouse cytochrome P450s are listed on the Internet directory of cytochrome P450s (www.icgeb.trieste.it/~p450srv) and of these, cytochromeP4504A1 and

---

[16] C. S. Giometti, S. L. Tollaksen, X. Liang, and M. L. Cunningham, *Electrophoresis* **19**, 2498 (1998).

[17] U. Edvardsson, M. Alexandersson, H. Brockenhuus von Lowenhielm, A. Nystrom, B. Ljung, F. Nilsson, and B. Dahllof, *Electrophoresis* **20**, 935 (1999).

[18] K. Schoonjans, J. Peinado-Onsurbe, A.-M. Lefebvre, R. A. Heyman, M. Briggs, S. Deeb, B. Staels, and J. Auwerx, *EMBO J.* **15**, 5336 (1996).

[19] P. Ellinghaus, C. Wolfrum, G. Assmann, F. Spener, and U. Seedorf, *J. Biol. Chem.* **274**, 2766 (1999).

[20] A. W. Zomer, G. A. Jansen, B. Van Der Burg, N. M. Verhoeven, C. Jakobs, P. T. Van Der Saag, R. J. Wanders, and B. T. Poll-The, *Eur. J. Biochem.* **267**, 4063 (2000).

[21] K. D. Setchell, M. Schwarz, N. C. O'Connell, E. G. Lund, D. L. Davis, R. Lathe, H. R. Thompson, R. Weslie Tyson, R. J. Sokol, and D. W. Russell, *J. Clin. Invest.* **102**, 1690 (1998).

FIG. 2. Quantification of PP and PPARα-dependent relative expression using cyanine dyes. Apolipoprotein E expression is suppressed, and hydroxyphytanoyl-CoA lyase is induced by the PP DEHP in a PPARα-dependent manner. (A) Proteins from three mice for each treatment group, pooled by treatment, were labeled with cyanine dyes; 50 μg labeled protein sample was loaded per 2D gel image and was scanned on an APB fluorescent scanner. Triplicate gels are run; typical gel images are shown. (B) Average relative expression. For each comparison, one sample was labeled with Cy3 and one with Cy5. The two samples were loaded on a single IEF strip and, after 2D electrophoresis, the gels were scanned at the two wavelengths appropriate for Cy3 and Cy5. Spot fluorescence quantification was carried out using PDQuest (Bio-Rad). Average Cy5 : Cy3 ratios of the spots highlighted in (A) for three repeats of each comparison are shown ($\pm$SD). Ratios: 1 = equal expression in each group, >1 = induction by PP and/or PPARα, <1 = suppression by PP and/or PPARα. ApoE and hydroxyphytanoyl-CoA lyase identified by MALDI-TOF and Q-TOF analysis, respectively, of spots excised from parallel Coomassie-stained 2D gels loaded with 420 μg pooled proteins.

4A3 mRNAs are induced by PPs in a PPARα-dependent manner.[5] This apparent discrepancy may be attributed to the relatively low abundance of P450s compared with structural and/or metabolic proteins. In addition, their relatively high pl means that they focus in a region of 2D gels that often exhibit poor resolution.[22,23] Enrichment may be necessary and could be achieved by preparing microsomes to look at cytochrome P450s or nuclear preparations to look at transcription factors. Further technical challenges are presented by membrane-bound proteins such as cytochrome P450s—how best to release them from the membrane? Choice of protein solubilization buffer prior to IEF can have a profound effect on the final outcome (see Fig. 3), and some initial experiments are recommended in order to optimize the ratios of urea to thiourea and to determine which zwitterionic detergent is most suited to the solubilization of particular tissues.[24]

Summary and Future Directions

Posttranslational modifications are clearly important in response to xenobiotics. The study of changes in pl and mass are informative, but Western blots of 2D gels transferred to PVDF can highlight even low abundance spots showing altered intensity and can focus on specific alterations. Cans et al.[25] highlighted the emergence of evidence for a network of tyrosine phosphorylation associated with nuclear events affecting growth control, cell cycle, apoptosis, and transcription, processes that are also regulated by PPs in rodents. Theoretically, treatment-dependent alterations in events such as these should be quantifiable using 2D Western blots probed with an antiphosphotyrosine antibody[26]; however, if MS analysis of the spot was required, some form of enrichment of phosphorylated spots would still be needed.

A variety of alternative technologies that strive to overcome some of the problems associated with 2D gels are available. These include protein affinity chips and antibody arrays. Affinity chips, such as those sold by Ciphergen, aim to speed up the process of identifying differentially expressed proteins by separating proteins using their binding affinity to a variety of specific chemical substrates laid down on a MALDI-TOF target (the chip). The protein sample is allowed to hybridize to the substrates, and after washing away nonspecifically bound proteins, the remaining

[22] Xiaogin Ye, Liming Lu, and Sarjeet S. Gill, *Biochem. Biophys. Res. Commun.* **239**, 660 (1997).

[23] A. Butt, M. Davison, G. Smith, J. Young, S. Gaskell, S. Oliver, and R. Benyon, *Proteomics* **1**, 42 (2001).

[24] T. Rabilloud, T. Blisnick, M. Heller, S. Luche, R. Aebersold, J. Lunardi, and C. Braun-Breton, *Electrophoresis* **20**, 3603 (1999).

[25] C. Cans, R. Mangano, D. Barila, G. Neubauer, and G. Superti-Furga, *Biochem. Pharmacol.* **60**, 1203 (2000).

[26] H. Kaufman, J. E. Bailey, and M. Fussenegger, *Proteomics* **1**, 194 (2001).

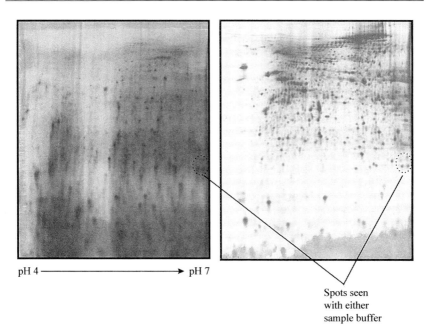

pH 4 ⟶ pH 7

Spots seen
with either
sample buffer

FIG. 3. Comparison of protein solubilization buffers. Choice of sample solubilization buffer influences the pattern observed on the final 2D gel. The two gels represent the same protein sample (hepatocyte cultures from three rats, pooled, 420 μg total protein loaded per gel) run on Immobiline pH 4 to 7 linear 18-cm IEF strips followed by PAGE on a 12% Tris–glycine gel. (Left) Samples were solubilized in a urea/CHAPS solubilization buffer. (Right) A urea/thiourea/CHAPS solubilization buffer was used. Both were solubilized in 370 μl of buffer at room temperature for 2 hr with shaking, followed by centrifugation to remove any insoluble proteins. Gels were silver stained as described previously.[3] Circled region shows spots common to both gels.

proteins are identified by MALDI-TOF MS. This allows the quantification of proteins that are poorly resolved on 2D gels, such as hydrophobic or small molecular weight proteins, but is less successful at resolving proteins larger than 30 kDa. An additional advantage of this technology is that extremely small quantities of protein can be examined, allowing affinity chips to be combined with laser microdissection to examine very specific groups of cells within a tissue sample.[27] Identification of the individual proteins requires a Q-TOF step, which can be integrated with Ciphergen's chip technology. Other companies, such as Biosite and Large Scale Biology Corporation, are investigating an antibody chip approach, creating specific antibodies for all members of entire protein libraries, attaching

[27] L. C. Wright, Jr., S.-M. Leung, S. Nasim, B.-L. Adam, T.-T. Yip, P. F. Schellhammer, L. Gong, and A. Vlahou, *Prostate Cancer Prostatic Dis.* **2**, 264 (1999).

those antibodies to a support substrate and using this in much the same way as a cDNA array to quantify the expression of proteins. These technologies may supersede 2D gels or may come to occupy complementary niches in the analysis of the proteome.

In summary, proteomics is an evolving group of technologies that harness the power of genomics data to identify and quantify functional changes in protein expression and posttranslational modification associated with cellular processes, toxicant treatment, or disease states. By marrying the new technology with classical cell biology and protein biochemistry, it is possible to use proteomics to profile changes in specific organelles or subgroups of proteins within the total proteome. By applying these technologies to systems such as PP-induced rodent hepatocarcinogenesis, we will increase our understanding of the molecular mechanisms underlying their effects and the species specificity of those effects.

Section IV

Metabolism

[25] Kinetic Analysis for Multiple Substrate Interaction at the Active Site of Cytochrome P450

By MAGANG SHOU

An enzyme reaction that involves one or multiple substrates consists of a sequence of reversible conversions between various enzyme species. The steady-state velocity equation for the given scheme of an enzyme reaction of any complexity can be derived by the method of King and Altman.[1] Cytochrome P450 enzymes (CYPs) are a superfamily of enzymes that play a prominent role in the metabolism of a vast array of drugs and xenobiotics.[2–8] Cytochrome P450 isoforms possess an identical prosthetic group but different apoprotein structures that are responsible for the broad and overlapping substrate specificities. Most P450-mediated reactions follow simple Michaelis–Menten kinetics from which kinetic constants (K_m and V_{max}) are easily derived.[9] If the addition of an effector (inhibitor or activator) to a P450 reaction results in an inhibition or activation of the enzyme, a value for K_i or K_A can be determined. These parameters are widely accepted to predict *in vivo* pharmacokinetic and pharmacodynamic consequences caused by exposure to one or multiple drugs.[10–12] However, some P450s display unusual kinetics that cannot be solved by Michaelis–Menten kinetics.[13–20] Most of the examples reported

[1] E. L. King and C. J. Altaman, *J. Phys. Chem.* **60,** 1375 (1976).

[2] D. R. Nelson, L. Koymans, T. Kamataki, J. J. Stegeman, R. Feyereisen, D. J. Waxman, M. R. Waterman, O. Gotoh, M. J. Coon, R. W. Estabrook, I. C. Gunsalus, and D. W. Nebert, *Pharmacogenetics* **6,** 1 (1996).

[3] S. A. Wrighton and J. C. Stevens, *Crit. Rev. Toxicol.* **22,** 1 (1992).

[4] J. H. Lin and A. Y. H. Lu, *Pharmacol. Rev.* **49,** 403 (1997).

[5] F. J. Gonzalez, *Pharmacol. Rev.* **40,** 243 (1988).

[6] F. G. Guengerich, *in* "Cytochrome P450: Structure, Mechanism, and Biochemistry" (P. R. Ortiz de Montellano, ed.), p. 473. Plenum Press, NY, 1995.

[7] F. P. Guengerich, *Mol. Pharmacol.* **33,** 500 (1988).

[8] F. P. Guengerich, E. M. J. Gillam, M. V. Martin, T. Baba, B. R. Kim, K. D. Raney, and C. H. Yun, *in* "Assessment of the Use of Single Cytochrome P450 Enzymes in Drug Research" (M. R. Waterman and M. Hildebrand, eds.), p. 161. Springer, Berlin, 1994.

[9] I. H. Segel, "Behaviour and Analysis of Rapid Equilibrium and Steady State Enzyme Systems." Wiley, New York, 1975.

[10] A. D. Rodrigues, G. A. Winchell, and M. R. Dobrinska, *Clin. Pharmacol.* **41,** 1 (2001).

[11] K. Ito, T. Iwatsubo, S. Kanamitsu, K. Ueda, H. Susuki, and Y. Sugiyama, *Pharmacol. Rev.* **50,** 465 (2000).

[12] T. D. Leemann and P. Dayer, *in* "Advances in Drug Metabolism in Man" (G. M. Pacifici and G. N. Fracchia, eds.), p. 784. European Commission, Brussels, 1995.

[13] M. Shou, J. Grogan, J. A. Mancewicz, K. W. Krausz, F. J. Gonzalez, H. V. Gelboin, and K. R. Korzekwa, *Biochemistry* **33,** 6450 (1994).

to date have been attributed to CYP3A4, which makes the Michaelis constants difficult to estimate. CYP3A4 is characterized by a broad substrate specificity. Substrates range from generally small to large in size (up to 1200 Da) and typically are highly lipophilic. K_m values for substrates vary markedly between 1 and 1500 μM. To date, many reports have shown that CYP3A4 can hold a large active site with a well-defined access channel for substrates, in keeping with the known preference of this enzyme for large, structurally diverse substrates.[21,22] Even the cyclosporin molecule, which is of considerable size relative to substrates for other P450s, is accommodated by the CYP3A4 active site. As a result, it is assumed that the active site of CYP3A4 can accommodate two or more intermediate-sized substrate molecules for metabolism. The hypotheses are supported extensively by numerous observations, which include enzyme steady-state kinetics,[14] CO rebound kinetics,[18] site-directed mutagenesis,[15,19] and binding spectral analysis.[23] The hypotheses propose that two (or more) substrate molecules of the same or different chemical entity can bind simultaneously to two (or more) distinct portions within the active site, and either of the two substrates can be metabolized through interaction with the reactive oxygen, which is bound to the heme iron. When two molecules coexist within the active site, some interactions may occur due to steric, electronic, and/or allosteric effects, leading to an alteration of kinetic properties (K_m and V_{max}) from that of singly bound complexes. Therefore, appropriate kinetic models for various unusual kinetics need to be developed, according to the observed changes in apparent kinetic constants, to adequately fit experimental data and yield accurate estimates of kinetic parameters. Multiple binding site-mediated kinetics involving CYP3A4 have been found to include mainly sigmoidicity or substrate activation,[17,19,24] partial inhibition,[25–27] substrate inhibition,[28–31] activation,[13,15,32]

[14] K. R. Korzekwa, N. Krishnamachary, M. Shou, A. Ogai, R. A. Parise, A. E. Rettie, F. J. Gonzalez, and T. S. Tracy, *Biochemistry* **37,** 4137 (1998).
[15] T. L. Domanski, Y.-A. He, G. R. Harlow, and J. R. Halpert, *J. Pharmacol. Exp. Ther.* **293,** 585 (2000).
[16] M. L. Schrag and L. C. Wienkers, *Drug Metab. Dispos.* **391,** 49 (2001).
[17] Y. G. Ueng, T. Kuwabara, Y. J. Chun, and F. P. Guengerich, *Biochemistry* **36,** 370 (1997).
[18] A. P. Koley, J. T. M. Buters, R. C. Robinson, A. Markowitz, and F. K. Friedman, *J. Biol. Chem.* **270,** 5014 (1995).
[19] G. R. Harlow and J. R. Halpert, *Proc. Natl. Acad. Sci. U.S.A.* **95,** 6636 (1998).
[20] J. B. Houston and K. E. Kenworthy, *Drug Metab. Dispos.* **28,** 246 (2000).
[21] B. A. Swanson, D. R. Dutton, J. M. Lunetta, C. S. Yang, and P. R. Ortiz de Montellano, *J. Biol. Chem.* **266,** 19258 (1991).
[22] D. F. V. Lewis, *in* "Cytochromes P450: Metabolic and Toxicological Aspects" (C. Ioannides, ed.), p. 355. CRC Press, Boca Raton, FL, 1996.
[23] N. A. Hosea, G. P. Miller, and F. P. Guengerich, *Biochemistry* **39,** 5929 (2000).
[24] M. Shou, Q. Mei, J. R. Ettore, R. Dai, T. A. Baillie, and R. H. Rushmore, *Biochem. J.* **340,** 845 (1999).

and differential kinetics.[17,33] In addition to CYP3A4, CYP2C9 and CYP1A2 have been confirmed to contain two binding domains at the active site and to exhibit unusual kinetics.[34,35] This suggests that cytochrome P450s may comprise somewhat common features capable of holding multiple substrates at the active site that is rarely observed for other enzyme systems. The software programs that are used to perform nonlinear regression and statistical analysis include Axum 5.0, Mathematica 4.0, SigmaPlot 6.0, Enzyme Kinetic Module 1.1 (SigmaPlot), and KaleidaGraph 3.09.

Substrate Activation: Sigmoidal Kinetics

Some CYP3A4-mediated reactions yield characteristics of sigmoidal velocity saturation curve, e.g., the metabolism of aflatoxin B_1,[17] steroids,[19,36] carbamazepine,[14,37] amitriptyline,[38] triazolam,[28] and diazepam.[24] Sigmoidicity of velocity is due to the allosteric effect of the substrate on the enzyme, resulting in a positive cooperativity. This could be a result of multiple substrate interaction with the enzyme that gives rise to an altered affinity and rate of product formation for other substrate-binding sites. The Hill equation [Eq. (1)] is usually employed

[25] R. W. Wang, D. J. Newton, T. D. Scheri, and A. Y. H. Lu, *Drug Metab. Dispos.* **25,** 502 (1997).

[26] K. E. Kenworthy, J. C. Bloomer, S. E. Clarke, and J. B. Houston, *Br. J. Clin. Pharmacol.* **48,** 716 (1999).

[27] M. Shou, Y. Lin, P. Lu, C. Tang, Q. Mei, D. Cui, W. Tang, J. S. Ngui, C. Lin, R. Singh, B. Wong, J. A. Yergey, J. H. Lin., P. G. Pearson, T. A. Baillie, A. D. Rodrigues, and T. H. Rushmore, *Curr. Drug Metab.* **2,** 17 (2001).

[28] M. L. Schrag and L. C. Wienkers, *Drug Metab. Dispos.* **29,** 70 (2001).

[29] Y. Lin, P. Lu, C. Tang, Q. Mei, A. D. Rodrigues, T. H. Rushmore, and M. Shou, *Drug Metab. Dispos.* **29,** 368 (2001).

[30] K. Venkatakrishnan, L. L. Von Moltke, and D. J. Greenblatt, *J. Pharmacol. Exp. Ther.* **297,** 326 (2001).

[31] M. D. Perloff, L. L. Von Moltke, M. H. Court, T. Kotegawa, R. I. Shader, and D. J. Greenblatt, *J. Pharmacol. Exp. Ther.* **292,** 618 (2000).

[32] J. S. Ngui, W. Tang, R. A. Sterrns, M. Shou, R. R. Miller, Y. Zhang, J. H. Lin, and T. A. Baillie, *Drug Metab. Dispos.* **28,** 1043 (2000).

[33] M. Shou, R. Dai, D. Cui, K. R. Korzekwa, T. A. Baillie, and T. H. Rushmore, *J. Biol. Chem.* **276,** 2256 (2001).

[34] J. M. Hutzler, M. J. Hauer, and T. S. Tracy, *Drug Metab. Dispos.* **29,** 1029 (2001).

[35] G. P. Miller and F. P. Guengerich, *Biochemistry* **40,** 7262 (2001).

[36] G. E. Schwab, J. L. Raucy, and E. F. Johnson, *Mol. Pharmacol.* **33,** 493 (1998).

[37] B. M. Kerr, K. E. Thummel, C. J. Wurden, S. M. Klein, D. L. Kroetz, F. J. Gonzalez, and R. H. Levy, *Biochem. Pharmacol.* **47,** 1969 (1994).

[38] J. Schmider, D. J. Greenblatt, L. L. von-Moltke, J. S. Harmatz, and R. I. Shader, *J. Pharmacol. Exp. Ther.* **275,** 592 (1995).

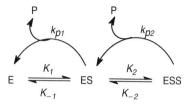

SCHEME 1. Kinetic scheme for a P450 enzyme with two binding domains within the active site (substrate activation).

to measure the number of binding sites (n) by data fitting.[9] The K' is a constant comprising the interaction factors and no longer equals K_m except when $n = 1$.

$$v = \frac{V_{max}[S]^n}{K' + [S]^n} \quad (1)$$

To characterize sigmoidal kinetics, the two-site model was proposed to describe simultaneous two substrate binding or substrate–effector binding to a P450 active site.[14] Accordingly, velocity equations were derived by Korzekwa et al.[14] to fit sigmoidal kinetic observations and, therefore, to generate kinetic parameters (e.g., K_{m1}, K_{m2}, V_{max1}, and V_{max2}). Kinetic expression is shown in Scheme 1 and Eq. (2), respectively.

$$v = \frac{\dfrac{V_{max1}[S]}{K_{m1}} + \dfrac{V_{max2}[S]^2}{K_{m1}K_{m2}}}{1 + \dfrac{[S]}{K_{m1}} + \dfrac{[S]^2}{K_{m1}K_{m2}}} \quad (2)$$

Shou et al.[24] later proposed a more detailed model (Scheme 2) that can discriminate all possible equilibration and rate steps of enzyme–substrate complexes on the two binding sites, resulting in four pairs of dissociation and rate constants (K_{S1}, K_{S2}, K_{S3}, K_{S4}, k_α, k_β, k_δ, and k_γ). Its kinetic predictions can be determined quantitatively by Eq. (3).

$$\frac{v}{[E]_{total}} = \frac{\dfrac{k_\alpha}{K_{S1}} + \dfrac{k_\beta}{K_{S2}} + \dfrac{[S]}{K_{S1}K_{S3}}(k_\delta + k_\gamma)}{\dfrac{1}{[S]} + \dfrac{1}{K_{S1}} + \dfrac{1}{K_{S2}} + \dfrac{[S]}{K_{S2}K_{S4}}} \quad (3)$$

Diazepam 3-hydroxylation by cDNA-expressed CYP3A4 is given as an example in Fig. 1. In Scheme 2, results indicate that K_{S1} and K_{S2} for the two single molecule-bound enzymes (SE and ES) were 13- to 32-fold larger than K_{S3} and K_{S4} for the two molecule-bound enzyme (SES), but rate constants k_α and k_β were

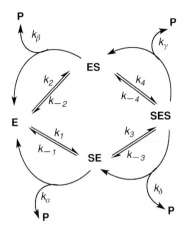

SCHEME 2. Allosteric kinetic scheme for a P450 enzyme with two binding domains in the active site (substrate activation).

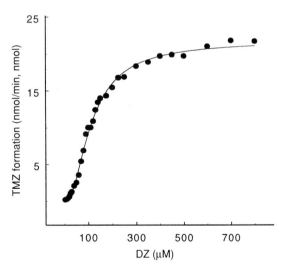

FIG. 1. Sigmoidal saturation curve of v vs [S] for the recombinant CYP3A4-catalyzed conversion of diazepam (DZ) to temazepam (TMZ). The curve was fitted with Eq. (3). Kinetic parameters were estimated by Eq. (2) ($K_{m1} = 356\,\mu M$, $V_{max1} = 2.1$ nmol/min/nmol; $K_{m2} = 11.4\,\mu M$, $V_{max 2} = 24.2$ nmol/min/nmol; $RSS = 1.220$ and $R^2 = 0.998$) and Eq. (3) ($K_{S1} = 170\,\mu M, k_\alpha = 1.8\,min^{-1}$; $K_{S2} = 421\,\mu M$, $k_\beta = 49.6\,min^{-1}$; $K_{S3} = 39.6\,\mu M$, $k_\delta = 392\,min^{-1}$; $K_{S4} = 13.1\,\mu M$, $k_\gamma = 808\,min^{-1}$; $RSS = 0.761$ and $R^2 = 0.999$), respectively.

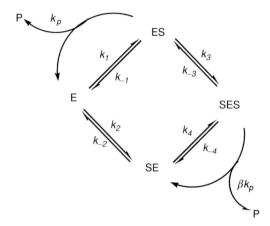

SCHEME 3. Kinetic model for substrate inhibition. ES, product-forming species; SE, inhibitory species.

8- to 449-fold less than k_δ and k_γ, respectively. Similarly, using the model of Scheme 1, the K_{m1} value for ES is 15-fold greater than that of the K_{m2} for ESS ($K_{m1} > K_{m2}$), and V_{max2} for ESS was 11.5-fold greater than V_{max1} ($V_{max2} > V_{max1}$). These values demonstrate that access and binding of the first molecule to either site of CYP3A4 could enhance cooperatively the binding affinity and reaction rate of the second molecule. Thus, the cooperative nature of the enzyme that gives a sigmoidal response can be described by the two-site model.

Substrate Inhibition

A substrate that causes a decrease in the rate of product formation as its concentration increases leads to a reaction that displays substrate inhibition kinetics. Of the P450-mediated reactions, substrate inhibition is commonly observed and has been recognized for decades.[14,28–31] Although the mechanism of P450-mediated substrate inhibition remains unknown, ignoring the phenomenon and truncating data can result in substantial errors in the values derived for critical kinetic parameters.[29] To address this issue, a two-site model was proposed, as shown in Scheme 3. In this kinetic scheme, ES denotes substrate binding to the site of the enzyme that is productive (k_p). SE denotes substrate binding to the site that reduces the rate of enzyme catalysis. The inhibitory site is assumed to be less productive or nonproductive ($k_p \rightarrow 0$ for SE). SES is a complex in which the two sites are occupied with substrates. Once the inhibitory site is bound at the excess substrate concentration, the turnover for SES no longer equals to that of ES and is reduced by factor β (usually $0 < \beta < 1$), which determines the potency of the inhibition. Equation (4) is a mathematic expression of the kinetics:

$$v = \frac{V_{max}\left(\dfrac{1}{K_S} + \dfrac{\beta[S]}{\alpha K_i K_S}\right)}{\dfrac{1}{[S]} + \dfrac{1}{K_S} + \dfrac{1}{K_i} + \dfrac{[S]}{\alpha K_S K_i}} \tag{4}$$

where $V_{max} = k_p[E]_{total}$, $K_m \approx K_S = k_{-1}/k_1$ ($k_p \ll k_{-1}$), and $K_i = k_{-2}/k_2$. K_S and K_i are dissociation constants of substrate binding to the catalytic site (ES) and to the inhibitory site (SE), respectively. α and β represent the factors by which the dissociation (K_S and K_i) of substrate at both sites and the maximal velocity (V_{max}) at the catalytic site change when a second substrate is bound. Dextromethorphan O-demethylation by CYP2D6 (Fig. 2A) and progesterone 6β-hydroxylation by CYP3A4 (Fig. 2B) exhibit substrate inhibition at a high concentration of the substrates. According to the present study and previous reports, some common features in the kinetics can be observed: (1) K_i values (substrate inhibition) are generally greater than the respective K_S values, suggesting that the substrate has binding affinity for the active site greater than the inhibitor; and (2) both K_S and K_i may be affected by the interaction of the two bound substrates within the enzyme, exhibited by a factor α (usually $\alpha > 1$). This suggests that the affinity of the substrate for the doubly bound complex is less than that for singly bound complexes. (3) The maximal potency of substrate inhibition at an excess concentration of the

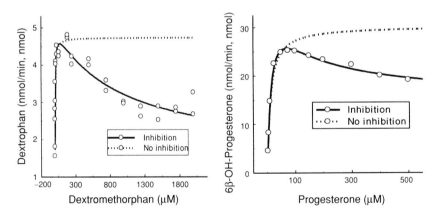

FIG. 2. Substrate inhibition of CYP-catalyzed reactions. Dotted lines represent hyperbolic saturation curves fitted to the Michaelis–Menten equation after truncating the inhibited rates at high substrate concentrations, and solid lines represent substrate inhibition curves fitted to Eq. (4). (A) CYP2D6-catalyzed O-demethylation of dextromethorphan ($\alpha = 23.8$, $K_S = 4.8\,\mu M$, $K_i = 48.4\,\mu M$, $\beta = 0.12$, $V_{max} = 4.3$ nmol/min/nmol, $RSS = 1.83$ and $R^2 = 0.976$). (B) CYP3A4-catalyzed 6β-hydroxylation of progesterone ($\alpha = 13.2$, $K_S = 16.2\,\mu M$, $K_i = 96\,\mu M$, $\beta = 0.41$, $V_{max} = 37$ nmol/min/nmol, $RSS = 2.69$, and $R^2 = 0.966$).

substrates is estimated by factor β $(0 < \beta < 1)$, which is shown to be less capable of metabolizing substrate than ES. Thus, the magnitude of substrate inhibition at a given concentration is dependent on the ratio of [SES] to [ES].

Partial Inhibition

Partial inhibition reflects incomplete inhibition, in which the enzyme is saturated with the inhibitor. This phenomenon is commonly encountered in P450-catalyzed reactions[25–27] and is due to the formation of the substrate–inhibitor–enzyme complex (ESI, Scheme 4) that is productive (ESI → P). The kinetics can be understood in terms of simultaneous binding of substrate and inhibitor to an enzyme in which the inhibitor incompletely inhibits the metabolism of a given substrate. Partial inhibition is similar to mixed-type inhibition in nature and is a result of either an inhibitor-induced conformational change of the enzyme that alters the binding affinity for substrate or steric hindrance that hampers the metabolism of the substrate.

$$v = \frac{V_{\max}\left(\dfrac{[S]}{K_S} + \dfrac{\beta[S][I]}{\alpha K_S K_i}\right)}{1 + \dfrac{[S]}{K_S} + \dfrac{[I]}{K_i} + \dfrac{[S][I]}{\alpha K_S K_i}} \tag{5}$$

In Eq. (5), factors α and $\beta(0 < \beta < 1)$ in the model represent the change in K_S and K_i and $V_{\max}$ for the substrate when the enzyme is bound with the inhibitor. In

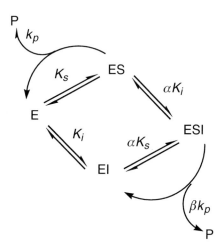

SCHEME 4. Kinetic model for partial inhibition.

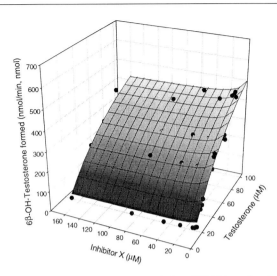

FIG. 3. Partial inhibition kinetics of inhibitor X for CYP3A4-catalyzed testosterone 6β-hydroxylation in human liver microsomes. Surface plots are predicted results, and scatter plots are experimental observations. Predicted kinetic parameters are calculated by Eq. (5) ($V_{max} = 0.958$ nmol/min/nmol, $K_S = 95\,\mu M$, $K_a = 9.7\,\mu M$, $\alpha = 0.97$, $\beta = 0.69$, $RSS = 78$, and $R^2 = 0.957$).

this case, K_i (K_i and αK_i) and β values are important estimates of P450 inhibition. If $\alpha \neq 1$, the two K_i values can be obtained through data fitting. As a matter of fact, the potency of P450 inhibition relies on the inhibitor and substrate interaction. For example, Fig. 3 shows a partial inhibition of CYP3A4-catalyzed testosterone 6β-hydroxylation by inhibitor X (inhibitor X, $K_i = 9.7\,\mu M$) that inhibits maximally the reaction by 31% ($\beta = 0.69$). This is due to the ESI complex that forms the product(s). The extent of inhibition is dependent on substrate and inhibitor interaction.

Activation

An enzyme activity toward one substrate can be increased in the presence of a second substrate.[13–16,18,19,23,32,39] Because both substrate and activator are substrates for the enzyme and do not replace each other, the two-site model can be adapted to interpret activation kinetics. Korzekewa et al.[14] have developed kinetic schemes for enzyme activation to explain the effect of α-naphthoflavone

[39] J. S. Ngui, Q. Chen, M. Shou, R. W. Wang, R. A. Stearns, T. A. Baillie, and W. Tang, Drug Metab. Dispos. 29, 877 (2001).

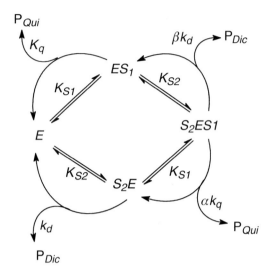

SCHEME 5. Proposed activation kinetics for CYP3A4 with two substrate-binding sites in the diclofenac–quinidine interaction.

on CYP3A4-catalyzed phenanthrene metabolism. The kinetic model for activation can also be projected in many ways in terms of the change of apparent K_m and V_{max} observed for each substrate. Ngui et al.[32] reported in Scheme 5 that (i) quinidine increased CYP3A4 activity in the metabolism of diclofenac by an increase in apparent V_{max} (~5-fold) but had no effect on the apparent K_m for diclofenac, and (ii) diclofenac had no effect on apparent K_m but slightly decreased the V_{max} for quinidine. Accordingly, substrate (S) and enzyme (E) in the model can be arranged in different orientations, e.g., ES_1, S_2E, and S_1ES_2. Because apparent K_m values for the two singly bound species (ES_1 and S_2E) remain unchanged in the presence of the second substrate, K_{S1} (K_{S2}) can be used in the equilibria for both ES_1 (S_2E) and S_2ES_1. Therefore, velocity equations are expressed as Eqs. (6) and (7).

$$v_{Qui} = \frac{V_{max\,Q}[S_1](K_{S2} + \alpha[S_2])}{K_{S1}K_{S2} + K_{S2}[S_1] + K_{S1}[S_2] + [S_1][S_2]} \tag{6}$$

$$v_{Dic} = \frac{V_{max\,D}[S_2](K_{S1} + \beta[S_1])}{K_{S1}K_{S2} + K_{S2}[S_1] + K_{S1}[S_2] + [S_1][S_2]} \tag{7}$$

Data fitting to the equations are shown in Figs. 4A and 4B ($R^2 = 0.931$ and 0.969, respectively). Two sets of K_{S1} values (1.32 and 1.42 μM) and K_{S2} (63.3 and

TABLE I

CALCULATED KINETIC PARAMETERS FOR QUINIDINE-MEDIATED ACTIVATION
OF DECLOFENAC METABOLISM

Substrate	Product	$K_{S1} \pm SE^a$ (μM)	$K_{S2} \pm SE^b$ (μM)	$V_{max} \pm SE^c$ (min^{-1})	Factor^d	Equation^e
Quinidine	3-OH-Q (P_{Qui})	1.32 ± 0.19	63.33 ± 9.60	5.92 ± 0.28 (V_{maxQ})	0.61 ± 0.38 (α)	(6)
Diclofenac	5-OH-D (P_{Dic})	1.42 ± 0.18	54.19 ± 4.78	10.49 ± 1.45 (V_{maxD})	5.87 ± 0.76 (β)	(7)

[a] K_{S1} is the dissociation constant for $ES_1 \rightleftharpoons E$ and $S_2ES_1 \rightleftharpoons S_2E$ equilibria (see Scheme 5).
[b] K_{S2} is the dissociation constant for $S_2E \rightleftharpoons E$ and $S_2ES_1 \rightleftharpoons ES_1$ equilibria.
[c] V_{max} is the maximal velocity for the conversion of quinidine to 3-OH-quinidine ($ES_1 \rightarrow P_{Qui}$) or of diclofenac to 5-OH-declofenac ($S_2E \rightarrow P_{Dic}$).
[d] Factors by which V_{max} values change when the second substrate binds to the enzyme.
[e] Equation used for the calculation of kinetic parameters.

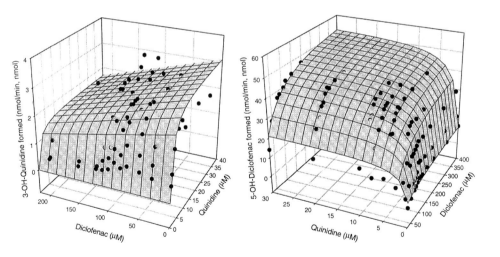

A   B

FIG. 4. Activation kinetics for the interaction between quinidine and diclofenac. Surfaces represent predicted results to experimental data (scatter plots). All parameters are given in Table I [Eq. (6) and (7)]. (A) Effect of diclofenac on the CYP3A4-mediated formation of 3-hydroxyquinidine and (B) effect of quinidine on the CYP3A4-mediated formation of 5-hydroxydiclofenac.

54.2 μM), calculated from Eqs. (6) and (7) (Table I), respectively, were found to be consistent and represent the binding affinities for quinidine and diclofenac either in the presence or in the absence of the second molecule. In contrast, K_S for quinidine is lower than that for diclofenac ($K_{S1} < K_{S2}$). Interestingly, quinidine increases V_{max} for diclofenac metabolism by 5.9-fold ($\beta = 5.9$); diclofenac, however, decreases V_{max} for the formation of 3-OH-quinidine by 39% ($\alpha = 0.61$). Results suggest that CYP3A4 has two distinct binding sites, which can be occupied separately by quinidine and diclofenac. The two substrates cannot be displaced kinetically by one another, as exhibited by the unchanged K_m. However, the site for quinidine appears to be allosteric, such that the binding of quinidine to the enzyme may change the kinetic characteristics of the second site, which preferred diclofenac for metabolism. The kinetic changes at the site could be due largely to the allosteric effect of quinidine at the site that catalyzes diclofenac 5-hydroxylation. More recently, Ngui et al.[39] postulated another model for the quinidine and warfarin interaction that is slightly different from the interaction of quinidine and diclofenac. Thus, two-site models can be postulated in several ways, based on changes in apparent kinetic constants, that interpret various consequences of the metabolic fate of the two substrate interaction.

Differential Kinetics

Differential kinetics occurs when an effector (also a substrate) activates the metabolism of a given substrate at one position but inhibits the metabolism at another position. The kinetic model is more complicated than any of those discussed earlier, as more product-forming substrate–enzyme complexes exist. Ueng et al.[17] reported that α-naphthoflavone (α-NF) modulates CYP3A4 activity in the regioselective metabolism of carcinogen aflatoxin B$_1$ by decreasing the 3α-hydroxylation pathway but increasing the 8,9-epoxidation reaction. The latter is a metabolic activation of the aflatoxin B$_1$ carcinogen. Similar results were observed in our laboratory[33] and showed that (1) α-NF increases the formation of losartan carboxylic acid (P$_2$, Fig. 5B) and decreases the formation of ω-3-hydroxylosartan (P$_1$, Fig. 5A) by CYP3A4; (2) losartan inhibits the CYP3A4-mediated formation of α-NF 5,6-epoxide (Q, Fig. 5C); and (3) changes in apparent kinetic constants, e.g., K_m and V_{max}, when two substrates coexist do not conform Michaelis–Menten assumptions for the drug–drug interaction. We proposed a two-site model to address the effect of α-NF on the differential metabolism of losartan.[33] Scheme 6 depicts the model and binding orientations of the substrates to the enzyme that are defined on the basis of the changes in apparent kinetic constants and the metabolic fates of each substrate observed. Due to interaction of the two-bound substrates, the apparent V_{max} and K_m of the substrate on each site vary with an increase in the concentration of the second substrate. Kinetic constants can be solved by the velocity equations derived from the

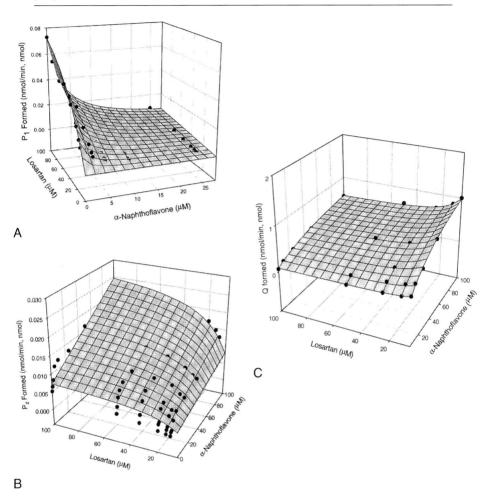

FIG. 5. Kinetics of the CYP3A4-mediated interaction between losartan and α-NF. Data (scatter plots) are predicted from Eqs. (8), (9), and (10), respectively. All calculated kinetic parameters are shown in Table. II. (A) Effect of α-NF on the CYP3A4-catalyzed formation of ω-3-hydroxylosartan (P_1); (B) effect of α-NF on the CYP3A4-catalyzed formation of the carboxylic acid derivative of losartan (P_2); and (C) effect of losartan on the CYP3A4-catalyzed formation of the 5,6-epoxide of α-NF (Q).

model in Scheme 6 [Eqs. (8)–(10)], which reflect the relationship between the substrate(s) and the enzyme in equilibria.

$$
v_{P1} = \frac{V_{\max P1}\left(\dfrac{1}{K_{S1}} + \dfrac{\alpha[S]}{K_{S1}K_A}\right)}{\dfrac{1}{[S]} + \dfrac{1}{K_{S1}} + \dfrac{1}{K_{S2}} + \dfrac{[F]}{K_F[S]} + \dfrac{[S]}{K_{S1}K_A} + \dfrac{[F]}{K_{S2}K_{SF}}} \tag{8}
$$

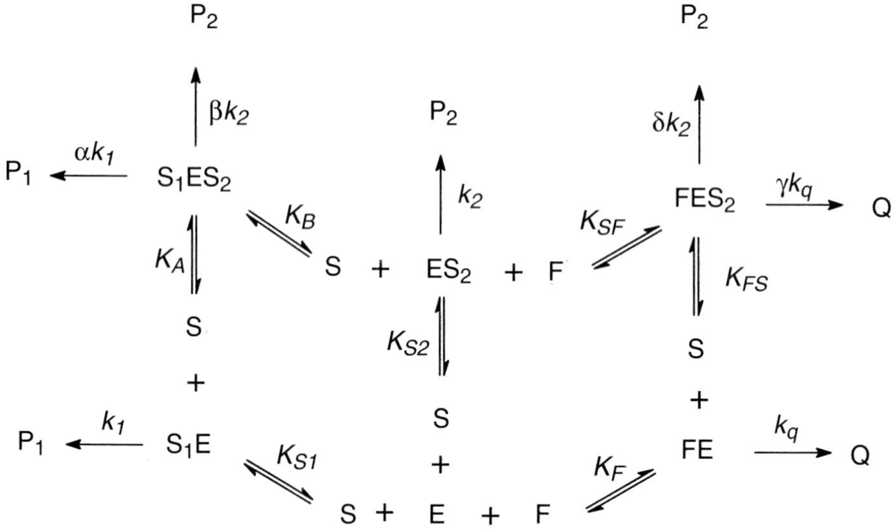

SCHEME 6. Proposed model of differential kinetics. S, losartan; F, α-NF; P_1, ω-3 hydroxyl losartan; P_2, losartan carboxylic acid; Q, α-NF-5,6-epoxide. S_1 and S_2 are losartan molecules, which bind to the two separate domains, leading to P_1 and P_2 formation, respectively.

$$v_{P2} = \frac{V_{\max P2}\left(\dfrac{1}{K_{S2}} + \dfrac{\beta[S]}{K_{S1}K_A} + \dfrac{\delta[F]}{K_{S2}K_{SF}}\right)}{\dfrac{1}{[S]} + \dfrac{1}{K_{S1}} + \dfrac{1}{K_{S2}} + \dfrac{[F]}{K_F[S]} + \dfrac{[S]}{K_{S1}K_A} + \dfrac{[F]}{K_{S2}K_{SF}}} \qquad (9)$$

$$v_Q = \frac{V_{\max Q}\left(\dfrac{[F]}{K_F[S]} + \dfrac{\gamma[F]}{K_F K_{FS}}\right)}{\dfrac{1}{[S]} + \dfrac{1}{K_{S1}} + \dfrac{1}{K_{S2}} + \dfrac{[F]}{K_F[S]} + \dfrac{[S]}{K_{S2}K_B} + \dfrac{[F]}{K_F K_{FS}}} \qquad (10)$$

In Eqs. (8)–(10), [S] and [F] are the concentrations of losartan and α-NF, and $V_{\max P1}$, $V_{\max P2}$, and $V_{\max Q}$ refer to the maximum velocities for the formation of each metabolite at its singly occupied site, respectively. K values are dissociation constants for individual enzyme–substrate complexes, whereas α, β, δ, and γ, respectively, represent factors that determine the change in $V_{\max}$ at one site when the second site is occupied by the substrate molecule. Table II lists all predicted kinetic parameters from Eqs. (8)–(10). The kinetic parameters indicate that α-NF increases $V_{\max P2}$ for the conversion of losartan to P_2 (FES$_2 \to$ FE + P$_2$) by 8.3-fold greater ($\delta = 8.3$) than $V_{\max P2}$ for ES$_2$ (losartan singly bound species), implying that the addition of α-NF accelerates the formation of FES$_2$ to losartan

TABLE II
PREDICTED KINETIC PARAMETERS FOR DIFFERENTIAL KINETICS OF α-NAPHTHOFLAVONE
ON METABOLISM OF LOSARTAN

Equilibria[a]	Product[b]	$K \pm SE$[c]	Factor[d]	$V_{max} \pm SE$[e]	Eq.
$S_1E \rightleftharpoons E$	P_1	K_{S1} 97 ± 12[f]		V_{maxP1} 0.16 ± 0.04	(8)
		K_{S1} 67 ± 23			(9)
		K_{S1} 107 ± 30			(10)
$S_1ES_2 \rightleftharpoons ES_2$	P_1	K_B 435 ± 108	α 0.38 ± 0.12	αV_{maxP1} 0.06	(8), (10)
$ES_2 \rightleftharpoons E$	P_2	K_{S2} 25.1 ± 5.6		V_{maxP2} 0.007 ± 0.001	(9)
		K_{S2} 40.5 ± 1.1			(8)
		K_{S2} 36.1 ± 3.2			(10)
$S_1ES_2 \rightleftharpoons S_1E$	P_2	K_A 250 ± 29	β 0.43 ± 0.21	βV_{maxP2} 0.003	(9)
		K_A 150 ± 45			(8)
$FES_2 \rightleftharpoons ES_2$	Q	K_{SF} 203 ± 20	γ 0.29 ± 0.11	γV_{maxQ} 1.19	(8), (10)
		K_{SF} 182 ± 44			(9)
$FE \rightleftharpoons E$	Q	K_F 183 ± 51		V_{maxQ} 4.1 ± 0.5	(10)
		K_F 203 ± 89			(9)
		K_F 158 ± 67			(8)
$FES_2 \rightleftharpoons FE$	P_2	K_{FS} 15.1 ± 0.8	δ 8.3 ± 2.1	δV_{maxP2} 0.058	(9), (10)

[a] Enzyme–substrate complexes and their equilibria shown in Scheme 6.
[b] Product formed from the specific enzyme–substrate complex.
[c] K is the dissociation constant of each substrate–enzyme complex in the model, calculated by the equations shown.
[d] Factor that determines the change in V_{max} for the first site when the second site is occupied.
[e] V_{max} values for that maximum velocity at the specific site of each product-forming complex in the model, calculated by the equations, respectively. RSS (residual sum of squares) and R^2 are 0.0213 and 0.992 for Eq. (8), 0.101 and 0.998 for Eq. (9), and 0.0153 and 0.9899 for Eq. (10), respectively.

carboxylate acid (P_2). This suggests that α-NF binding to the one site changes the kinetic properties, exhibited by K_{FS} and factor δ, of the vacant site for losartan, which converts losartan preferentially to P_2. However, α-NF inhibits the conversion of losartan to P_1 by a significant increase in $K_{mP1(app)}$ and a decrease in $V_{maxP1(app)}$. The differential effect of α-NF suggests that (1) α-NF (F) competes with S_1 at the one site by decreasing S_1E (and S_1ES_2) and increasing FE (and FES_2), leading to less production of P_1, and (2) α-NF stimulates the other site to which S_2 binds and increases the formation of P_2 ($FES_2 \rightarrow FE + P_2$). Conversely, losartan had similar, but opposite, effects on the kinetic properties of α-NF, such that α-NF metabolism was inhibited by 71% ($\gamma = 0.29$, Table II) for $FES_2 \rightarrow$ Q. These findings illustrate that the two-bound sites are capable of interacting with each other in different manners (i.e., allosteric effect and steric hindrance), resulting in the appreciable changes in binding affinity and reaction rate.

Drug–drug interactions have become an important clinical issue due to the effects of one drug on efficacy, toxicity, or disposition of another drug.

Cytochromes P450 are recognized to play an important role in clinically relevant drug–drug interactions as a result of which new chemical entities (NCEs) typically are subjected to high-throughput CYP inhibition screening.[4,40–42] A key cause of drug–drug interactions is the effect of one drug on the activities of critical enzymes involved in the metabolism of a second drug. If unusual enzyme kinetics occur in screening of the compounds on P450 activity, a quantitative relationship between IC_{50} (EC_{50}) and K_i (K_A) cannot be maintained in terms of the Michaelis–Menten concept and IC_{50} (EC_{50}) values no longer are an estimate of K_i (K_A). Therefore, the development of appropriate kinetic models and the approximation of constants, which can adequately describe the nature of the interaction between the substrates and the enzyme, are of considerable importance in the prediction of metabolic outcome, drug clearance, and the likelihood of clinically important drug–drug interactions.[4,10,40,41]

[40] C. L. Crespi, *Curr. Opin. Drug Disc. Dev.* **2**, 15 (1999).
[41] A. D. Rodrigues and J. H. Lin, *Curr. Opin. Chem. Biol.* **5**, 396 (2001).
[42] G. C. Moody, S. J. Griffin, A. N. Mather, D. F. McGinnity, and R. J. Riley, *Xenobiotica* **9**, 53 (1999).

[26] Design and Application of Fluorometric Assays for Human Cytochrome P450 Inhibition

By CHARLES L. CRESPI, VAUGHN P. MILLER, and DAVID M. STRESSER

The commercial success of a new drug entity (NDE) depends on its pharmacological activity and safety. One important aspect of drug safety is the extent to which a NDE causes metabolism-based pharmacokinetic interactions with coadministered medications. If the coadministered medication has a narrow therapeutic index, toxicity can result. In addition, if the coadministered medication is a prodrug, efficacy can be impaired. The inhibition of metabolism by cytochrome P450 enzymes (CYP) is a principal mechanism for such interactions.[1] It is well established that appropriate *in vitro* assays are useful predictors of the potential for *in vivo* interactions.

The rationale for *in vitro* assays is based on the single-enzyme paradigm for drug–drug interactions. In this paradigm, it is assumed that if a NDE inhibits the metabolism of one probe substrate for an enzyme, then it similarly inhibits all substrates of that enzyme. Therefore, potential drug–drug interactions can be tested for each enzyme of interest with a suitable probe substrate. The validity of these

[1] S. Rendic and F. J. Di Carlo, *Drug Metab. Rev.* **29**, 413 (1997).

assays depends on the use of human enzymes and the probe substrate/assay conditions measuring the activity of the subject enzyme with a high degree of specificity. Enzyme specificity is always assured when individual, cDNA-expressed enzymes are used. Therefore, the use of expressed enzymes provides considerable flexibility in the choice of probe substrates. However, if enzyme mixtures, such as human liver microsomes, are to be used, probe substrate choices are more limited, as substrates with multiple pathways of metabolism are usually inappropriate.

Classical enzyme kinetic analyses can be applied to measure the ability of NDEs to inhibit the major drug-metabolizing CYP enzymes (generally agreed to be CYP1A2, CYP2C8, CYP2C9, CYP2C19, CYP2D6, and CYP3A4). Indeed, such studies utilizing classical drug molecules have been routinely conducted for more than a decade. However, quantitation of the amount of drug in each incubation typically required a time-consuming separation step (typically liquid chromatography). As tests for CYP inhibition began to be applied earlier in the drug discovery processes and the number of chemicals to be examined increased by several orders of magnitude, a bottleneck resulted. The development and application of direct fluorometric substrates for the major human drug metabolizing CYPs presented a way around this bottleneck. Such assays can be conducted in multiwell plates and analyzed directly, thus bypassing any separation steps and increasing assay throughput dramatically.

Suitable substrates for five of the principal drug metabolizing enzymes, CYP1A2, CYP2C9, CYP2C19, CYP2D6, and CYP3A4, were reported in 1997.[2] These assays are based on cytochrome P450-catalyzed O-dealkylation reactions for commercially available chemicals, which generate a readily detectable fluorescent product. The human cytochrome P450 enzymes were introduced into these assays as single, cDNA-expressed enzymes. This was necessitated by the fact that these substrates were not enzyme specific and hence inappropriate for use in enzyme mixtures. In addition, a number of the reactions were catalyzed slowly, and relatively large amounts of enzyme were needed to achieve an adequate signal in the assay. High enzyme concentrations have the potential to introduce artifacts due to the depletion of inhibitor by cytochrome P450-mediated metabolism.

Improvements to the substrates were achieved by several approaches. First, other commercially available or published CYP substrates were tested to determine if they offered improved signal-to-noise ratios (i.e., enabled a reduction in the amount of enzyme). 7-Methoxy-4-trifluoromethylcoumarin/CYP2C9, 7-benzyloxyquinoline/CYP3A4, and 3-O-methylfluorescein/CYP2C19 were identified by this approach. Second, knowledge of specific CYP active sites was used to design substrates. Onderwater et al.[3] used a computer model of the CYP2D6 active

[2] C. L. Crespi, V. P. Miller, and B. W. Penman, *Anal. Biochem.* **248**, 188 (1997).
[3] R. C. Onderwater, J. Venhorst, J. N. Commandeur, and N. P. Vermeulen, *Chem. Res. Toxicol.* **12**, 555 (1999).

site to develop 7-methoxy-4-(aminomethyl)coumarin, a relatively selective, fluorometric CYP2D6 substrate. Renwick *et al.*[4] used a similar approach to develop 2,5-bis(trifluoromethyl)-7-benzyloxy-4-trifluoromethylcoumarin, a CYP3A4 substrate that shows good CYP3A4 selectivity in human liver microsomes. Our laboratory used a simple overlay with dextromethorphan, a rigid, prototypical CYP2D6 substrate, to identify 3-[2-(*N*,*N*-diethyl-*N*-methylammonium)ethyl]-7-methoxy-4-methylcoumarin as a selective, fluorometric CYP2D6 substrate.[5] Clearly, this approach can also be applied to the CYP2C subfamily. Finally, knowledge of the regioselectivity of some enzymes was used to guide the development of new substrates. We capitalized on the tendency of CYP3A to oxidize positions of high electron density such as benzyl positions to develop two CYP3A substrates with benzyl ether substituents: dibenzylfluorescein and 7-benzyloxy-4-trifluoromethylcoumarin.[6]

Methodology

This section first discusses some general considerations for conducting fluorometric CYP inhibition assays and then provides a general method for conducting the assay for several of the drug-metabolizing CYPs.

In order to detect competitive inhibitors, a probe substrate concentration should be chosen that is close to or below the apparent K_m. For competitive inhibitors tested with a substrate present at the apparent K_m concentration, the IC_{50} is twice the apparent K_i. For noncompetitive inhibitors tested under the same conditions, the IC_{50} is the apparent K_i.

Assays should be conducted under initial rate conditions—formation of the metabolite should be linear with respect to enzyme concentration and incubation time, and the total consumption of the substrate should be less than 20%. Assays conducted at 37° in the plate scanner with continuous data acquisition assure assay linearity, but occupy the instrument during the entire incubation period (i.e., lowers throughput dramatically). However, coumarins without an electron-withdrawing substituent(s) require an elevated pH to exhibit fluorescence and thus are not suitable for direct assays. Similarly, the need for a post-assay incubation step excludes fluoroscein derivatives from this approach.

High enzyme concentrations or long incubation times can result in inhibitor depletion. In addition, some potential NDEs may bind extensively to microsomal

[4] A. B. Renwick, D. F. V. Lewis, S. Fulford, D. Surry, B. Williams, P. D. Worboys, B. G. Lake, and D. C. Evans, *Xenobiotica* **31**, 187 (2001).
[5] N. Chauret, B. Dobbs, R. L. Lackman, K. Bateman, D. A. Nicoll-Griffith, D. M. Stresser, J. M. Ackermann, S. D.Turner, V. P. Miller, and C. L. Crespi, *Drug Metab. Dispos.* **29**, 1196 (2001).
[6] D. M. Stresser, A. P. Blanchard, S. D. Turner, J. C. Erve, A. A. Dandeneau, and C. L. Crespi, *Drug Metab. Dispos.* **28**, 1440 (2000).

protein and thus the free inhibitor concentration may be less than the nominal concentration. Accordingly, it may be desirable to keep the microsomal protein concentration as low as possible and standardize the protein concentration by the addition of "control" microsomes (microsomes devoid of cytochrome P450). (High microsomal protein concentrations tend to quench the fluorescence of the metabolite. However, the magnitude of this effect is modest.) In addition, NDEs may bind nonspecifically to the wells of the plate and thus deplete the aqueous phase of inhibitor.

Some potential NDEs may interfere with the assay by either fluorescing at the wavelengths being used to monitor metabolite formation or quenching the fluorescence of the metabolite. In general, metabolites with higher excitation and emission wavelengths are less sensitive to interference. However, concurrent (or follow-up) testing for possible assay interference is often desirable.

Many organic solvents are inhibitory to cytochromes P450.[7] Therefore, organic solvent concentrations should be kept as low as possible. Use of up to 2% (v/v) acetonitrile, 1% (v/v) methanol, and 0.2% (v/v) dimethyl sulfoxide (DMSO) is compatible with these fluorometric assays. Some enzyme/substrate pairs (CYP1A2/CEC, CYP2D6/AMMC, and CYP3A4/BQ) permit the use of higher DMSO concentrations (up to 0.5%).

The choice of the type of plate and the quality of the excitation and emission filters can substantially affect the signal-to-noise ratio in the assay. In general, black plates provide better signal-to-noise ratios than white or clear plates. However, the differences are more pronounced with coumarins and becomes less pronounced as the excitation wavelength increases. Similarly, clear bottom plates will have a higher background than black bottom plates. Obviously, an opaque bottom requires the use of an instrument that can read fluorescence from the top of the well.

Finally, it is now well established that with CYP3A4 the extent of inhibition is commonly substrate dependent. For some inhibitors, qualitative differences and quantitative differences of more than 20-fold in IC_{50} values can be obtained, depending on the probe substrate being used in the assay. This property is not fully understood, but may be related to the ability of CYP3A enzymes to accommodate multiple molecules simultaneously in the active site. As a practical matter, it is prudent to use multiple probe substrates in any test for CYP3A4 inhibition.

The method summarized here describes the determination of IC_{50} values using an eight-point curve. In any given application, more abbreviated curves or more extended curves may be appropriate. The highest inhibitor concentration is an experimental variable, but it is typically $\geq 100 \ \mu M$ if aqueous solubility and compatability with organic solvents permit. Table I provides the assay parameters for

[7] W. F. Busby, Jr., J. M. Ackermann, and C. L. Crespi, *Drug Metab. Dispos.* **27,** 246 (1999).

TABLE I
FLUOROMETRIC SUBSTRATE/ENZYME PAIRS AND THEIR METABOLITE EXCITATION
AND EMISSION WAVELENGTHS

Substrate	Enzyme(s) assayed	Metabolite	Excitation (nm)	Emission (nm)
3-[2-(N,N-Diethyl-N-methylamino) ethyl]-7-methoxy-4-methylcoumarin	CYP2D6	(3-[2-(N,N-Diethyl-N-methylammonium)ethyl]-7-methoxy-4-methylcoumarin)[a]	390	460
7-Ethoxy-3-cyanocoumarin	CYP1A2, CYP2C9, CYP2C19	7-Hydroxy-3-cyanocoumarin	410	460
7-Methoxy-4-trifluoromethylcoumarin	CYP2C9	7-Hydroxy-4-trifluoromethylcoumarin	410	538
7-Benzyloxy-4-trifluoromethylcoumarin	CYP3A4, CYP2C19	7-Hydroxy-4-trifluoromethylcoumarin	410	538
Dibenzylfluorescein	CYP3A4, CYP2C8	Fluorescein	485	538
3-O-Methylfluorescein	CYP2C19	Fluorescein	485	538
7-Benzyloxyquinoline	CYP3A4	7-Hydroxyquinoline	410	538
Resorufin benzyl ether	CYP3A4	Resorufin	530	590

[a] 3-[2-(N,N-Diethyl-N-methylamino)ethyl]-7-hydroxy-4-methylcoumarin was used as an external standard for quantification of the O-demethyl metabolite (3-[2-(N,N-diethyl-N-methylammonium)ethyl]-7-methoxy-4-methylcoumarin).

the enzyme and substrate pairs as used routinely in our laboratory. Figure 1 contains the structures of the substrates. The assay is conducted as a series of reagent additions, which are easily compatible with robotic liquid-handling systems. For the serial dilutions in step 1, the inhibitor solution is pipetted several times to assure proper mixing. Mixing for the initiation and stopping of the incubations is accomplished by rapid reagent addition.

1. The 12 wells in the row of a 96-well plate can be used for one inhibition curve. Wells 1 to 8 contain serial 1 : 3 dilutions of the inhibitors (50 μl transferred between wells containing 100 μl). Wells 9 and 10 contain no inhibitor, and wells 11 and 12 are blanks for background fluorescence (stop solution is added before the enzyme) or, alternatively, tests for interference or quenching. Assays should contain a positive control. Enzyme-specific positive control inhibitors (with the highest inhibitor concentration as added to well 1) for the different enzymes are as follows: CYP1A2, furafylline (100 μM); CYP2C9, sulfaphenazole (10 μM); CYP2C19, tranylcypromine (500 μM); CYP2D6, quinidine (0.5 μM); and CYP3A4, ketoconazole (5 μM). Alternatively, the number of positive control preparations needed may be reduced by the use of nonspecific inhibitors such as miconazole.

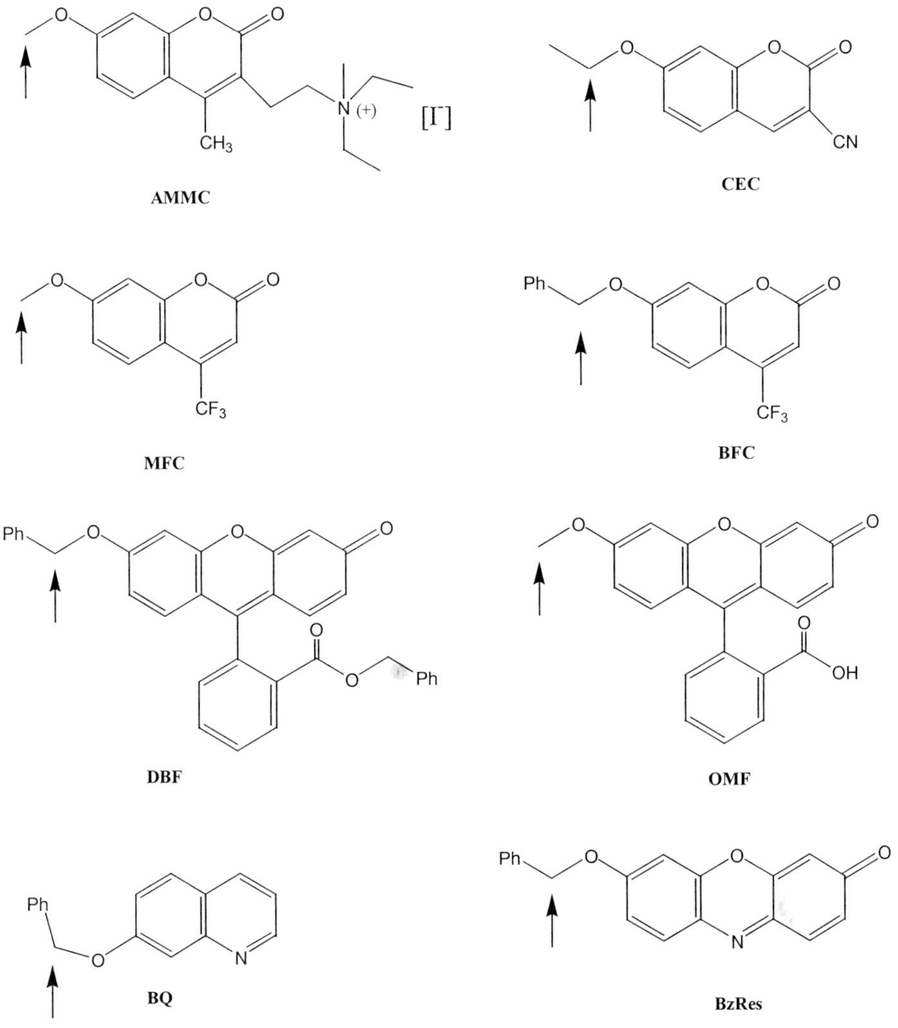

FIG. 1. Structures of fluorometric substrates and sites of oxidation. Structures are provided for 3-[2-(N,N-diethyl-N-methylamino)ethyl]-7-methoxy-4-methylcoumarin (AMMC), 7-ethoxy-3-cyanocoumarin (CEC), 7-methoxy-4-trifluoromethylcoumarin (MFC), 7-benzyloxy-4-trifluoromethylcoumarin (BFC), dibenzylfluorescein (DBF), 3-O-methylfluorescein (OMF), 7-benzyloxy-quinoline (BQ), and resorufin benzyl ether (BzRes). Arrows mark the site of oxidation that results in the fluorescent metabolite.

2. Carry out serial dilution of the inhibitors in potassium phosphate buffer (pH 7.4, the optimal concentration varies depending on the enzyme) containing the NADPH generating system. CYP3A4 is most active at concentrations above 100 mM, CYP1A2 and CYP2D6 at 100 mM, and CYP2C enzymes at 25 to 50 mM. For all enzyme/substrate pairs except CYP2D6/ AMMC, the final cofactor concentrations are 1.3 mM NADP$^+$, 3.3 mM glucose 6-phosphate, and 0.4 U/ml glucose-6-phosphate dehydrogenase. For CYP2D6, the final cofactor concentrations are 0.0081 mM NADP$^+$, 0.41 mM glucose 6-phosphate, and 0.4 U/ml glucose-6-phosphate dehydrogenase. The lower NADP$^+$ concentration is required in order to reduce interference from the fluorescence of NADPH, which can be substantial with 390-nm excitation and 460-nm emission wavelengths. Indeed, for the other substituted coumarins (410-nm excitation wavelength), NADPH may still contribute significantly to background fluorescence. A modest reduction in NADP$^+$ concentration may result in an improved signal-to-background ratio and also reduce reagent cost.

3. Incubations are initiated by the addition of 0.1 ml prewarmed enzyme (37°) and substrate. Incubations are carried out for 10 min (0.5 pmol CYP3A4, 1 μM DBF), 20 min (0.5 pmol CYP1A2, 5 μM CEC), 30 min (4 pmol CYP2C8, 1 μM DBF; 0.5 pmol CYP2C19, 25 μM CEC; 1 pmol CYP2C19, 2 μM OMF; 1 pmol CYP3A4, 50 μM BFC; and 3 pmol CYP3A4, 40 μM BQ), or 45 min (1 pmol CYP2C9, 75 μM MFC; 1.5 pmol CYP2D6, 1.5 μM AMMC; and 3 pmol CYP3A4, 50 μM BzRes). The substrates are prepared in acetonitrile.

4. Incubations are stopped by the addition of 0.075 ml of 80% acetonitrile–20% 0.5 M Tris base or 2 N NaOH (fluorescein derivatives only). For most substrates, the plates can be analyzed for fluorescence signal immediately. However, for fluoresceins, the plates should be incubated further for at least 2 hr (up to overnight) to allow the signal to develop. Excitation and emission wavelengths are contained in Table I.

5. Data can be exported and analyzed using a variety of software tools, such as an Excel spreadsheet. IC$_{50}$ values are calculated by linear interpolation or curve fitting.

Results

Figure 2 contains a representative inhibition curve for diclofenac and sulfaphenazole inhibition of CYP2C9 with the fluorometric substrates CEC, MFC, and DBF. In general, with CYP1A2, CYP2C9, CYP2C19, and CYP2D6 enzymes, we have observed a good concordance between IC$_{50}$ values obtained among and between different classical substrates and different fluorometric substrates. However, such concordance does not generally apply to CYP3A4. This enzyme demonstrates marked substrate specificity in IC$_{50}$ values among different probe

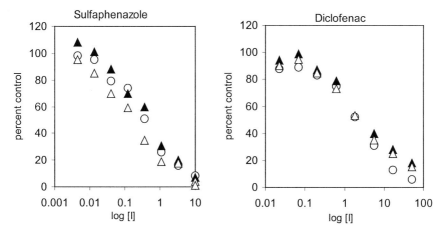

FIG. 2. Inhibition of CYP2C9 activity by sulfaphenazole and diclofenac. Eight-point inhibition curves were generated using 3-cyano-7-ethoxycoumarin (○), 7-methoxy-4-trifluoromethylcoumarin (▲), and dibenzylfluorescein (△).

substrates and this effect is observed from both traditional substrates and fluorometricsubstrates. This poor concordance between substrates cannot be attributed to artifacts due to differences in experimental conditions.

Human CYP3A4 has been demonstrated to simultaneously bind and metabolize multiple compounds in its active site.[8] Figure 3 shows example curves for itraconazole and nimodipine inhibition of CYP3A4 using four different fluorometric substrates. With itraconazole, clear substrate-dependent inhibition is evident and IC_{50} values varied more than 10-fold. With nimodipine, the response was similar across substrates, with IC_{50} values differing by less than 4-fold. Results with nimodipine suggest that for some compounds, inhibition of CYP3A4 is not *always* substrate dependent. At this time the full extent of substrate dependence in CYP3A4 inhibition is an area of active research. A prudent approach to evaluate CYP3A4 inhibition potential would be to use multiple probe substrates in a screening mode and follow-up studies with likely comedications.

Some cytochrome P450 substrates can covalently modify and inactivate the enzymes. Such mechanism-based inhibitors may not be potent competitive inhibitors and thus may not be detected in assays for competitive inhibition with short incubation times (less than 10 min). Different experimental approaches (typically time- and NADPH-dependent inhibition) are needed to identify this class of inhibitors. Two aspects of the experimental design facilitate the detection

[8] M. Shou, R. Dai, K. R. Korzekwa, T. A. Baille, and T. H. Rushmore, *J. Biol. Chem.* **276,** 2256 (2001).

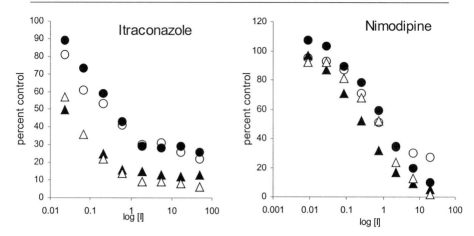

FIG. 3. Inhibition of CYP3A4 activity by itraconazole and nimodipine. Eight-point inhibition curves were generated using 7-benzyloxy-4-trifluoromethylcoumarin (▲), 7-benzyloxyquinoline (●), dibenzylfluorescein (△), and resorufin benzyl ether (○).

of mechanism-based inhibitors. First, because the incubation times are relatively long, there is an opportunity for time- and NADPH-dependent inactivation to occur. Second, because most of the assays are direct, the plates can be scanned at multiple time points and temporal changes in IC_{50} analyzed. Mechanism-based inhibitors demonstrate a substantial, time-dependent decrease in IC_{50} values. Such an observation would suggest that more detailed studies may be appropriate. In contrast, time-dependent increases in IC_{50} values imply that the inhibitor is being depleted by metabolism.

Further refinements in fluorometric CYP inhibition assays are possible. In particular, a higher fluorescence output of the metabolite and more rapid substrate turnover would further reduce the time and materials needed for conducting these assays. The complexity of CYP3A4 inhibition analysis indicates that the use of a single substrate is inappropriate. Therefore, it is desirable to identify additional fluorometric CYP3A4 substrates and an appropriate approach for decision making based on the results of experiments with multiple substrates.

[27] Automated Quantitative and Qualitative Analysis of Metabolic Stability: A Process for Compound Selection during Drug Discovery

By STEPHEN A. WRING, IVIN S. SILVER, and COSETTE J. SERABJIT-SINGH

Introduction

The number of compounds entering drug discovery programs has increased beyond the capacity of conventional *in vivo* techniques to evaluate their metabolism and pharmacokinetics. Consequently, many pharmaceutical companies have established *in vitro* high-throughput metabolism assays to avoid progressing metabolically labile compounds.[1–7] Compounds that are most stable when incubated in hepatic subcellular fractions are generally selected for further study.

This chapter describes three stages of development leading to a fully automated high-throughput assay of metabolic stability that has assayed over 10,000 compounds. The assay takes advantage of robotics and high-throughput high-performance liquid chromatography (HPLC)-tandem mass spectrometry to overcome the two key bottlenecks to successful and timely assay of a large number of diverse compounds. These are (1) the labor-intensive liquid transfer steps and (2) large numbers of samples for bioanalysis. Other bottlenecks, compound procurement, and automated data processing have been addressed.[8] The final software tools for converting LC-MS/MS peak area data to the extent of metabolism will be published elsewhere.

Our challenge was to increase throughput while maintaining the success rate of the assay. To improve throughput, we used automation that could perform all liquid handling procedures with the versatility to allow incubations with enzymes from either single or multiple mammalian species. The percentage of compounds detected successfully was improved over conventional LC-MS or HPLC-UV by the introduction of high-throughput LC-MS/MS and new software tools that improved sensitivity and selectivity.

---

[1] P. Eddershaw, A. Beresford, and M. Bayliss, *Drug Disc. Today* **5,** 409 (2000).

[2] R. White, *Annu. Rev. Pharmacol. Toxicol.* 133 (2000).

[3] P. Eddershaw and M. Dickins, *Pharm. Sci. Technol. Today* **2,** 13 (1999).

[4] W. Korfmacher, C. Palmer, C. Nardo, K. Dunn-Meynell, and D. Grotz, *Rapid Commun. Mass Spectrom.* **13,** 901 (1999).

[5] M. Tarbit and J. Berman, *Curr. Opin. Chem. Biol.* **2,** 411 (1998).

[6] D. Rodrigues, *Pharm. Res.* **14,** 1504 (1997).

[7] T. N. Thompson, *Curr. Drug Metab.* **1,** 215 (2001).

[8] L. W. Frick, D. M. Higton, S. A. Wring, and K. J. Wells-Knecht, *Med. Chem. Res.* **8,** 472 (1998).

This chapter presents the automated and bioanalytical procedures used; the design criteria we selected as being most important for high-throughput assays using enzymes from either single or multiple species; data that demonstrate the capacity, success rate, and reproducibility achieved; and metabolic stability data typified by a diverse set of 1000 compounds, and more focused results that exemplify a species-dependent structure–activity relationship. Finally, we describe early work using a new application of time-of-flight (TOF) mass spectrometry to address the instance when all compounds within a series are metabolically labile. Identifying the labile sites of a molecule helps chemists to design compounds that achieve adequate metabolic stability. In this application, LC-TOF/MS provides accurate mass data that are used to propose molecular formula of metabolites generated during the *in vitro* incubations.

Methods

Incubation Conditions

The optimized and automated method employs an eight-probe Tecan Genesis 150 robotic system (Tecan US, Research Triangle Park, NC) to perform all liquid handling steps. Reaction mixtures for metabolic stability studies consist of pooled liver S9 homogenates (5 mg/ml final protein concentration, XenoTech LLC, Kansas City, KS), candidate drug (1 μM final concentration), 5 mM UDPGA, 2 mM NADPH, and 10 mM MgCl$_2$ in a final volume of 0.5 ml of 100 mM potassium phosphate buffer, pH 7.4. Microsomes (0.5 mg/ml final protein concentration, XenoTech LLC) instead of the S9 homogenate have also been used to measure metabolic stability where a similar protocol to the just described was used without the addition of UDPGA, and the concentration of MgCl$_2$ was decreased from 10 to 5 mM. Samples are preincubated for 5 min, with shaking, at 37° in 96-well polypropylene (2 ml) plates prior to the addition of NADPH. An initial time point (t_0) is collected immediately after preincubation, and subsequent aliquots (preprogrammed into the Tecan software) are taken automatically at predetermined time points. For S9 homogenates, samples are collected at 0, 15, 30, and 60 min, and for liver microsomes, samples are collected at 0, 5, 10, 15, 20, 25, and 30 min. Each reaction is terminated by transferring aliquots of the incubation mixture (0.1 ml) into acetonitrile (0.2 ml), maintained at 4°. Precipitated protein is removed by centrifugation (3800 rpm for 15 min), and the resultant supernatant is transferred to a new 96-well polypropylene deep well plate for subsequent analysis by LC-MS/MS. Drug-free matrix control samples and QC standards (e.g., bufuralol, oxazapam, and verapamil, Table I) are included in each run and treated in the same manner as unknowns.

For metabolite identification work, higher substrate concentrations (5–25 μM) may be required to yield higher levels of metabolites; however, the concentration of dimethyl sulfoxide (DMSO) should be maintained at $\leq 1\%$ (v/v).

TABLE I

INTERASSAY PRECISION FOR ANALYSIS OF PROBE SUBSTRATES[a]

Incubation time (min)	Bufuralol (mean metabolized ± SD)	Oxazepam (mean metabolized ± SD)	Verapamil (mean metabolized ± SD)
15	2 ± 5	4 ± 5	39 ± 9
30	3 ± 5	9 ± 6	75 ± 6
60	4 ± 5	25 ± 7	95 ± 2

[a] Incubated with human S9 liver homogenates in the stage III assay. Replicate analysis ($n = 46$ independent assays over 6 months) for three probe substrates, incubated at 1 μM concentration, for 15, 30, and 60 min.

Quantitative Bioanalysis by High-Throughput LC-MS/MS

All analyses for the stage II or III assays employ HPLC with tandem mass spectrometry (LC-MS/MS). Mass spectrometry is performed on Applied Biosystems (Foster City, CA) API 300, 2000, or 3000 series triple quadrupole instruments equipped with either a turbo ion spray (TISP) source for electrospray ionization or a heated nebulizer (HN) source for atmospheric pressure chemical ionization (APCI). Polarity is selected for optimum sensitivity. Detection by tandem mass spectrometry is based on precursor ion transitions to the strongest intensity product ions. Key instrumental conditions are optimized to yield best sensitivity. Source temperature is typically 450°. Instrument optimization may be automated for high-throughput analysis with software tools such as PROMS,[9] Autotune, or Automaton (Applied Biosystems).

Chromatography has been performed on several instruments. A suitable system for the stage III assay employs a Hewlett Packard (Agilent) 1100 HPLC equipped with a switching valve to allow unretained materials to be diverted to waste (0–2 min). The sample volume (10 to 20 μl) is loaded on the column by means of a Gilson 215 autosampler. Fast gradient chromatography is performed on short Phenomonex Aqua or Luna C$_{18}$ columns (30 × 2 mm i.d., 3 μm) at a flow rate of 0.6 ml min^{-1}. The mobile phase consists of two solvents: (A) either 10 mM ammonium formate, pH 3.5, for positive ion electrospray or 10 mM ammonium acetate, pH 6.8, for positive or negative mode APCI, or negative mode electrospray, and (B) 100% acetonitrile. The typical gradient profile is 0–2 min 1% (v/v) B, 2.0–3.0 min linear gradient to 95% (v/v) B, 3.0–3.9 min 95% (v/v) B, 3.9–4.0 min linear gradient to 1% (v/v) B; and 4.0–4.6 min 1% (v/v) B.

Analytical throughput is enhanced by fast gradient chromatography and pooling of samples postincubation (cassette analysis).[9,10] For cassette analysis, test

[9] D. M. Higton, *Rapid Commun. Mass Spectrom.* **15,** 1922 (2001).

[10] S. A. Wring, L. S. Birkemo, J. W. Polli, D. G. Morgan, J. B. Morgan, L. W. Frick, and C. S. Serabjit-Singh, "Proceedings of the 48th ASMS Conference on Mass Spectrometry and Allied Topics," p. 779. Long Beach, CA, 2000.

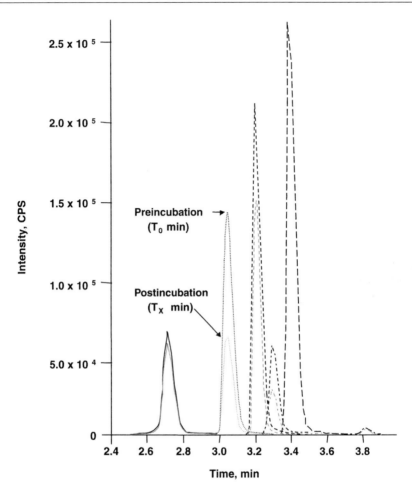

FIG. 1. Tandem mass chromatograms representing peaks for test compounds before (solid line) and after (dashed line) incubation. Peak area ratios were calculated as described in Methods to reflect the degree of metabolism.

compounds are assigned to cassettes to confer the largest difference in m/z (mass : charge) ratio between analytes. Mass differences that may have corresponded to metabolites, such as 16 m/z, are also avoided. This removes isobaric clashes between the different parent ions and their putative metabolites.[9] Cassettes typically contain three to five compounds (Fig. 1); ultimately, the size of cassettes is limited by analytical sensitivity and any constraints required to avoid isobaric clashes. Fast gradient LC and cassette analysis reduce the total number of injections required and shorten the chromatographic run time with no loss in peak shape or selectivity. In most instances, this allows analytical work to be completed within an overnight run following incubation.

Combined Quantitative and Qualitative Accurate Mass Analysis by LC-TOF/MS

Combined quantitative and qualitative bioanalysis is performed on an Applied Biosystems Mariner TOF mass spectrometer operated in the low mass range mode ($m/z < 1000$) to maximize sensitivity. Spectra are obtained over the mass range 100–800 m/z at an acquisition rate of 20 MHz and summed to yield 0.2 sec per spectrum. The mass range corresponds to the compounds under investigation. All data acquisition and reduction are performed using Mariner Systems Software V.4.0. Data mining for metabolite peaks is performed using a custom macro developed by Applied Biosystems. The instrument is calibrated externally using the manufacturer's recommended low mass calibration standards. All bioanalytical procedures are followed as described earlier for LC-MS/MS, except discrete analysis replaces cassette analysis. This increases total analytical run times; however, the cost of the extra analysis time is offset by the availability of a sensitive assay providing accurate mass data.

Calculations for Metabolic Stability

Metabolic stability by LC-MS/MS and LC-TOF/MS, expressed as the amount of compound metabolized, is calculated from the peak area ratio (%) of compound remaining after incubation compared to the preincubation sample (Fig. 1).

$$\text{Compound metabolized (\% remaining)}$$
$$= \frac{\text{Peak area of postincubation } (t_x \text{ min}) \text{ sample}}{\text{Peak area of preincubation } (t_0 \text{ min}) \text{ sample}} \times 100$$

Results and Discussion

Assay Design and Implementation

Before our stage I assay, automation technologies existed for increasing the throughput of complex liquid handling procedures but had not been applied to *in vitro* metabolic stability. Over a period of 4 years our approach has evolved with the flexibility and throughput of the assay being increased in three stages (Fig. 2). Throughput is expressed as compound equivalents (the number of compounds multiplied by the number of sources of enzyme studied). The success rate represents the percentage of compounds assayed with duplicates agreeing within 20% and where peaks for parent compound were resolved from other components within the sample matrix.

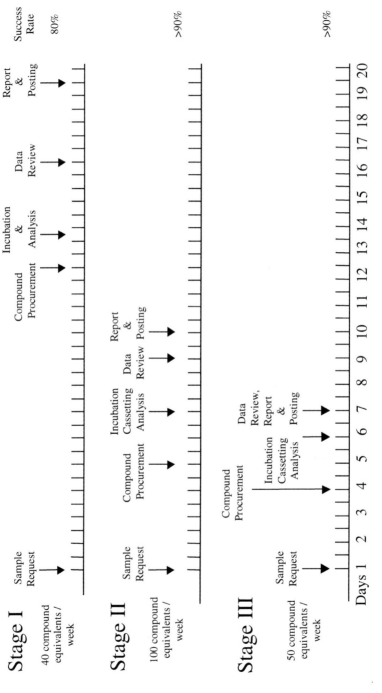

FIG. 2. Development stages of the *in vitro* metabolic stability screen. Stage I, II, and III assays represent the stepwise development of the high-throughput screen. Arrows indicate completion times for each phase of the overall process and illustrate how the turnaround time was shortened from 20 to 7 days.

Stage I. This stage was an automated single time point (30 min) assay with only human hepatic microsomes that provided the capability to study metabolic stability for up to 50 compounds (incubated at 5 μM) per week using an LC-MS end point. We assayed an average of 40 compound equivalents per week sustained over 1 year. The success rate was approximately 80% with an average processing time (turnaround) of 14 days from compound receipt to report. Two full-time employees (FTEs) could provide the needed throughput.

Stage II. Tandem mass spectrometry (LC-MS/MS) with cassette analysis to enhance analytical specificity, sensitivity, and throughput was applied. The assay was expanded to include a single time point (30 min) assay with either the microsomal or the S9 fraction from multiple species (typically rat, human, dog, and/or monkey) or multiple tissues (typically liver and intestine). The nominal capacity of the assay was increased to 125 compounds per week. We assayed an average of 100 compound equivalents per week sustained over 2 years. A typical week would comprise work from 5 to 10 project teams, each supplying 10 to 15 compounds to be studied in two to three species. The success rate was >90% with an average turnaround of 10 days. The equivalent of three FTEs could provide the needed throughput.

Stage III. Changes were made to the automated liquid handling protocol that allowed on-line incubations and collection of multiple time points (0, 15, 30, and 60 min). A more sensitive mass spectrometer (API3000) allowed substrate concentration to be reduced to 1 μM. The nominal capacity was maintained at 125 compound equivalents per week, but the number of data points was increased; this reflected the need of project teams for more detailed information. Despite increasing the number of time points, we maintained the same compound throughput by dropping duplicate analyses. To date, we have assayed an average of 50 compound equivalents per week, and our success rate has remained at >90% with a typical turnaround of 6 days.

The stepwise development of this high-throughput assay involved three major aspects. We increased, yet balanced, the capacity of different steps in the process; we worked within the constraints of resources (instrumentation, software development, and staff); and we responded to the changing needs of drug discovery by broadening the scope of the assay.

In stage I, automation of incubations quickly made bioanalysis rate limiting. The commercial availability of liver microsomes and robust automation of liquid handling in a 96-well plate format made incubations facile, fast, and reproducible. The analytical end point, a standard LC-MS method (not optimized for each test compound), did enhance sensitivity compared to HPLC with UV detection, but had only a minor effect on throughput because selected ion monitoring lacked the selectivity required to shorten chromatographic run times. Calculation of metabolic stability from MS output (data reduction) was a time-consuming manual process. Success rates were typically 80%, with failures being due to inadequate chromatographic separation of the test compound from other components in the incubation matrix.

In stage II the most marked improvement in capacity (Fig. 2) was achieved by a concerted use of tandem mass spectrometry (LC-MS/MS) with automated optimization of instrumental parameters and cassette analysis. Enhanced capacity allowed for incubations with enzyme from multiple species and additional quality control samples (QCs). Table I presents quality control data for three drugs representing low and high metabolism. As expected, imprecision was high in the instance of low turnover; however, this was unimportant as compounds were reliably classified as stable.

Overall, in stage II we enhanced the sensitivity and selectivity of our assay (success rates increased to >90%), albeit at the cost of a more expensive triple quadrupole instrument. We increased throughput and reduced compound receipt-to-report time from 14 to 10 days.

The stage II assay had a nominal capacity of 125 compound equivalents per week, which was greater than demand; requests were rarely >100 compound equivalents per week. Instead, there was a need for more detailed information. Consequently, during stage III we exploited the versatility of the automation to assay at multiple time points and allow calculation of an *in vitro* half-life $(t_{1/2})$.[11,12]

The enhanced analytical sensitivity also allowed us to lower substrate concentrations to below the likely K_m. Microsomes or S9 homogenates from different species or tissues were used to evaluate diverse sets of compounds, up to 100 per week.

Quantitative Metabolic Stability

Figure 3 presents the typical distribution of metabolic stability in human liver S9 homogenates for approximately 1000 diverse compounds across 19 different project teams. Thirty-eight percent of compounds possessed good *in vitro* metabolic stability (<25% metabolized at 60 minutes) and would have been considered for further *in vitro* and/or *in vivo* studies.

The automated screen allows elucidation of structure–metabolic stability relationships within a set of compounds. For instance, the six compounds presented in Table II represent a small subset of a series studied for metabolic stability. GSK A, GSK B, and GSK C were metabolized extensively in incubations with rat, human, and dog liver S9 fractions. Modifications at R_1, R_2, and R_3, yielding GSK D and GSK E, enhanced *in vitro* stability with a concomitant increase in bioavailability in the rat, indicating concordance between *in vitro* and *in vivo* data. *In vivo* data were not available for dog or human. The modification of the R_2 moiety on GSK F yielded

---

[11] S. Obach, *Drug Metab. Dispos.* **27**, 1350 (1999).
[12] S. Obach, *Curr. Opin. Drug Disc. Dev.* **4**, 36 (2001).

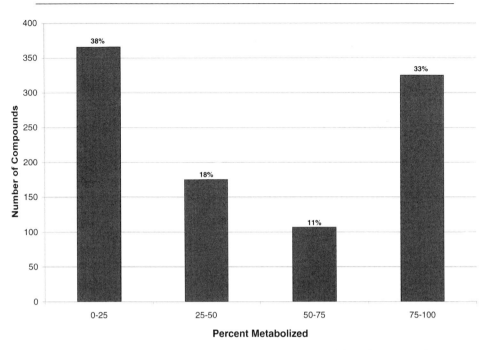

FIG. 3. Distribution of program compounds assayed for metabolic stability using the stage III assay with human liver S9 homogenates. Representative data from 60-min incubations of 973 program compounds (1 μM incubation concentration) from 19 different programs. The label above each bar represents the percentage of compounds in each band of metabolic stability.

a mixed result, revealing the inherent differences among species: GSK F was stable to rat and human metabolism (similar to GSK D) yet labile to metabolism by the dog (similar to that of GSK A, GSK B, and GSK C). *In vitro* data contributed to selection of compound GSK D with greater confidence that rat and dog are reasonable models of human metabolic stability and that the stability is probably sufficient to get adequate human exposure.

Combined Quantitative and Qualitative Accurate Mass Metabolic Stability

The stage III assay evaluates quantitative metabolic stability and could be adapted to generate metabolite profiles, but would require that incubation samples be reanalyzed. TOF mass spectrometers can generate quantitative and qualitative information with no appreciable loss in sensitivity in a single injection.[13] A study comparing quantitative data from the stage III assay and the same incubation samples assayed by LC-TOF/MS yielded general agreement for a diverse

[13] H. Zhang, K. Heinig, and J. Henion, *J. Mass Spectrom.* **35,** 423 (2000).

TABLE II

In Vitro METABOLIC STABILITY MEASURED IN STAGE III ASSAY, AND RELATIVE BIOAVAILABILITY FOR SUBSET OF TEST COMPOUNDS

Compounds	R_1	R_2	R_3	Liver metabolism (% metabolized)[a]			Bioavailability[b] (%F in Rat)
				Rat	Human	Dog	
GSK A	—[c]	—[c]		96	66	99	14
GSK B	F	—[c]		88	63	99	3
GSK C	—[c]	—[c]		99	94	99	7
GSK D	F	CF_3		7	5	4	40
GSK E	F	F		15	43	46	58
GSK F	F	F		7	21	90	—[c]

[a] Liver metabolism is defined as percentage of parent compound metabolized by human S9 at 37° in 1 hr. Each compound was incubated at μM concentration.

[b] F, Bioavailability in rats is defined as AUC_{po}/AUC_{iv}.

[c] — Data not available.

set of marketed central nervous system (CNS) and non-CNS compounds ($R^2 = 0.844$, $n = 75$, range 1–100% metabolism). This indicates that quantitative data by LC-MS/MS and LC-TOF/MS are likely to be similar.

The advantage of LC-TOF/MS is the ability to deduce molecular formulas from the accurate mass of the metabolites generated in the samples incubated at $t_{60\,min}$ compared to control samples at $t_{0\,min}$. Table III presents accurate mass data for verapamil and four major *in vitro* metabolites resolved to four decimal places (versus to typically only unit resolution by the quantitative LC-MS/MS described earlier). Accurate mass data are in concordance with the reported N-dealkylated, O-dealkylated, and hydroxylated metabolites. This demonstrates that while accurate mass data alone do not allow definitive structural elucidation,

TABLE III
MOLECULAR FORMULAS FOR VERAPAMIL AND ITS MAJOR METABOLITES DETERMINED
FROM LC-TOF/MS ACCURATE MASS DATA[a]

Input m/z	Calculated mass	Error		Formula	Isotope match score
		mDa	ppm		
291.2095	291.2067	2.8	9.5	$C_{17}H_{27}N_2O_2$	0.961029
427.2611	427.2591	1.9	4.5	$C_{25}H_{35}N_2O_4$	0.991883
441.2783	441.2748	3.5	7.9	$C_{26}H_{37}N_2O_4$	0.998159
471.2905	471.2854	5.2	11.0	$C_{27}H_{39}N_2O_5$	0.907187
455.2923	455.2904	1.8	4.0	$C_{27}H_{39}N_2O_4$	0.99784

[a] Data on samples generated in the stage III assay. Calculated masses correspond to the following metabolic processes for verapamil: 291.21073, N-dealkylation; 427.25913, 2-O-dealkylation; 441.27478, O-dealkylation; 471.28535, N-oxidation or hydroxylation[b]; and 455.29044, parent verapamil. Limits used for formula assignment: maximum error, ±15 ppm; charge, +1; C (0–27); H (0–45); O (0–6); N (0–2); all other elements (0).

[b] Most likely hydroxylation because chromatographic retention was shorter than that of the parent and N-oxides are typically retained longer than parent by reversed-phase HPLC.

they can suggest molecular formulas for major metabolites and contribute to the design of more metabolically stable candidate drugs.

Conclusions

We have applied automation and high-throughput bioanalysis to *in vitro* metabolic stability screens that, during the three stages of development, have assayed over 10,000 diverse compounds in 4 years. The screen has supported, on average, 24 drug discovery programs per year. Our current fully automated assay has provided metabolic stability data (with *in vitro* half-life) on the equivalent of >1500 compounds in the past year. In the majority of instances, these studies included enzyme from two to three species.

Acknowledgments

The authors acknowledge colleagues, past and present, from the Drug Metabolism and Pharmacokinetic Departments at GlaxoSmithKline who contributed to the methods and experimental work described. In particular, we thank Cindy Rewerts and Lindsey Tyler for developing the automation and performing incubations; Dr. Daniel Morgan, Steven Cook, and Dr. Mary Wells-Knecht for bioanalysis; and David Reynolds for *in vivo* data presented in Table II. We also thank Dr. Charles Liu and Brian Boucher of Applied Biosystems for the loan of the Mariner TOF mass spectrometer and custom software employed for data mining in the qualitative studies.

[28] Characterization of Covalent Adducts to Intact Cytochrome P450s by Mass Spectrometry

By LUKE K. LIGHTNING and WILLIAM F. TRAGER

Introduction

Cytochrome P450s (P450s) comprise a superfamily of heme-containing monooxygenases that have molecular masses (MMs) of approximately 50 kDa. P450 genes have been isolated from a variety of organisms,[1] and the amino acid sequence for each of these isoforms has been established. One of the functions of mammalian P450s is to catalyze the first oxidative step in the elimination pathway of most drugs. Although the secondary structure for these enzymes is known, crystal structure data for mammalian P450s are lacking, with the notable exception of a rabbit P450.[2] The identity and catalytic role of critical active site residues for a particular human P450, therefore, can be determined only through the use of various indirect methods. One such method involves the use of mechanism-based inactivators (MBIs) to identify active site peptides. A MBI of a P450 is a compound that is oxidized to some reactive intermediate that, instead of escaping into the bulk medium, leads to covalent modification of the P450 active site. Our studies focused on (1) characterizing MBIs that attach covalently to an active site amino acid and (2) developing an analytical technique that would allow us to accomplish our goals in an efficient and accurate manner. To this end, we chose electrospray ionization (ESI)-mass spectrometry (MS) as our method of analysis; this chapter describes our use of a rapid, sensitive, and accurate LC/ESI-MS assay to determine the molecular mass of intact P450s prior to and following mechanism-based inactivation by several different compounds.[3–5]

Assay Methods

Instrumentation

LC/ESI-MS is performed on a Fisons VG Quattro II triple quadrupole mass spectrometer fitted with a Megaflow Electrospray source and controlled by a Digital DECpcLPx 466d2 computer running Fisons Mass Lynx 0.1 software. Instrument

[1] D. R. Nelson, L. Koymans, T. Kamataki, J. J. Stegeman, R. Feyereisen, D. J. Waxman, M. R. Waterman, O. Gotoh, M. J. Coon, R. W. Estabrook, I. C. Gunsalus, and D. W. Nebert, *Pharmacogenetics* **6**, 1 (1996).

[2] P. A. Williams, J. Cosme, V. Sridhar, E. F. Johnson, and D. E. McRee, *Mol. Cell* **5**, 121 (2000).

[3] L. L. Koenigs, R. M. Peter, A. P. Hunter, R. L. Haining, A. E. Rettie, T. Friedberg, M. P. Pritchard, M. Shou, T. H. Rushmore, and W. F. Trager, *Biochemistry* **38**, 2312 (1999).

[4] L. L. Koenigs and W. F. Trager, *Biochemistry* **37**, 13184 (1998).

[5] L. K. Lightning and W. F. Trager, submitted for publication.

settings are as follows: source temperature 100°; N_2 drying gas, 150 liter/hr; neb-ulizing gas, 20 liter/hr; probe voltage, 3.8 kV; cone voltage, 42 kV, 30–35 kV; probe position, on axis. Data acquisition is carried out from m/z 200 to 2000 Da in the CONTINUUM scanning mode, with unit resolution up to at least 1500 Da based on calibration and resolution optimization using a mixture of poly(ethylene glycol) 300/600/1000/1500. The LC interface to the mass spectrometer consists of Shimadzu LC10AD solvent delivery modules and a Shimadzu SPD-10AV mod-ule UV–VIS spectrophotometric detector with wavelength detection at 214 and 400 nm. The LC is equipped with a POROS R2 perfusion column (4.6×100 mm) from Perseptive Biosystems (Cambridge, MA).

P450 Expression and Purification

P450s 1A2, 2A6, 2C9, 2C9 C175R, and 3A4 are expressed using a baculovirus/ insect cell system as described previously.[3] Polyhistidine-tagged (HT) P450s cam (P450 cam-HT) and P450 3A4-HT are expressed using an *Escherichia coli* system as described previously.[3] Proteins are purified from rat liver microsomes (P450s 1A1 and 2B1), insect cell paste (P450s 1A2, 2A6, 2C9, 2C9 C175R, and 3A4), or *E. coli* bacterial cultures (P450 reductase, cytochrome b_5, and P450s cam-HT and 3A4-HT) to electrophoretic homogeneity as described previously.[3] Analysis of purified P450s using the LC/ESI-MS system is performed without any further work-up.

Reconstitution and Mechanism-Based Inactivation of P450s

The relative molar amount of P450, P450 reductase, and cytochrome b_5 used in the reconstituted systems is 1 : 2 : 1, 1 : 2 : 1, and 1 : 3 : 1 for P450 2A6, 2B1, and 2C9, respectively. The different ratios of P450:P450 reductase:cytochrome b_5 used are optimized as described previously.[3] The desired amounts of P450, P450 reduc-tase, and cytochrome b_5 are combined and dialyzed overnight against at least 100 volumes of buffer (50 mM potassium phosphate, pH 7.4). This initial dialysis step is performed in an effort to enhance P450 activity by removing any residual glyc-erol and/or detergent present in the purified protein samples. The amount of lipid (L-α-dilaurylphosphatidylcholine) and catalase added to the reconstituted P450 system is 50 μg/nmol P450 and 2000 U/mL final incubation volume, respectively, and this mixture is set on ice for 1 hr. Catalase is added to prevent inactivation of P450 by reactive oxygen species formed during P450 uncoupling.

The amount of P450 used in each mechanism-based inactivation experiment is 1 nmol, and the reaction is performed in potassium phosphate buffer (50 mM, pH 7.4). The reconstituted P450 2A6 and 2B1 enzyme mixtures are preincu-bated at 30° for 2 min with 5-MOP (5-methoxypsoralen) (100 nmol), 8-MOP (8-methoxypsoralen) (100 nmol), or P (psoralen) (100 nmol). The reconstituted P450 2C9 enzyme mixture is preincubated at 30° for 2 min with tienilic acid (500 nmol) and glutathione (GSH, where indicated). Each MBI is added as a

TABLE I
LC/ESI-MS ANALYSIS OF CYTOCHROME P450s[a]

Protein	Molecular mass			
	Experimental	Predicted	Difference	Error (%)
E. coli P450 cam-HT	47366.6 ± 2.0	47361.0	5.6	0.012
Rat P450 1A1	59382.2 ± 14.1	59393.4	11.2	0.019
Bac. P450 1A2	58290.9 ± 5.5	58294.5	−3.6	0.006
Bac. P450 2A6	56544.4 ± 3.2	56541.6	2.8	0.005
Rat P450 2B1	55935.7 ± 7.3	55933.8	1.9	0.003
Bac. P450 2C9	55578.6 ± 0.5	55575.1	3.5	0.006
Bac. P450 2C9 C175R	55633.9 ± 3.9	55628.1	5.8	0.010
Bac. P450 3A4	57313.9 ± 7.6	57299.3	14.6	0.025
E. coli P450 3A4-HT	57137.7 ± 0.3	57143.1	−5.4	0.009

[a] Adapted with permission from L. L. Koenigs, R. M. Peter, A. P. Hunter, R. L. Haining, A. E. Rettie, T. Friedberg, M. P. Pritchard, M. Shou, T. H. Rushmore, and W. F. Trager, *Biochemistry* **38**, 2312 (1999). Copyright © 1999. American Chemical Society.

solution in methanol, but the final concentration of methanol in the incubation is less than 1%. The reactions are initiated by the addition of NADPH (500 nmol) and allowed to proceed for 1 hr at 30° to ensure complete enzyme inactivation. No further treatment is necessary, and the inactivated enzyme mixtures are kept on ice prior to injection onto the LC/ESI-MS system for determination of the molecular mass of the intact P450.

LC/ESI-MS Analysis of Intact P450s

For analysis of intact P450s prior to and following mechanism-based inactivation (see Tables I and II), 300 pmol and 1 nmol of spectrally detectable P450, respectively, are injected onto the LC/ESI-MS system. There is no need to pretreat the purified P450 samples or reconstituted enzyme mixtures prior to injection on the LC column. The flow rate of the LC is 3 ml/min with approximately 50 μl (2%) directed to the mass spectrometer using a splitter. The solvent gradient consists of buffer A [0.05% CF$_3$COOH (TFA, pH 3.0)] and buffer B [0.05% TFA in 95 : 5 acetonitrile (ACN) : H$_2$O (pH 2.5)]. The proteins and other components present in the purified P450 samples and reconstituted enzyme mixtures are separated using a gradient elution of 35–50% buffer B from 0 to 5 min, an isocratic elution at 50% buffer B from 5.0 to 6.5 min, and a gradient elution of 50–100% buffer B from 6.5 to 10 min. Under these conditions, heme elutes at 2.4 min, P450 reductase elutes at 4.0 (truncated form) and 5 min (full-length form), and P450 elutes at approximately 8 min. Data acquisition is carried out from m/z 200 to 2000 Da in 4.5 sec, and the scans across the HPLC peak are summed to give an ESI mass

TABLE II

LC/ESI-MS ANALYSIS OF COVALENTLY MODIFIED CYTOCHROME P450s[a]

	Experimental molecular mass			
P450 + MBI	P450-MBI	P450	Difference	Predicted difference[c]
P450 2A6 + 8-MOP	56774.1 ± 7.9	56538.4 ± 10.0	235.6 ± 3.8	232.2
P450 2A6 + 5-MOP	56780.3 ± 7.3	56542.5 ± 8.3	237.8 ± 4.0	232.2
P450 2A6 + P	56749.0 ± 4.6	56542.7 ± 4.4	206.5 ± 8.2	202.2
P450 2B1 + 8-MOP	56163.6 ± 7.4	55925.7 ± 2.2	237.9 ± 9.6	232.2
P450 2B1 + 5-MOP	56164.1 ± 6.7	55934.8 ± 4.3	240.9 ± 6.2	232.2
P450 2B1 + P	56123.9 ± 3.4	55919.1 ± 8.6	204.8 ± 11.8	202.2
P450 2C9 + tienilic acid	55923.0 ± 1.1[b]	55578.6 ± 0.5	344.4 ± 1.1	349.0
	56273.0 ± 4.4		694.4 ± 4.2	698.0

[a] Adapted with permission from L. L. Koenigs, R. M. Peter, A. P. Hunter, R. L. Haining, A. E. Rettie, T. Friedberg, M. P. Pritchard, M. Shou, T. H. Rushmore, and W. F. Trager, *Biochemistry* **38,** 2312 (1999); L. L. Koenigs, and W. F. Trager, *Biochemistry* **37,** 13184 (1998); and L. K. Lightning and W. F. Trager, submitted for publication.

[b] The two experimentally determined MMs for the covalently modified P450 2C9 reflect the formation of a monoadduct and diadduct of P450 2C9 after incubation with tienilic acid. Formation of the diadduct was not observed in the presence of GSH (10 mM).

[c] The predicted MM difference was calculated by assuming the addition of a furanoepoxide and thiophene epoxide on incubation of the P450 with furanocoumarin and tienilic acid, respectively.

spectrum that is subsequently deconvoluted using the MaxEnt computer program. These experiments are performed on at least three different days and are highly reproducible. The intraday variability in the determination of the molecular masses of intact P450s and P450 reductase is found to range from ±0.002 to 0.035%. The standard deviation of the calculated MM based on the charge state distribution observed is less than ±0.09%. The average predicted MMs are calculated based on the amino acid sequences of the P450s, and the MM range for the eight mammalian P450s is 3818.3 Da.

Remarks

The dialysis step performed prior to exposure of the P450 to the mechanism-based inactivator was necessary to achieve an optimal level of P450 activity and, as a consequence, a binding stoichiometry of >0.5 : 1. We suspect that this might be due to the removal of various detergents or glycerol that remained after purification or the increased time that the enzymes were allowed to preincubate. However, because we were never able to achieve a 1 : 1 binding stoichiometry, it was necessary to inject a substantial amount (1 nmol) of P450 after inactivation in order to produce mass spectra with acceptable signal-to-noise ratios. We found that the R2 POROS column was significantly better than standard C$_4$ HPLC columns in separating the components of the reconstituted system in the relatively short periods of time (<10 min) we desired. Due to the high flow rates required for the fast separation of the proteins in the reconstituted mixture, it was necessary to split the flow so that only 2% of the entire flow was delivered to the mass spectrometer. Thus, a significant portion of the P450 injected was discarded during each run. It is likely that this experimental necessity can now be avoided by using the recently introduced capillary POROS column.[6] This column allows for all of the flow to be directed to the mass spectrometer without any effect on resolution or HPLC run time. Therefore, the amount of P450 necessary for this type of analysis with use of this column can be decreased significantly (100-fold). As stated earlier, residual amounts of glycerol or detergent from our purified P450s remained despite repeated dialysis steps. The POROS column was found to be both necessary and effective in removing these contaminants prior to the introduction of the protein sample into the mass spectrometer. Presumably, the residual impurities led to the decreased sensitivity we observed if the purified P450 sample was injected directly into the mass spectrometer without an initial POROS column cleanup. Therefore, we highly recommend use of this LC column (or its capillary counterpart) to achieve the highest level of sensitivity.

Conclusions

An LC/ESI-MS procedure has been developed to analyze small amounts of intact purified P450 (P450s cam-HT, 1A1, 1A2, 2A6, 2B1, 2C9, 2C9 C175R, 3A4, 3A4-HT) in a short period of time. The difference between experimentally determined (based on experiments performed on at least three different days) and predicted (based on the amino acid sequence) MMs of the P450s ranged from 0.002 to 0.025%. Each experimentally determined MM possessed a standard deviation of less than 0.09% (based on the charge state distribution). In addition, this technique proved useful for identifying the monoadducts of P450 2A6 and 2B1 and 8-MOP, 5-MOP, or P, as well as the mono- and diadducts of P450 2C9 and tienilic acid following mechanism-based inactivation. To summarize, this analytical method was found to be sensitive, rapid, accurate, and one that could be used effectively for the analysis of intact P450s prior to and following mechanism-based inactivation by various compounds. Because it does possess these properties, this procedure should prove useful in high-throughput studies involving covalent adducts of P450s and the characterization of mutant forms.

Acknowledgments

This work was supported by the National Institutes of Health Grant GM32165 to W.F.T. and a Dorothy Danforth Compton graduate student fellowship to L.K.L.

[6] U. Hoch and P. R. Ortiz de Montellano, *J. Biol. Chem.* **276**, 11339 (2001).

[29] Mutagenesis Testing Based on Bacterial Expression of Human P450s

By TETSUYA KAMATAKI and KEN-ICHI FUJITA

Introduction

Many promutagens, present in our environment, require metabolic activation by so-called drug-metabolizing enzymes, including cytochrome P450 (CYP), to exert their genotoxicity.[1] The hepatic CYP system consists of a number of CYPs[2]; each form of CYP shows some substrate specificity to activate promutagens to their reactive intermediates capable of causing an inheritable alteration in the DNA of a cell, namely, a mutation.[3] Thus, it is important to clarify the roles of the CYP form(s) in the mutagenic activation of promutagens because the catalytic property and the population of CYP determine the formation of reactive metabolites in the body. Because the catalytic properties of CYP even in the same family vary among animal species, it is necessary to use human CYP to predict the bioactivation of chemicals in humans.

The Ames test has received the greatest attention as one of the short-term assay systems to elucidate the mutagenicity of chemicals. However, in the classical Ames test, the $9000g$ supernatant fraction of homogenates (S9) prepared from rodent liver is added to the reaction mixture as a source of enzymes activating promutagens.[4] To overcome the species difference in the properties of rodent CYP and human CYP, it may be possible to use human S9 in the Ames test, even though the use of the human preparation is limited for the following reasons: the population of each form of CYP varies according to the medical background of donor subjects and the level of CYPs in these materials is sometimes too low to examine the mutagenic activation of chemicals.

Another source of human CYP has been considered to be the heterologous expression system. Efforts have been made to establish genetically engineered *Escherichia coli* cells expressing human CYP because of an advantage of a high yield of CYP protein. Subsequent to the study demonstrating the successful expression of bovine CYP17A in *E. coli* cells,[5] the use of *E. coli* cells has become

[1] J. A. Miller, *Cancer Res.* **30,** 559 (1970).
[2] D. R. Nelson, L. Koymans, T. Kamataki, J. J. Stegeman, R. Feyereisen, D. J. Waxman, M. R. Waterman, O. Gotoh, M. J. Coon, R. W. Estabrook, I. C. Gunsalus, and D. W. Nebert, *Pharmacogenetics* **6,** 1 (1996).
[3] F. P. Guengerich, *Cancer Res.* **48,** 2946 (1988).
[4] D. M. Maron and B. N. Ames, *Mutat. Res.* **113,** 173 (1983).
[5] H. J. Barnes, M. F. Arlotto, and M. R. Waterman, *Proc. Natl. Acad. Sci. U.S.A.* **88,** 5597 (1991).

popular to express human CYP and NADPH-CYP reductase (OR).[6–10] The genetically engineered *E. coli* cells were used successfully for studies of drug metabolism. Therefore, we expected that the technology to express the human CYP and the OR in *E. coli* cells could also be applied to the *Salmonella typhimurium* cells used for the Ames test to examine the mutagenic activation of chemicals by human CYP.[11,12] *S. typhimurium* tester strains YG7108 and TA1538 sensitive to base pair substitutions and frameshift mutations, respectively, are employed to express each form of human CYP and the OR. The *S. typhimurium* TA1538 tester strain co-expressing human CYP1A2, OR, and *S. typhimurium* *O*-acetyltransferase (OAT) has also been developed to detect the activation of promutagens such as heterocyclic amines,[12] as these promutagens undergo activation via *N*-hydroxylation by CYP1A2, followed by OAT.[13]

The activation of promutagens by human CYP was detectable with the genetically engineered *S. typhimurium* cells expressing human CYP with a high sensitivity. For example, the mutagenic activation of *N*-nitrosamine by human CYP was detectable with the established *S. typhimurium* YG7108 cells at a micromolar level.[12] The metabolic activation of heterocyclic amines was detectable with the genetically engineered *S. typhimurium* TA1538 expressing human CYP1A2 and OR at a concentration lower than 1 μM. Furthermore, the coexpression of OAT together with CYP1A2 and OR in *S. typhimurium* TA1538 cells resulted in the detection of the mutagenicity of heterocyclic amines at the picomolar level.[13] One of the reasons for the high sensitivity has been thought to be that the incubation of promutagens with genetically engineered *S. typhimurium* cells expressing a high level of CYP produced a high amount of reactive metabolite(s) inside the bacterial cells. The reactive intermediate(s) thus formed might bind to the bacterial DNA easily, inducing a higher level of gene mutation.

The purpose of this article is to introduce the methods to establish the genetically engineered *S. typhimurium* tester strain expressing human CYP and OR

[6] C. W. Fisher, M. S. Shet, D. L. Caudle, A. Cheryl, and C. A. Martin-Wixtrom, *Proc. Natl. Acad. Sci. U.S.A.* **89,** 10817 (1992).

[7] J. Dong and T. D. Porter, *Arch. Biochem. Biophys.* **327,** 254 (1996).

[8] J. A. R. Blake, M. Pritchard, S. Ding, G. C. M. Smith, B. Burchell, C. R. Wolf, and T. Friedberg, *FEBS Lett.* **397,** 210 (1996).

[9] H. Iwata, K. Fujita, H. Kushida, A. Suzuki, Y. Konno, K. Nakamura, A. Fujino, and T. Kamataki, *Biochem. Pharmacol.* **55,** 1315 (1998).

[10] M. P. Pritchard, M. J. Glancey, J. A. R. Blake, D. E. Gilham, B. Burchell, R. C. Wolf, and T. Friedberg, *Pharmacogenetics* **8,** 33 (1998).

[11] H. Kushida, K. Fujita, A. Suzuki, M. Yamada, T. Nohmi, and T. Kamataki, *Mutat. Res.* **471,** 135 (2000).

[12] A. Suzuki, H. Kushida, H. Iwata, M. Watanabe, T. Nohmi, K. Fujita, F. J. Gonzalez, and T. Kamataki, *Cancer Res.* **58,** 1833 (1998).

[13] R. Kato, T. Kamataki, and Y. Yamazoe, *Environ. Health Persp.* **49,** 21 (1983).

(and the OAT) and to describe the procedure of mutation assay with the established
S. typhimurium tester strain.

Expression Plasmids

The pCW vector (Dr. Rick Dahlquist, University of Oregon, OR), containing
two *tac* promoters induced by isopropyl-β-D($-$)-thiogalactopyranoside (IPTG)
at the upstream of an *Nde*I restriction enzyme cloning site coincident with the
initiation ATG codon, is used to construct the expression plasmid carrying cDNAs
of human CYP and OR. The vector also contains a *trpA* transcription terminator
sequence to prevent a read-through, a phage M13 origin of DNA replication, and
the *lacI*^{q} gene encoding the *lac* repressor that inhibits transcription from the *tac*
promoters in the absence of inducing agents.

Prior to the insertion of CYP cDNA into the pCW expression vector, the 5'
terminus of CYP cDNAs is modified to achieve a high level of expression following
the methods described previously.[9,14,15] Each modified CYP cDNA contains the
*Nde*I restriction enzyme site at the initiation ATG codon to conveniently ligate
into the cloning site of the pCW vector. A restriction enzyme site (*Hind*III, *Sal*I,
or *Xba*I) is also introduced into the 3'-franking region of the CYP cDNA close
to the stop codon (within 10 bp) to insert the CYP cDNA into the pCW vector,
as the expression vector possesses these cloning sites downstream of the *Nde*I
cloning site. All modifications are introduced by polymerase chain reaction (PCR)
mutagenesis.

Modified N-terminal amino acid sequences and cDNA sequences for each form
of CYP are shown in Fig. 1. Each CYP cDNA thus modified is ligated into the
pCW vector. The blunt-ended human OR cDNA is inserted into the *Msc*I cloning
site of the pSE420 expression vector (Invitrogen, Carlsbad, CA). Then the plasmid is digested with *Ssp*I to prepare a DNA fragment carrying a *trc* promoter,
the OR cDNA, and a terminator. The cDNA fragment thus obtained is ligated
into pCW carrying human CYP at the *Bst*1107I cloning site to construct a coexpression plasmid carrying human CYP and OR (pCYP/OR). The structure of the
coexpression plasmid is shown is Fig. 2.

The OAT cDNA is inserted into the *Eco*RV site of a tetracycline-resistant
gene in a pACYC184 vector carrying tetracycline- and chloramphenicol-resistant
genes.[16]

---

[14] T. Shimada, R. M. Wunsch, I. H. Hanna, T. R. Sutter, F. P. Guengerich, and E. M. J. Gillam, *Arch.
Biochem. Biophys.* **357**, 111 (1998).
[15] E. M. J. Gillam, Z. Guo, Y. F. Ueng, H. Yamazaki, I. Cock, P. E. B. Reilly, W. D. Hooper, and F. P.
Guengerich, *Arch. Biochem. Biophys.* **317**, 374 (1995).
[16] A. C. Y. Chang and S. N. Cohen, *J. Bacteriol.* **134**, 1141 (1978).

1A1: Native   M L F P I S M S A T
      Modified M A F P I S M S A T
      atg gct ttt cca att tca atg tca gca acg

1A2: Native   M A L S Q S V P F S A T E L L L A S A I F C L
      Modified M A L   L L A V F L F C L
      atg gct ctg   tta tta cga gtt ttt ctg ttc tgc ctg

1B1: Native   M G T S L S P N D P W
      Modified M L S P N D P W P L N
      atg ctt tct tca aat gat ccg tgg ccg cta aac

2A6: Native   M L A S G M L L V A L L V C L T V M V L M S V W Q Q R K S K G K L P P G
      Modified M A   R Q V H S S W N L P P G
      atg gct   cgt caa gtt cat tct tct tgg aat ctg cct ccg gga

2C8: Native   M E P F V V L V L C L S F M L L F S L W R Q S C R R R K L P P G
      Modified M A   R Q V H S S W N L P P G
      atg gct   cgt caa gtt cat tct tct tgg aat ctg cct cct ggc

2C9: Native   M D S L V V L V L C L S C L L L L S L W R Q S S G R G K
      Modified M A   R Q S S G R G K
      atg gca   acg caa tct tct gga cga gga aaa

2C19: Native   M D P F V V L V L
      Modified M A L L L A V F L
      atg gct ctg tta tta gca gtt ttt ctc

2D6: Native   M G L E A L V P L A V I V A I F L L L V D L M H R R Q R W A A R Y
      Modified M A   R Q V H S S W N L
      atg gct   cgt caa gtt cat tct tct tgg aat ctg

2E1: Native   M S A L G V T V A L L V W A A F L L L V H S M W R Q V S S W N L P
      Modified M A   R Q V S S W N L P
      atg gct   cgt caa gtt cat tct tct tgg aat ccc

3A4: Native   M A L I P D L A M E T W L L L A V S L
      Modified M A L   L L A V S L
      atg gtc ctg   tta tta gca gtt ttt ctg

3A5: Native   M D L I P N L A V E T W L L L A V S L
      Modified M A L   L L L A V F L
      atg gct ctc   ctg tta tta gca gtt ttt ctg

FIG. 1. Modification of N-terminal amino acid and nucleotide sequences for the expression of each form of human CYP in *S. typhimurium* cells.

Salmonella typhimurium Tester Strains

Base pair substitutions are detectable with *S. typhimurium* YG7108.[17] The *S. typhimurium* YG7108, derived from TA1535, lacks two O^6-methylguanine-DNA methyltransferase genes, *ada* and *ogt*, showing a high sensitivity to alkylating agents such as *N*-nitrosamines.

Salmonella typhimurium TA1538 was developed by Ames *et al.*[18] and is used to detect frameshift mutations in the classical Ames test.

[17] M. Yamada, B. Sedgwick, T. Sofuni, and T. Nohmi, *J. Bacteriol.* **177,** 1511 (1995).
[18] B. N. Ames, F. D. Lee, and W. E. Durston, *Proc. Natl. Acad. Sci. U.S.A.* **70,** 782 (1973).

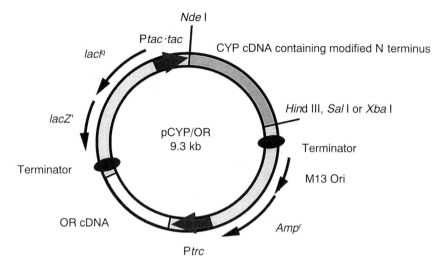

FIG. 2. Structure of the coexpression plasmid for human CYP and OR.

Transfection Procedure

The expression plasmid pCYP/OR is first modified by introduction into *S. typhimurium* LB5000 (R^-M^+) cells to prevent digestion of the plasmid by restriction enzymes present in the *S. typhimurium* YG7108 or TA1538.[19] The *S. typhimurium* cells were transfected with the plasmid using the electroporation method. The competent cells of *S. typhimurium* LB5000 are prepared as follows. Bacterial cells stored at −80° are cultured in a Luria–Bertani medium (10 ml) at 37° for 12 hr. Twenty five microliters of the culture is inoculated into a magnesium-free SOB medium and incubated at 37°.[20] The growth of the culture is measured by light scattering at 600 nm, namely absorbance at 600 nm (A_{600}). When the A_{600} reaches a value of 0.5, the bacterial cells are pelleted by centrifugation at 2150g for 5 min at 4°, and the supernatant is decanted. The precipitation is suspended into 1 ml of 10% (v/v) glycerol (Merck, Darmstadt, Germany). The suspension thus obtained is again centrifuged at 2150g for 5 min at 4°, and the bacterial cells are resuspended into 100 μl of the 10% (v/v) glycerol.

A cell porator system Gene-Pulser (Bio-Rad, Hercules, CA) is used for electroporation. Twenty microliters of the suspension of *S. typhimurium* LB5000 competent cells is mixed with 2 μl of the 50-ng/μl pCYP/OR plasmid solution. The

[19] L. R. Bullas and J. Ryu, *J. Bacteriol.* **156**, 471 (1983).
[20] J. Sambrook, E. F. Fritsch, and T. Maniatis, "Molecular Cloning: A Laboratory Manual," 2nd Ed. Cold Spring Harbor Laboratory Press, Cold Spring Harbor, NY, 1989.

mixture is loaded into a disposable electroporation chamber (Bio-Rad), and then electroporation is performed at 1.5 kV/mm at 4°. The mixture is transferred into an SOC medium (1 ml) and incubated at 4° for 70 min with vigorous shaking.[20] After incubation, a 2-μl portion of the mixture is plated onto a Luria–Bertani plate supplemented with ampicillin (25 μg/ml). The pCYP/OR plasmids thus modified in *S. typhimurium* LB5000 cells are extracted from *S. typhimurium* cells and then introduced into *S. typhimurium* YG7108 or TA1538 cells by electroporation according to the following methods. Competent cells of the *S. typhimurium* YG7108 and TA1538 are prepared according to the method described earlier. The conditions of electrotransformation are the same as those for the introduction of pCYP/OR into *S. typhimurium* LB5000 competent cells, except that the electric field is 1.0 kV/mm for the transformation of *S. typhimurium* YG7108 competent cells. Introduction of the pACYC184 carrying OAT into *S. typhimurium* TA1538 cells is performed by the same procedure as mentioned earlier, except that the *S. typhimurium* cells are plated onto a Luria–Bertani plate supplemented with chloramphenicol (10 μg/ml) after electroporation.

Expression Procedure

Expression of CYP and OR in S. typhimurium YG7108 or TA1538

Expression of each form of CYP and OR in genetically engineered *S. typhimurium* YG7108 or TA1538 cells is performed as follows. The culture conditions vary depending on the form of CYP to achieve the high levels of expression. Twenty microliters of a bacterial stock suspension is inoculated into 5 ml of a nutrient broth (Difco, Detroit, MI) supplemented with ampicillin (50 μg/ml). Cultures are incubated overnight with shaking at 37°. Two hundred microliters of the culture is used to inoculate 200 ml of a modified Terrific Broth medium reported by Sandhu *et al.*[21] containing 0.5 mM δ-aminolevulinic acid in a 500-ml flask, except for the expression of CYP2C8 in *S. typhimurium* YG7108 cells. For the expression of CYP2C8 in *S. typhimurium* YG7108 cells, 2.0 ml of the culture is inoculated into 200 ml of the modified Terrific Broth.

Optimal culture conditions to express each form of human CYP and OR in genetically engineered *S. typhimurium* YG7108 cells are as follows. The bacteria are grown with shaking at 25° (to express CYP1A1, CYP1B1, CYP2A6, CYP2C9, CYP2C19, CYP3A4, or CYP3A5) or 30° (to express CYP1A2, CYP2C8, or CYP2D6) for 8 hr prior to induction with 1.5 mM IPTG. Recombinant proteins are expressed following a further incubation. For the expression of CYP1A1 or CYP1B1, the culture is incubated at 30° for 18 hr with shaking. Bacteria are grown at 30° for 24 hr to express CYP2A6, CYP2C9, CYP2C19, CYP3A4, or CYP3A5.

---

[21] P. Sandhu, T. Baba, and F. P. Guengerich, *Arch. Biochem. Biophys.* **306,** 443 (1993).

The bacterial cell culture is performed at 25° for 18 hr to express CYP1A2 and for 24 hr to express CYP2C8 or CYP2D6, respectively. The expression of CYP2E1 is performed by an incubation of the culture at 37° for 2 hr prior to the induction by IPTG and a further incubation at 30° for 18 hr.

The genetically engineered *S. typhimurium* TA1538 cells are grown with shaking at 25° (to express CYP1A2, CYP2D6, and CYP2E1) or 30° (to express CYP1A1, CYP2A6, CYP2C8, CYP2C9, CYP2C19, CYP3A4, and CYP3A5) for 8 hr prior to the addition of 1.5 mM IPTG. Recombinant proteins are expressed following a further incubation for 18 hr, except for CYP2C19 (24 hr). The temperature of a culture is the same as that of a culture before the induction with IPTG.

The modified Terrific Broth medium consists of a potassium phosphate buffer (pH 5.5) and potassium phosphate buffer (pH 6.0) to express CYP2E1 and CYP2D6 in the *S. typhimurium* YG7108 and TA1538 cells, respectively. The Terrific Broth medium does not contain 0.4% (v/v) glycerol for the expression of CYP2A6 and CYP2E1 in the *S. typhimurium* YG7108 and TA1538 cells.

After the culture, the content of CYP in the bacterial cells is determined by $Fe^{2+} \cdot CO$ versus Fe^{2+} difference spectra, according to the method of Omura and Sato.[22] Five milliliters of bacterial suspension is recovered and centrifuged at 2150g at 4° for 10 min. The remainder of the bacterial culture is stored at 4° until use for the mutation assay. The pellets of *S. typhimurium* cells are resuspended in 2.5 ml of a 100 mM potassium phosphate buffer (pH 7.4) containing 0.2% (v/v) emulgen 911 (Kao, Tokyo, Japan) and 20% (v/v) glycerol (Kanto Chemical, Tokyo, Japan). Difference spectra are recorded by a UV–visible spectrophotometer (Model U-3000, Hitachi, Tokyo, Japan).

The OR activity in bacterial cells is measured with cytochrome c (Wako Pure Chemicals, Osaka, Japan) as an electron acceptor by measuring the absorbance change at 550 nm at 20° according to the method of Phillips and Langton.[23] Bacterial cells are disrupted by sonication prior to determination of the OR activity.[24] The unit of OR is defined as the amount of the enzyme that reduces 1 μmol of cytochrome c/min. Tables I and II show the typical expression levels of each form of human CYP and OR in *S. typhimurium* YG7108 and TA1538 cells, respectively.

Expression of CYP1A2, OR, and OAT in S. typhimurium TA1538

Twenty microliters of a bacterial stock suspension is inoculated into 10 ml of a nutrient broth supplemented with ampicillin (25 μg/ml) and chloramphenicol

[22] T. Omura and R. Sato, *J. Biol. Chem.* **239**, 2370 (1964).
[23] A. H. Phillips and R. G. Langton, *J. Biol. Chem.* **237**, 2370 (1962).
[24] P. Sandhu, Z. Guo, T. Baba, M. V. Martin, R. H. Turky, and F. P. Guengerich, *Arch. Biochem. Biophys.* **309**, 168 (1994).

TABLE I

EXPRESSION LEVELS OF HUMAN CYP AND OR IN GENETICALLY
ENGINEERED *S. typhimurium* YG7108 CELLS[a]

Source	CYP (nmole/liter culture)	OR[b] (units/liter culture)
CYP1A1/OR	93 ± 11	370 ± 140
CYP1A2/OR	140 ± 4	410 ± 56
CYP1B1/OR	200 ± 83	1500 ± 400
CYP2A6/OR	160 ± 20	540 ± 140
CYP2C8/OR	95 ± 5	1000 ± 200
CYP2C9/OR	140 ± 66	930 ± 100
CYP2C19/OR	62 ± 30	210 ± 15
CYP2D6/OR	120 ± 19	600 ± 110
CYP2E1/OR	28 ± 4	640 ± 88
CYP3A4/OR	170 ± 27	790 ± 250
CYP3A5/OR	120 ± 6	560 ± 160

[a] Mean $\pm$ SD ($n = 3$).
[b] The unit of OR is defined as the amount of the enzyme that reduced 1 μmol cytochrome c per minute.

(10 μg/ml) to confirm whether the genetically engineered *S. typhimurium* cells harbor pCYP/OR and pACYC184 carrying OAT. Cultures are grown with shaking at 37° for 12 hr. Two milliliters of the culture is inoculated into 200 ml of the modified Terrific Broth and grown with shaking at 25° for 8 hr prior to induction with 1 mM IPTG. The expression of recombinant proteins is achieved by a further

TABLE II

EXPRESSION LEVELS OF HUMAN CYP AND OR IN GENETICALLY
ENGINEERED *S. typhimurium* TA1538 CELLS[a]

Source	CYP (nmole/liter culture)	OR[b] (units/liter culture)
CYP1A1/OR	54 ± 31	290 ± 100
CYP1A2/OR	300 ± 70	410 ± 44
CYP1B1/OR	130 ± 31	390 ± 66
CYP2A6/OR	200 ± 40	630 ± 260
CYP2C8/OR	84 ± 39	570 ± 150
CYP2C9/OR	170 ± 49	590 ± 59
CYP2C19/OR	84 ± 40	570 ± 320
CYP2D6/OR	120 ± 89	430 ± 340
CYP2E1/OR	32 ± 5	410 ± 260
CYP3A4/OR	69 ± 36	310 ± 200
CYP3A5/OR	72 ± 61	290 ± 27

[a] Mean $\pm$ SD ($n = 3$).
[b] The unit of OR is defined as the amount of the enzyme that reduced 1 μmol cytochrome c per minute.

incubation at 25° for 12 hr with shaking. The expression levels of CYP1A2 and OR are determined by the methods described earlier. The level of OAT in the cytosol fraction of *S. typhimurium* is determined by measuring *N*-acetyltransferase activity with isoniazid as a substrate, as *S. typhimurium* OAT had both OAT and *N*-acetyltransferase activities.[25] The cytosol fraction is prepared according to the method of Sandhu *et al.*[24] The protein concentration of the cytosolic fraction is determined by the method of Lowry *et al.*[26] A typical *N*-acetyltransferase activity of the OAT expressed in the cytosol fraction of the *S. typhimurium* is 136 nmol/mg protein.[12]

Mutation Assay

The assay is carried out as described by Maron and Ames[4] with modifications. If a promutagen is insoluble in aqueous solution, dimethyl sulfoxide (DMSO) is used as a solvent to dissolve the chemical. In general, the amount of DMSO in a reaction mixture is restricted up to 1% (v/v) to prevent the inhibition of CYP activity by DMSO. One hundred microliters of a promutagen solution is mixed with 500 μl of 0.1 M sodium phosphate buffer (pH 7.4) in an incubation tube.

Before the mutation assay, we determine the expression level of CYP in the genetically engineered *S. typhimurium* YG7108 or TA1538 cells. According to our experience, the OR (and OAT) is expressed in the bacterial cells when CYP is expressed in the cells. Then, the number of *S. typhimurium* cells is adjusted to give $1-2 \times 10^9$ cells/ml by dilution with a nutrient broth. One hundred microliters of the cell suspension is added to each assay to start the activation reaction by human CYP. Bacterial cells expressing human CYP and OR are preexposed to a promutagen at 37° for 20 min. An NADPH-generating system is not added to a reaction mixture, as the mutagenic activation of promutagens is not affected by the addition of NADPH, probably because NADPH present in the bacterial cells is utilized as an electron donor. Immediately after preincubation, the reaction mixture is poured onto a minimal glucose plate with a top agar at 45°. The plates are incubated at 37° for 2 days before the number of His$^+$ revertant per plate is counted. Assays are carried out in duplicate at each concentration of a promutagen. Results are evaluated as mutagenic activation of a promutagen by human CYP. As an example, the result of a mutation assay with 2-amino-3,4-dimethylimidazo[4,5-*f*]quinoline (MeIQ) and *S. typhimurium* TA1538 cells coexpressing human CYP1A2, OR, and OAT is shown in Fig. 3. The mutagenic activation of MeIQ is detectable with *S. typhimurium* cells at the concentration of 0.3 pM.

[25] M. Watanabe, M. Ishidate, and T. Nohmi, *Mutat. Res.* **234,** 337 (1990).
[26] O. H. Lowry, N. J. Rosenbrough, A. L. Farr, and R. J. Randall, *J. Biol. Chem.* **193,** 265 (1951).

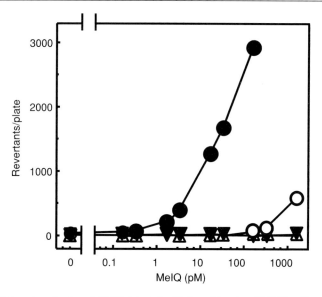

FIG. 3. Mutagenic activation of MeIQ detected with *S. typhimurium* TA1538 cells coexpressing human CYP1A2, OR, and OAT. ●, *S. typhimurium* TA1538 cells expressing CYP1A2, OR, and OAT; ○, *S. typhimurium* TA1538 cells expressing CYP1A2 and OR; ▼, *S. typhimurium* TA1538 cells expressing OAT; △, *S. typhimurium* TA1538 cells harboring control plasmids.

When test chemicals used are expected to be mutagenic and carcinogenic, extreme care should be taken to prevent human exposure.

Acknowledgments

We thank Dr. Rick Dahlquist (University of Oregon, Oregon) for supplying the pCW vector, and we also thank Dr. Takehiko Nohmi (National Institute of Health Sciences, Tokyo, Japan) for providing the *Salmonella typhimurium* LB5000, TA1538, and YG7108 strains. This study was supported in part by a Grant-in-Aid from the Ministry of Education, Science, Sports and Culture of Japan, by a grant (No. 99-2) from the Organization for Pharmaceutical Safety and Research (OPSR), by Grants-in-Aid for Cancer Research from the Ministry of Health and Welfare of Japan, and by a fund under a contract with the Environment Agency of Japan, and in part by a grant from the Smoking Research Foundation.

[30] Use of Long-Term Cultures of Human Hepatocytes to Study Cytochrome P450 Gene Expression

By Lydiane Pichard-Garcia, Sabine Gerbal-Chaloin, Jean-Bernard Ferrini, Jean-Michel Fabre, and Patrick Maurel

Introduction

Prediction of drug-mediated cytochrome P450 (CYP) enzyme induction has been rendered mandatory by drug regulation agencies because of the possible impact of this process on pharmacokinetics, side effects, and toxicity of drugs.[1] Due to species specificity, animal studies are not fully reliable in this respect.[2] Instead, primary cultures of normal adult human hepatocytes have been shown to represent the best *in vitro* model to investigate CYP gene expression.[3–5] However, the limited and unpredictable availability of human liver necessitates efficient storage conditions, allowing the whole cellular functions of these cells to be maintained. Cryopreservation of freshly isolated hepatocytes has been developed.[6–9] However, despite significant recent improvements, available data suggest that the normal phenotype of cryopreserved hepatocytes is not fully conserved after thawing.[10]

As an alternative to cryopreservation, we have developed long-term primary cultures (LTPC) of human hepatocytes.[11,12] Indeed, such cultures, if successful, would reduce the need for liver tissue, would allow repeating experiments with the

[1] G. T. McInnes and M. J. Brodie, *Drugs* **36**, 83 (1988).

[2] S. A. Kliewer, J. T. Moore, L. Wade, J. L. Staudinger, M. A. Watson, S. A. Jones, D. D. McKee, B. B. Oliver, T. M. Willson, R. H. Zetterstrom, T. Perlmann, and J. M. Lehmann, *Cell* **92**, 73 (1998).

[3] B. Clement, C. Guguen-Guillouzo, J. P. Campion, D. Glaise, M. Bourel, and A. Guillouzo, *Hepatology* **4**, 373 (1984).

[4] A. P. Li, P. Maurel, M. J. Gomez-Lechon, L. C. Cheng, and M. Jurima-Romet, *Chem. Biol. Interact.* **107**, 5 (1997).

[5] P. Maurel, *Adv. Drug Deliv. Rev.* **22**, 105 (1996).

[6] G. de Sousa, F. Nicolas, M. Placidi, R. Rahmani, M. Benicourt, B. Vannier, G. Lorenzon, K. Mertens, S. Coecke, A. Callaerts, V. Rogiers, S. Khan, P. Roberts, P. Skett, A. Fautrel, C. Chesne, and A. Guillouzo, *Chem. Biol. Interact.* **121**, 77 (1999).

[7] A. Guillouzo, L. Rialland, A. Fautrel, and C. Guyomard, *Chem. Biol. Interact.* **121**, 7 (1999).

[8] H. G. Koebe, B. Muhling, C. J. Deglmann, and F. W. Schildberg, *Chem. Biol. Interact.* **121**, 99 (1999).

[9] A. P. Li, C. Lu, J. A. Brent, C. Pham, A. Fackett, C. E. Ruegg, and P. M. Silber, *Chem. Biol. Interact.* **121**, 17 (1999).

[10] A. P. Li, P. D. Gorycki, J. G. Hengstler, G. L. Kedderis, H. G. Koebe, R. Rahmani, G. de Sousas, J. M. Silva, and P. Skett, *Chem. Biol. Interact.* **121**, 117 (1999).

[11] J. B. Ferrini, J. C. Ourlin, L. Pichard, G. Fabre, and P. Maurel, "Cytochrome P450 Protocols," Vol. 107, p. 341. Humana Press, Totowa, NJ, 1998.

[12] J. B. Ferrini, L. Pichard, J. Domergue, and P. Maurel, *Chem. Biol. Interact.* **107**, 31 (1997).

same cells, thus increasing the number of tested molecules and decreasing the problems due to interindividual variability, and would permit chronic toxicological studies. Several hepatocyte phenotypic markers have been evaluated previously in LTPC of human hepatocytes. These include the expression of functional C/EBPα and β transcription factors,[13] production of plasmatic proteins such as albumin, α_1-antitrypsin, and apolipoproteins (apoA1 and apoB100),[12] the production of urea, and the permissivity to hepatitis C virus replication.[14] The full maintenance of these markers indicates that hepatocytes exhibit a highly differentiated phenotype in LTPC. The aim of this chapter is to describe the use of these cultures to study CYP gene induction.

Cell Cultures

Human hepatocytes are prepared as described,[15,16] and cells are plated and maintained in LTPC conditions.[11,12] The long-term culture medium described previously[17] has been used with modifications. It consists of Williams' E medium supplemented with the following additives: 7.2 μM linoleic and linolenic acid, 200 mg/liter bovine serum albumin (BSA), 10 mg/liter insulin, 5 mg/liter transferrin, 0.1 μM selenium acid, 20 μg/liter liver growth factor, 5 μM dibutyryl cyclic-AMP, 10 IU/liter prolactin, 1.25 μM ethanolamine, 1 mg/liter glucagon, 50 μg/liter epidermal growth factor, 2 mM glutamine, 0.1 μM dexamethasone, 100,000 IU/liter penicillin/streptomycin, and 0.75 mg/liter Fungizone. To investigate CYP gene induction, two different protocols have been used (Fig. 1). In protocol 1, hepatocytes are exposed for 4 days before analysis either to dimethyl sulfoxide (DMSO) alone (inducer solvent) or to various prototypic enzyme inducers: 1 nM dioxin (TCDD), 10 μM rifampicin (RIF), or 500 μM phenobarbital (PB). Inducer treatments start at days 5, 12, and 26, while the cells are cultured in the absence of inducer (washout) during the intervening periods. In protocol 2, hepatocytes are exposed to an inducer cocktail (TCDD + RIF + PB, same concentrations as in protocol 1) at day 5 and for the entire duration of culture. At selected times, including time 0 (T0, 12 hr postplating) and days 9, 16, and 30, cells are analyzed for CYP and nuclear receptor expression.

[13] J. B. Ferrini, E. Rodrigues, V. Dulic, L. Pichard-Garcia, J. M. Fabre, P. Blanc, and P. Maurel, *J. Hepatol.* **35,** 170 (2001).
[14] C. Fournier, C. Sureau, J. Coste, J. Ducos, G. Pageaux, D. Larrey, J. Domergue, and P. Maurel, *J. Gen. Virol.* **79,** 2367 (1998).
[15] L. Pichard, I. Fabre, M. Daujat, J. Domergue, H. Joyeux, and P. Maurel, *Mol. Pharmacol.* **41,** 1047 (1992).
[16] L. Pichard, I. Fabre, G. Fabre, J. Domergue, B. Saint Aubert, G. Mourad, and P. Maurel, *Drug Metab. Dispos.* **18,** 595 (1990).
[17] R. E. Lanford, K. D. Carey, L. E. Estlack, G. C. Smith, and R. V. Hay, *In Vitro Cell Dev. Biol.* **25,** 174 (1989).

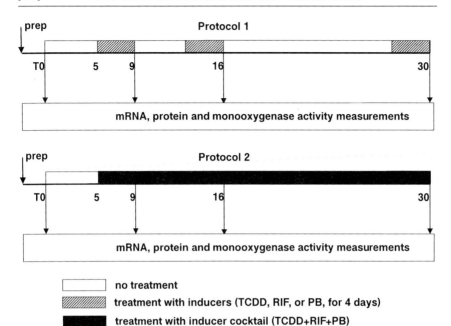

FIG. 1. Induction protocols.

CYP Gene Expression

The results on the induction of CYP proteins, including CYP1A2, CYP2B6, CYP2C9, and CYP3A4, assessed by immunoblotting, are shown in Fig. 2. These proteins remain inducible for 30 days. The levels observed after induction are very close, irrespective of the age of culture (compare data at days 9, 16, and 30). In addition, levels obtained in cells following either protocol 1 or 2 for induction are not significantly different (compare TCDD and cocktail for CYP1A2 or RIF, PB, and cocktail for CYP2C9 and CYP3A4). Furthermore, the levels reached after induction at day 30 are close to those measured at T0, except for CYP2B6, which exhibits a significantly lower expression.

In parallel experiments, CYP-related monooxygenase activities, including acetanilide 4-hydroxylase (CYP1A2), production of nirvanol from S-mephenytoin (CYP2B6), tolbutamide 4-hydroxylase (CYP2C9), and cyclosporin A oxidase (CYP3A4), are measured directly in the cultures by HPLC analysis of the extracellular medium. Data are presented in Table I. As observed with proteins, these activities remain inducible for the entire duration of cultures, irrespective of the induction protocol. The extent of induction varies from one culture to another. Interestingly, activities are greater (4- to 10-fold for acetanilide 4-hydroxylase and 2-fold for tolbutamide 4-hydroxylase and cyclosporin A oxidase) after induction

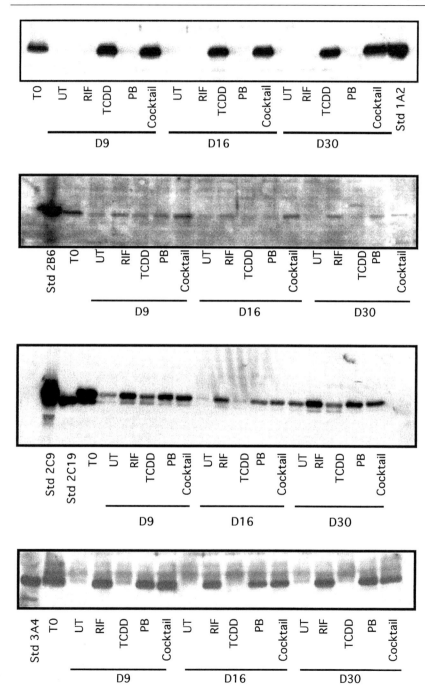

TABLE I

MONOOXYGENASE ACTIVITIES IN LTPC OF HUMAN HEPATOCYTES[a]

Time[b]	Treatment[b]	Acetanilide hydroxylase[c]		Nirvanol[d]		Tolbutamide hydroxylase[d]		Cyclosporin A oxidase[c]	
		166[e]	180	166	180	166	180	166	180
T0		1.2	4.0	274	950	1.1	20.8	1.8	1.5
Day9	UT	0.1	0.01	un	un	0.2	3.1	0.05	0.08
	RIF	0.1	0.03	—	140	0.3	21.0	0.15	2.5
	PB	—	0.04	—	40	—	13.0	—	1.9
	TCDD	8.1	19.9	14	20	0.2	2.9	0.1	0.1
	Cocktail	10.1	20.0	30	240	0.42	16.7	0.2	2.7
Day16	UT	0.1	0.02	un	un	0.02	0.6	0.08	0.04
	RIF	0.3	0.02	1	8	1.1	8.0	4.4	1.2
	PB	—	0.03	—	6	—	4.7	—	1.1
	TCDD	22.7	14.4	un	un	0.08	1.1	0.15	0.07
	Cocktail	25.0	17.3	10	9	1.8	6.8	4.3	0.9
Day30	UT	0.2	0.02	un	un	0.12	5.1	0.1	0.2
	RIF	0.4	0.1	5	170	2.3	20.8	4.3	3.2
	PB	—	0.06	—	30	—	16.5	2.2	2.4
	TCDD	19.0	16.2	un	20	0.1	3.3	0.2	0.2
	Cocktail	20.1	17.1	34	80	2.1	16.9	4.2	1.8

[a] Human hepatocytes were isolated, plated, and cultured under LTPC conditions. un, Undetectable; — not determined.

[b] T0, time 0 of experiments, i.e., 12 hr after plating. Protocol 1. At days 5, 12, and 26, hepatocytes were treated with DMSO (solvent, UT) or inducers: 10 μM rifampicin (RIF), 500 μM phenobarbital (PB), or 1 nM 2,3,7,8-tetrachlorodibenzo-p-dioxin (TCDD) for 96 hr. Protocol 2. Hepatocytes were treated continuously with inducer cocktail (10 μM RIF + 500 μM PB + 1 nM TCDD) from T0 to the time of analysis. Monoxygenase activities were measured at days 9, 16, and 30 by HPLC analysis of extracellular medium with the following substrates: 25 μM acetanilide 26.4 μM tolbutamide, 33.6 μM S-mephenytoin, and 5 μM cyclosporin A.

[c] Acetanilide hydroxylase, tolbutamide hydroxylase, and cyclosporin A oxidase in micromolar of metabolites released in extracellular medium (four million cells) after 6, 8, and 4 hr of incubation, respectively.

[d] Production of nirvanol from S-mephenytoin in picomoles released in extracellular medium (four million cells) after 8 hr incubation.

[e] Two different cultures from two different patients (FT166 andn FT180) are compared.

FIG. 2. Inducible expression of CYP proteins. Human hepatocytes isolated from patient FT180 were plated and maintained under LTPC conditions. At days 5, 12, and 26, some cells were treated with solvent (DMSO, UT) or inducers: 10 μM rifampicin (RIF), 500 μM phenobarbital (PB), or 1 nM 2,3,7,8-tetrachlorodibenzo(p)dioxin (TCDD) for 96 hr (protocol 1, Fig. 1). In other plates, cells were treated continuously with inducer cocktail (10 μM RIF + 500 μM PB + 1 nM TCDD) from day 5 to the time of analysis (protocol 2, Fig. 1). At days 9, 16, and 30, cells were harvested and microsomes were prepared and analyzed by immunoblot. T0, time 0 of experiments, i.e., 12 hr after plating. Std refers to authentic protein markers using cDNA expressed CYPs (Gentest).

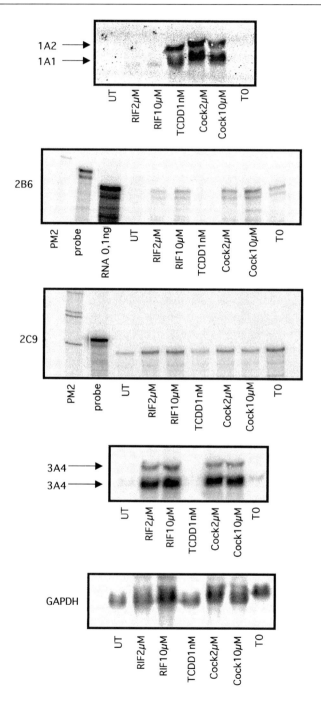

at day 30 than at T0, except for the production of nirvanol (CYP2B6), which only reaches approximately 10% of its T0 level in induced cells, in agreement with protein levels. These data show that the original levels of expression of CYP1A2, CYP2C9, and CYP3A4 proteins and related activities are maintained for at least 30 days in LTPC exposed to prototypical CYP inducers.

In order to further characterize the nature of induction, expression of the respective mRNAs is evaluated by Northern blot (CYP1A1, CYP1A2, and CYP3A4) and the RNase protection assay (CYP2B6, CYP2C9) after 30 days of culture in both induction protocols. Results are shown in Fig. 3. CYP1A1 and CYP1A2 mRNAs are not detectable in freshly plated cells (at T0), but exhibit high levels at day 30 in LTPC exposed to TCDD or inducer cocktail. Similar observations are made with CYP3A4 mRNA. In contrast, CYP2B6 and CYP2C9 mRNAs are expressed at T0. CYP2B6 mRNA is not detectable in untreated cells but returns to its T0 level in LTPC exposed to RIF or cocktail, in sharp contrast to the protein (see Fig. 2 and Table I). CYP2C9 mRNA remains detectable for the entire duration of culture, notably in untreated cells. Its levels are increased as expected from protein analysis in response to RIF, PB, and in LTPC exposed continuously to the inducer cocktail. These data suggest that the induction of these four CYP genes is of transcriptional or posttranscriptional origin and that induction of the CYP1A family does not interfere significantly with the induction of CYP2 and CYP3A families and *vice versa* (data with inducer cocktail).

The expression of other CYP genes has been examined (not shown). CYP2C19 responds moderately to RIF and PB, whereas CYP2D6 and CYP2E1 do not. In terms of protein and enzyme activity at day 30: (i) CYP2C19 (*S*-mephenytoin hydroxylase) roughtly behaves as CYP2B6 with levels $\leq$10% of T0 in RIF-, PB-, or inducer cocktail-treated cells; (ii) CYP2D6 (debrisoquine hydroxylase) is expressed constitutively and exhibits levels approximately 10% of T0; and (iii) CYP2E1 protein expression is not detectable beyond day 10, presumably because of the absence of inducer of this protein in our culture medium.

Nuclear Receptor Expression

It has been shown that induction of CYP3As, CYP2Bs, and CYP2C8-9 in response to xenobiotics is mediated primarily through the activation of two members

---

FIG. 3. Inducible expression of CYP mRNAs. Human hepatocytes isolated from patient FT166 were plated and cultured under LTPC conditions. At day 26, some cells were treated with solvent (DMSO, UT) or inducers: 10 μM rifampicin (RIF) or 1 nM 2,3,7,8-tetrachlorodibenzo-*p*-dioxin (TCDD) for 96 hr (protocol 1, Fig. 1); in other plates, cells were treated continuously with inducer cocktail (2 or 10 μM RIF + 500 μM PB + 1 nM TCDD protocol 2, referring to cock2 and cock10, respectively, Fig. 1) from day 5 to the time of analysis. At day 30, cells were harvested and total RNA was extracted and analyzed by Northern blot (1As, 3A4, GAPDH) or RNase protection assays (2B6, 2C9). T0, time 0 of experiments, i.e., 12 hr after plating. PM2, size marker. Probe: undigested riboprobe; note that this probe migrates slightly above the actual protected probe due to the presence of the plasmid sequence.

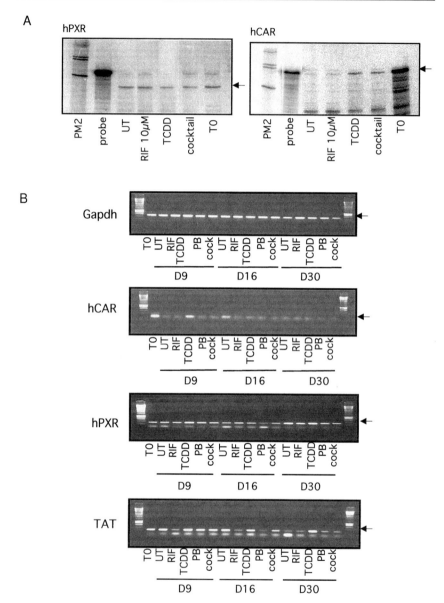

FIG. 4. Expression of nuclear receptor mRNAs. Human hepatocytes isolated from patient FT166 (A) or FT180 (B) were plated and cultured under LTPC conditions. (A) At day 26, some cells were treated with solvent (DMSO, UT) or inducers: 10 μM rifampicin (RIF) or 1 nM 2,3,7,8-tetrachlorodibenzo-p-dioxin (TCDD) for 96 hr (protocol 1, Fig. 1); in other plates, cells were treated continuously with inducer cocktail (10 μM RIF + 500 μM PB + 1 nM TCDD protocol 2, Fig. 1) from day 5 to the time of analysis. At day 30, cells were harvested and total RNA was extracted and analyzed by RNase protection assays. T0, time 0 of experiments, i.e., 12 hr after plating. PM2, size marker. Probe:

of the steroid receptor family : the pregnane X receptor (PXR)[18] and the constitutive androstane receptor (CAR).[19] In addition, both PXR and CAR have been shown to be controlled by the glucocorticoid receptor (GR).[20,21] We therefore decided to evaluate the expression of thyrosine aminotransferase (TAT as a marker of GR activity), PXR, and CAR in our LTPC by RT-PCR and RNase protection assay. The results are reported in Fig. 4. It is remarkable that these mRNAs are expressed at significant levels in LTPC with respect to T0.

Discussion

This chapter showed that the cellular machinery necessary for CYP1-3 gene induction is functional in human hepatocytes in LTPC. First, TAT, a GR-driven gene, appears to be expressed constitutively in these cultures, suggesting a constitutive expression of GR. As this receptor has been shown to control PXR and CAR expression,[20,21] it is not surprising that these two nuclear receptors are significantly expressed and functional, and this is finally consistent with CYP2B-2C and CYP3A gene induction. Moreover, the constitutive expression of CYP2C9,

[18] J. M. Lehmann, D. D. McKee, M. A. Watson, T. M. Willson, J. T. Moore, and S. A. Kliewer, *J. Clin. Invest.* **102**, 1016 (1998).
[19] T. Sueyoshi, T. Kawamoto, I. Zelko, P. Honkakoski, and M. Negishi, *J. Biol. Chem.* **274**, 6043 (1999).
[20] J. M. Pascussi, L. Drocourt, J. M. Fabre, P. Maurel, and M. Vilarem, *J. Mol. Pharmacol.* **58**, 361 (2000).
[21] J. M. Pascussi, S. Gerbal-Chaloin, J. M. Fabre, P. Maurel, and M. Vilarem, *J. Mol. Pharmacol.* **58**, 1441 (2000).

---

undigested riboprobe; note that this probe migrates slightly above the actual protected probe due to the presence of the plasmid sequence. (B) At days 5, 12, and 26, some cells were treated with solvent (DMSO, UT) or inducers: 10 μM RIF, 500 μM phenobarbital (PB), or 1 nM TCDD for 96 hr (protocol 1, Fig. 1). In other plates, cells were treated continuously with inducer cocktail (10 μM RIF + 500 μM PB + 1 nM TCDD) from day 5 to the time of analysis (protocol 2, Fig. 1). At days 9, 16, and 30, cells were harvested and total RNA was extracted and analyzed by RT-PCR. Sense and reverse primers (5′ to 3′) were, respectively, as follows: GAPDH, GGT CGG AGT CAA CGG ATT TGG TCG, and CAA AGT TGT CAT GGA TGA CC (position 15–485, size of fragment 470 bp); TAT, AGC AAA GGC AAC CTC CCC TC, and GGC ATC TTT CAT TGC CTG GG (position 28–300, size of fragment 273 bp); CAR, GCC ATG GCC CTC TTC TCT CC, and GCC CTG GAT GTG CTG GAT TT (position 790–1005, size of fragment 216 bp); and PXR, GGA AGA AGC TGC CAA GTG GAG C, and CCA GGT TCC AGT CTC CGC GTT G (position 882–1200, size of fragment 318 bp). One microliter of reverse transcription was amplified with a *Taq* DNA polymerase (Life Technologies) after mixing: 50 pmol of each primer 1.5 mM MgCl$_2$, 0.25 mM dNTPs, 1X polymerization buffer, and 2.5 U *Taq* in a final volume of 50 μl. Samples were submitted to 30 cycles of amplification: denaturation for 30 sec at 95°, annealing for 30 sec at 55° (GAPDH, CAR), 58° (TAT), or 65° (PXR), and elongation for 30 sec at 72°. Amplification products were analyzed on 1.2% agarose gel electrophoresis.

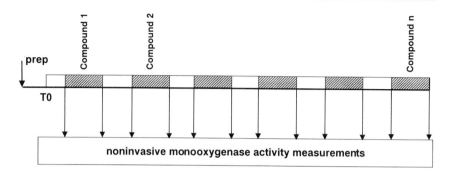

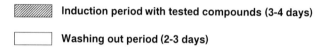

Induction period with tested compounds (3-4 days)

Washing out period (2-3 days)

FIG. 5. Experimental protocol for large-scale screening of enzyme inducers using LTPC of hepatocytes.

a primary glucocorticoid responsive gene,[22] is consistent with the constitutive expression of GR. The present study therefore shows that LTPC provides an invaluable means to investigate the inducible expression and functional activity of human CYPs in their natural cellular environment. Indeed, it is remarkable that the protein and mRNA levels, as well as the enzyme activity of human CYP genes, including CYP1A2, CYP2C9, and CYP3A4, are maintained for at least 1 month in these cultures. However, further medium modifications are currently being tested to increase the expression of other CYP forms, including CYP2C19, CYP2D6, and CYP2E1.

A first important application of these LTPC is their use for evaluating the putative enzyme-inducing effect of compounds, as described in Fig. 5. In this protocol, cells are treated for 3–4 days with a standard concentration of tested compounds. Induction of the enzyme under concern is assessed by measuring noninvasively (e.g., by HPLC analysis of extracellular medium) a specific enzyme activity before and after treatment. This is followed by a 2- to 3-day period of washout before starting a new treatment with another compound. We determined that by using 12-well plates (0.5 million hepatocytes per well), 600 compounds can be tested in duplicate for their inducing effect on a given CYP gene with 120 million hepatocytes. Such a protocol has been used successfully by other groups (not published).

[22] S. Gerbal-Chaloin, J. M. Pascussi, L. Pichard-Garcia, M. Daujat, F. Waechter, J. M. Fabre, N. Carrere, and P. Maurel, *Drug Metab. Dispos.* **29,** 242 (2001).

Another interesting aspect of these LTPC is that when exposed continuously to the cocktail of inducers (Fig. 1), cells remain metabolically competent, thus allowing evaluation of either the metabolic pathways or the putative hepatotoxicity of a large number of compounds. Finally, these LTPC permit chronic toxicological studies. In conclusion, this culture system represents a very pertinent *in vitro* tool for investigating CYP gene induction and drug metabolism in humans.

[31] Polarized Cell Cultures for Integrated Studies of Drug Metabolism and Transport

By CYNTHIA BRIMER-CLINE and ERIN G. SCHUETZ

A variety of approaches have been used to develop cellular model systems that express cytochromes P450 (CYPs). However, drug transport processes influence the intracellular concentration of CYP inducers, inhibitors, and substrates, and thus the pharmacological action of these drugs and their interactions with CYPs. The multidrug resistance gene MDR1 encodes the drug efflux transporter P-glycoprotein (Pgp), and because many drugs that interact with CYP3A are Pgp substrates, this transporter can influence the magnitude of CYP3A induction[1] and CYP3A metabolism.[2] The challenges for the future are to generate model systems in which the dynamic interplay between drug transporting proteins and drug metabolizing enzymes can be explored.

Data are only as good as the model. Caco-2 cells were selected because they express Pgp, polarize in culture, and retain functional characteristics of human small intestine and are used routinely to screen for intestinal permeability and oral drug bioavailability of compounds.[3] However, Caco-2 frequently lacks an important component of intestinal first-pass drug metabolism, CYP3A4.[4] LLC-PK1 cells and derivative cell lines stably expressing human Pgp were chosen because they polarize in culture and have been used extensively to characterize the transport function of Pgp.[5]

[1] E. G. Schuetz, A. H. Schinkel, M. V. Relling, and J. D. Schuetz, *Proc. Natl. Acad. Sci. U.S.A.* **93,** 4001 (1996).

[2] L. Lan, J. T. Dalton, and E. G. Schuetz, *Mol. Pharmacol.* **58,** 863 (2000).

[3] M.-C. Gres, B. Julian, M. Bourrie, V. Meunier, C. Roques, M. Berger, X. Boulenc, Y. Berger, and G. Fabre, *Pharmacol. Res.* **15,** 726 (1998).

[4] J. C. Kolars, W. M. Awni, R. M. Merion, and P. B. Watkins, *Lancet* **338,** 1488 (1991).

[5] A. H. Schinkel, E. Wagenaar, L. van Deemter, C. Mol, and P. Borst, *J. Clin. Invest.* **96,** 1698 (1995).

The expression system must have robust CYP3A4 activity. Transduction with recombinant adenovirus (AdV) offers several advantages.[6] Adenovirus infects most cell types, levels of the recombinant protein can represent 15–20% of total cellular protein, and multiple AdV can be introduced simultaneously. This article describes the development of polarized cellular systems coexpressing AdV with CYP3A4 (Ad3A4), P450 reductase (AdRed), and the MDR1/Pgp transporter. Because the experimental approach can be adapted to study the interplay of multiple enzyme/transporting systems, it may find significant application as a screening tool for the pharmaceutical industry and as a more basic research tool to study the kinetics of intestinal drug bioavailability.

Materials

Cell Lines and Media

Caco-2 (human colon adenocarcinoma cells) and LLC-PK1 (pig kidney epithelial cells) are available commercially (ATCC, Manassas, VA), and L-MDR1 cells are as described previously by Schinkel et al.[5] LLC-PK1 and L-MDR1 cells are grown in medium 199 and Caco-2 cells in (Dulbecco's modified Eagle's medium (DMEM), both media with L-glutamine and 10% (v/v) fetal calf serum (FCS). 293 cells (ATCC, CRL-1573) (embryo epithelial kidney cells transformed with adenovirus 5 DNA) are cultured in D2 medium composed of DMEM plus 2% (v/v) heat-inactivated FCS, 2 mM glutamine, and 1% (w/v) Fungizone. All media are supplemented with 50 units/ml penicillin and 50 μg/ml streptomycin (all reagents from Invitrogen, Carlsbad, CA). The 2× MEM(+) used for the 293 overlay contains 2× MEM, 15% heat-inactivated FCS, 4 mM glutamine, 100 units/ml penicillin, 100 μg/ml streptomycin, and 2% (w/v) Fungizone. All cells are grown at 37° in the presence of 5% (v/v) CO_2.

Adenovirus

Plaque pure recombinant adenoviruses Ad3A4 and AdRed are provided by Genotherapeutics, Inc. (Memphis, TN) and from Dr. Albert Li (University of Tennessee, Memphis) and have been described previously.[7]

Methods

Safe Handling of Adenovirus

Appropriate guidelines should be followed for the safe handling of adenovirus (see www.uth.tmc.edu/biosafety/html/adrenoviral2.html and web.

---

[6] T. Friedberg and C. R. Wolf, Adv. Drug Deliv. Rev. 22, 187 (1996).
[7] C. Brimer, J. T. Dalton, Z. Zhu, J. Schuetz, K. Yasuda, E. Vanin, M. V. Relling, Y. Lu, and E. Schuetz, Pharmacol. Res. 17, 803 (2000).

ncifcrf.gov/Campus/safety/safetygram/ism-193.pdf.) Adenovirus is a pathogen of the respiratory and gastrointestinal mucous and eye membranes and does not have to be replication competent to cause corneal and conjunctival damage. Briefly, all procedures using adenovirus are performed under biosafety level 2 (BSL-2), are never performed on an open bench, and areas where adenovirus is handled and stored are clearly designated. Protective clothing should be worn, samples should be centrifuged in closed containers using sealed rotors, and all contaminated materials must be treated with a virucide before autoclaving and all work surfaces must be decontaminated.

Recombinant Adenovirus Production from Purified AdV Amplification of Viral Particles and Purification by Double Cesium Chloride Gradients

Recombinant transfer vectors used for the creation of recombinant AdV have lost a required E1 region of the AdV genome on viral DNA homologous recombination, which makes the AdV vector replication deficient. This prevents the AdV from spurious replication in host cells. However, in order to grow stocks of AdV, 293 cells must be used, as they have been engineered with the entire E1 region of Ad5 integrated in its chromosome and thus provide the necessary E1 proteins *in trans*. 293 cells should be handled carefully, as they are sensitive to drying, cold temperatures, and excessive mechanical manipulation. Therefore, do not let cells sit for long after aspirating media, always use media that have been prewarmed to 37°, and add media to the side of the dish. Set up thirty 15-cm plates for adenoviral production using 293 cells in D2 medium. When the cells are 80–90% confluent, they are ready to be infected with adenovirus. Generally, use 1 μl of high-titer adenovirus stock [generally 10^{10} plaque-forming units (pfu)/ml] per 15-cm plate. There are approximately 10^7 cells per plate. Multiplicity of infection (MOI) is the number of virus particles/cell. Therefore, the MOI would be approximately 1. Remove adenovirus stock from −80° and thaw immediately before use. Dilute 1 μl of the adenovirus stock per 5 ml of D2. This is the infection media. Aspirate the media from 293 cells and add 5 ml of the infection media per 15-cm plate. Incubate the plates at 37°, 5% CO_2 for 1.5 hr. Gently rock plates every 15 min to assure that the media covers all parts of the plate. After 90 min, add 20 ml of DMEM with 10% FCS to each plate and incubate until the cells begin to show cytopathic effects (CPE; a rounded appearance) and start to detach from the plate. The cells will also appear as "grape-like clusters." Rounding of the cells begins after approximately 24 hr and is fully developed after 36–48 hr depending on the initial amount of virus added. The cells should be harvested within this time range. If CPE is evident before 24 hr, CPE could be due to the effect of viral proteins being produced. To harvest the cells, detach by pipetting fluid and cells up and down using a 25-ml pipette. Collect all cells (because virus is mostly intracellular) in 50-ml disposable polypropylene tubes. Pellet cells by centrifuging at 3000g for 15 min at 4°. Save 10 ml of the supernatant (saved supernatant) in a tube and aspirate the rest of the media from the pellet. Resuspend the pellet in the 250 μl of "saved supernatant" per original 15-cm plate. Because 30 plates were started with, the pellet will be resuspended in 7.5 ml. Treat the remaining supernatant with chlorox before discarding because it is biohazardous. Freeze/thaw the cells (cycling between dry ice and 37° water bath) five times to lyse and release the virus, vortexing between each freeze/thaw cycle. The last freeze can be done overnight and thawed the next day. Remove cellular debris from the crude viral lysate by centrifuging at 6500g at 4° for 30 min. Recover the viral supernatant and bring the volume to 4.25 ml (SW41 swinging bucket/Beckman or equivalent) or 5.0 ml (for SW40 swinging bucket/Beckman or equivalent) with sterile phosphate-buffererd saline (PBS).

To purify the virus, prepare ultraclear SW40 tubes (Beckman) by soaking in 95% (v/v) ethanol followed by sterile water and removal of all liquid. Prepare Cesium Chloride (CsCl) solutions in PBS as follows.

Density	Amount of solid CsCl (g)	Volume of solution (ml)
1.25	27	73
1.33	34	66
1.40	39	61

Density can be checked simply by weighing 1 ml of the CsCl solution.

The first ultracentrifugation involves purifying the AdV through a discontinuous (step) CsCl gradient. Depending on the rotor used, the gradients are made as indicated below. It is convenient to use ultraclear tubes for the remaining centrifugation steps as it is easier to visualize the adenoviral band.

Component	SW41 (ml)	SW40 (ml)	SW40 (ml)
CsCl (density 1.25 g/ml)	2.5	3.0	9.0
CsCl (density 1.40 g/ml)	2.5	3.0	9.0
Viral lysate	4.25	5.0	15

CsCl gradients are made by placing the lower density CsCl solution (1.25 g/ml) in the centrifuge tube first and then underlaying the higher density CsCl solution (1.40 g/ml) using a sterile 2-ml pipette. Carefully overlay the cleared viral lysate onto the CsCl gradient using a sterile 2-ml pipette, noting that the quality of the gradient affects the purity of the final virus. Perform all steps under a sterile hood. Centrifuge the tubes at 210,000g for 2 hr at 20°, with a deceleration rate of 0. Clean the tube with 95% ethanol prior to collection. Puncture the tube with a 3-ml syringe and 21-gauge needle below the *lower* opalescent (bluish white) band (infectious adenovirus) and collect the band.

The second ultracentrifugation step is a continuous CsCl gradient. In a sterile centrifuge tube, add 8 ml of CsCl solution (1.33 g/ml) and overlay with the solution from the previous centrifugation step. Centrifuge at 210,000g for 24–48 hr at 20°, with a deceleration rate of 0. Recover the opalescent adenoviral band as before. From this point onward, keep the solution at 4°. Cesium chloride must be removed before cell culture. Dialyze the adenoviral solution against sterile 10 mM Tris, pH 7.4, 1 mM MgCl$_2$, and 10% (v/v) glycerol. The dialysis solution should be changed three times in 12 hr, at which time the adenoviral solution can be aliquoted and stored at −70°.

Determining AdV Titer by Plaque Assay

Trypsinize one dish (150 mm) of confluent 293 cells onto five 100-mm dishes on day 1 so that cells are 80% confluent at the time of transduction on day 2. Set up enough plates to have duplicate plates for each dilution. Prepare 1 : 10 serial dilutions of AdV stock out to 10^{-11} in D2. Remove media from one plate at a time and add 2 ml of the adenovirus dilution to the appropriate plate by tilting the plate and adding fluid to the side of the plate. Allow the transduction to continue for 90 min at 37°. Rock plates every 15 min to ensure that the solution covers all cells. Carefully aspirate the infection media off each plate. Carefully overlay each plate with 10 ml of a 1 : 1 mixture of 2% Seaplaque agarose (Biowhittaker, Rockland, ME) and 2× MEM(+) prewarmed to 37°. The overlay is made by mixing an equal volume of soft agarose (SeaKem) that has been boiled and brought to 42° and 2× MEM that has been brought to 37°. Repeat the overlay (4 ml) after 5 days and again after 10 days. Plaques should be visible 7–14 days after transduction. Count plaques only after 21 days and determine the titer based on the number of plaques at a certain dilution. For example, if there are 75 plaques on the 10^{-9} plate, then the count is 75×10^9 pfu (plaque forming units) per 2 ml of adenoviral media added. Therefore, the count is $75 \times 10^9/2 = 37 \times 10^9$ pfu/ml.

Assessing Competency of Ad3A4 and AdRed Transduction

Although AdV will transduce almost every cell type, transduction efficiency varies between cell lines.[8] Therefore, the competency of AdV transduction of the host cells must first be determined. Although AdV does not integrate stably, the viral episome is stable in cells, so when terminally differentiated cells (or confluent monolayers, such as those in transwell culture for transport experiments) are infected, the viral genome will last essentially for the life of the cell. Because these cells will ultimately be used to assess CYP activity and transport function, it is most expeditious to simultaneously determine the optimal AdV MOI that results in the highest cell viability, maximal CYP and P450 reductase expression

---

[8] C. Hidaka, E. Milano, P. L. Leopold, J. M. Bergelson, N. R. Hackett, R. W. Finberg, T. J. Wickham, I. Kovesdi, P. Roelvink, and R. G. Crystal, *J. Clin. Invest.* **103,** 579 (1999).

and activity, integrity of membrane tight junctions, and efficient MDR1/Pgp transport function. LLC-PK1 and L-MDR1 cells are plated at day 0 on either 24.5-mm tissue culture plastic wells or transwells [Costar, 24.5 mm diameter, 3.0-μm pore size (Fisher, Suwanee, GA) (6 well plates)] at 2×10^6 cells/well and Caco-2 cells at 3×10^6 cells/ well. On day 1, trypsinize cells in one well and count the cell number. Transduce cells with AdV concentrations ranging from 1 to 1000 MOI, i.e., viral particles per cell. For example, for an MOI of 100, 100 virus particles/cell are needed, so for 3.0×10^6 cells, add 100 times 3.0×10^6 or 3.0×10^8 pfu/well. For a 24.5-mm well, add AdV in 0.5 ml of the serum-free media. Incubate the plates at $37°$, 5% CO_2 for 1.5 hr. Gently rock plates every 15 min to ensure that the medium covers all parts of the plate. After 90 min, add 1.5 ml of medium with 10% heat-inactivated fetal bovine serum. On day 3, change to fresh serum-containing medium. On day 5 cells, representative wells for each MOI are counted, cell viability is determined, and duplicate wells are assayed for CYP3A4, P450 reductase protein or activity or transport function.

Adenoviral Expression of CYP3A4 and P450 Reductase

Immunoblot analysis of cell lysates is carried out as described previously.[7] A dose-dependent increase in CYP3A4 protein was observed from an MOI of 10 to 1000 (Fig. 1A), with very high cellular toxicity evident at 1000. The expression of CYP3A4 was higher in LLC-PK1 compared to Caco-2 cells. Suboptimal expression of NADPH P450 reductase in many cell lines, including Caco-2, can limit maximal CYP catalytic activities.[9] Therefore, we determined whether cotransduction of a recombinant P450 reductase adenovirus (AdRed) would enhance CYP3A4 activity. Transduction of 1–25 MOI of AdRed resulted in a dose-dependent increase in P450 reductase protein expression (Fig. 1B). Testosterone 6β-hydroxylation assays were performed on microsomes prepared from LLC-PK1, L-MDR1, and Caco-2 cells,[2] as well as their counterparts transduced with either Ad3A4 or Ad3A4 plus AdRed (Table I). Activities were compared in microsomes using a single saturating concentration of testosterone (100 μM). Cell lines transduced with CYP3A4 alone had significantly lower activity than cell lines transduced with both CY3A4 and oxidoreductase. The kinetic constants (K_m and V_{max}) for testosterone 6β-hydroxylation were determined. The K_m values are slightly lower than those observed in insect cell microsomes[10] and human liver microsomes.[10,11] The V_{max} values are in the range of human liver microsomes.[11]

[9] M. Hu, L. Yiqi, C. M. Davitt, S.-M. Huang, K. Thummel, B. W. Penman, and C. L. Crespi, *Pharmacol. Res.* **16**, 1352 (1999).

[10] C. A. Lee, S. H. Kadwell, T. A. Kost, and C. J. Serabjit-Singh, *Arch. Biochem. Biophys.* **319**, 157 (1995).

[11] A. J. Draper, A. Madan, K. Smith, and A. Parkinson, *Drug Metab. Dispos.* **26**, 299 (1998).

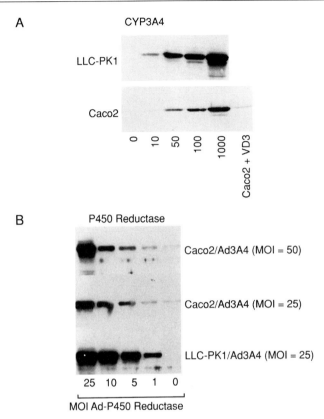

FIG. 1. Immunoblot analysis of CYP3A4 (A) and P450 reductase (B) in LLC-PK1, L-MDR1, and Caco2 cells transduced with various MOI (0–1000) of Ad3A4 alone (A) or cells cotransduced with Ad3A4 (MOI 25, 50) and various MOI (0–25) of AdRed (B). Lysate from Caco2 cells treated with 1,25-dihydroxyvitamin D_3(VD3) as a control. Adapted from C. Brimer, J. T. Datton, Z. Zhu, J. Schuetz, K. Yasuda, E. Vanin, M. V. Relling, Y. Lu, and E. Schuetz, *Pharmacol. Res.* **17,** 803 (2000), with permission.

Transport Function

Because the ultimate goal is to analyze functional interactions between CYP3A and Pgp, it needs to be determined that Pgp transport function is not affected by AdV transduction. Pgp transport function is determined by comparing the apical-to-basal and basal-to-apical transport of the xenobiotic across cells with little Pgp (LLC-PK1) to drug transport across cells expressing Pgp exclusively on the apical surface (L-MDR cells). Transport assays are performed as described previously.[5] On day 5 following AdV transduction of cells in 24.5-mm transwell dishes, the quality of cell monolayers is determined routinely by first measuring transepithelial

TABLE I
KINETIC PARAMETERS OF TESTOSTERONE 6β-HYDROXYLASE ACTIVITY
IN CELL MICROSOMES[a]

Cell line (microsomes)	Testosterone 6β-hydroxylase activity[b] (pmol/mg/min)	$K_m(\mu M)$	V_{max} (pmol/mg/min)
LLC-PK1	0	ND	ND
LLC-Ad3A4	182	14.8	196
LLC-Ad3A4/AdRed	956	33.0	1350
L-MDR1	0	ND	ND
L-MDR1-Ad3A4	188	30.0	259
L-MDR1-Ad3A4/AdRed	603	23.6	698
		15.9	
Caco-2	0	ND	ND
Caco-2/Ad3A4	61.9	ND	ND
Caco-2/Ad3A4/AdRed	338	18.0	414

[a] Adapted from C. Brimer, J. T. Datton, Z. Zhu, J. Schuetz, K. Yasuda, E. Vanin, M. V. Relling, Y. Lu, and E. Schuetz, *Pharmacol. Res.* **17**, 803 (2000), with permission. ND, not detectable.
[b] Testosterone 6β-hydroxylase activity at 100 μM (saturating concentration). Steady-state kinetic parameters were monitored by HPLC using isolated microsomal fractions as described (see reference in footnote *a*).

electrical resistance (TEER), which should range from 200 to 450 ohm · cm^2, indicating that the cells have produced a monolayer that has tight junctions to prevent leaking of media between the cells. Cells are washed, and the assay is started (time 1) by adding radiolabeled drug to either the apical or the basal compartment. Radiolabeled Pgp substrate [e.g., [^{3}H]vinblastine sulfate (11.7 Ci/mmol) (Moravek)] is added at a 2 or 5 μM final concentration containing ~0.25 μCi/ml. At 1, 2, 3, and 4 hr, 50-μl aliquots are sampled from the opposite compartment, radioactivity is counted, and results are expressed as the percentage radioactivity appearing in the opposite compartment relative to radioactivity added at time 0. If the drug is a Pgp substrate, compared to LLC-PK1 cells, basal-to-apical transport will increase and apical-to-basal transport of the drug will decrease in L-MDR1 cells, resulting in a net basal to apical transport in L-MDR1 cells (Fig. 2). The flux of drugs across LLC-PK1 cells and Pgp transport kinetics in cells with and without Ad3A4 and AdRed transduction was identical. These findings support the feasibility of using Ad3A4-transduced LLC-PK1, L-MDR1, and Caco-2 cells for coupled transport/metabolism analysis.

Conclusion

    Model cellular systems have been established previously to determine the way drugs interact singly with either CYP3A4 or Pgp. However, these

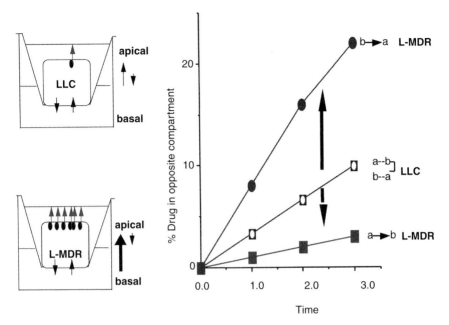

FIG. 2. Schematic representation of transepithelial transport in LLC-PK1 and L-MDR1 cells of a Pgp substrate. Drug is applied to one compartment [basal (b) or apical (a)], and the percentage of drug appearing in the opposite compartment at $t = 1, 2$, and 3 hr is plotted.

drug-metabolizing enzymes and transporters do not exist in isolation, but are part of an integrated detoxification system. LLC-PK1, L-MDR1, and Caco-2 polarized cells expressing Pgp and transduced with Ad3A4 and AdRed can be used to study the dynamic interplay between drug-transporting proteins and drug-metabolizing enzymes.

Acknowledgments

We thank Dr. Albert Li for recombinant AdV and Dr. Jim Dalton for measurements of CYP3A activity.

Section V

Invertebrate P450s

[32] Use of Methylotropic Yeast *Pichia pastoris* for Expression of Cytochromes P450

By Mette Dahl Andersen and Birger Lindberg Møller

Introduction

Cytochromes P450 (P450s) occur in all kingdoms. Genome sequencing projects have provided numerous P450 gene sequences. These have been used to construct multiple sequence alignments and phylogenetic trees. Likewise, intron positions and phases have been determined.[1] In most cases, the function of a particular P450 cannot be predicted solely from its primary sequence but needs to await biochemical analysis of the catalytic properties of the corresponding protein. However, purification of P450s from their natural sources is not trivial. Plant P450s are most often only present in minute amounts in specific tissues or developmental states and are often quite labile in plant extracts.[2,3] Therefore, many P450 genes are available for which the function remains unknown. The almost complete genome of *Arabidopsis thaliana* contains 247 P450 genes and 26 pseudogenes (www.biobase.dk/P450: The Arabidopsis Genome Initiative[4]), but the catalytic activity is only known for a very small number of these. Accordingly, expression in heterologous organisms is an essential method to obtain the corresponding P450 protein and functional knowledge of P450 genes. Heterologous expression of P450s has been established in several organisms (e.g., reviewed in Gonzalez and Korzekwa[5]). The suitability of the different expression systems is highly dependent on the individual P450 gene sequence and may, in some cases, require recoding to optimize for the codon usage of the host.[6]

Pichia pastoris is a methylotrophic yeast capable of utilizing methanol as its sole carbon and energy source.[7,8] The *AOX1* gene responsible for alcohol oxidase activity is tightly regulated, and the gene product is not detectable when glucose, ethanol, or glycerol constitutes the carbon source. In contrast, culturing *P. pastoris*

[1] S. M. Paquette, S. Bak, and R. Feyereisen, *DNA Cell Biol.* **19,** 307 (2000).
[2] F. Durst and D. P. O'Keefe, *Drug Metab. Drug Interact.* **12,** 171 (1995).
[3] G. P. Bolwell, K. Bozak, and A. Zimmerlin, *Phytochemistry* **37,** 1491 (1994).
[4] The Arabidopsis Genome Initiative, *Nature* **408,** 796 (2000).
[5] F. J. Gonzalez and K. R. Korzekwa, *Annu. Rev. Pharmacol. Toxicol.* **35,** 369 (1995).
[6] Y. Batard, A. Hehn, S. Nedelkina, M. Schalk, K. Pallett, H. Schaller, and D. Werck-Reichhart, *Arch. Biochem. Biophys.* **379,** 161 (2000).
[7] J. M. Cregg, J. F. Tschopp, C. Stillman, R. Siegel, M. Akong, W. S. Craig, R. G. Buckholz, K. R. Madden, P. A. Kellaris, G. R. Davis, B. L. Smiley, J. Cruze, R. Torregrossa, G. Velicelebi, and G. P. Thill, *Bio/Technology* **5,** 479 (1987).
[8] J. M. Cregg, T. S. Vedvick, and W. C. Raschke, *Bio/Technology* **11,** 905 (1993).

on methanol as the sole carbon source results in high levels of *AOX1* mRNA and protein. AOX1 levels up to 5% of total mRNA and up to 35% of total protein in the yeast cell have been reported (reviews on *P. pastoris* expression are reported elsewhere[9,10]). The *AOX1* promoter has been used to construct vectors for heterologous protein expression in *P. pastoris*, and the expression system is available commercially from Invitrogen (Leek, The Netherlands). A number of soluble proteins have been expressed in *P. pastoris* with expression levels of several grams per liter.[11,12] In contrast, reports on membrane protein expression in *P. pastoris* are few. Only two P450s from spiny dogfish shark and spiny lobster have been expressed previously in *P. pastoris*.[13,14] The expressions were verified by detection of the new expected P450 enzyme activity in the transformed cells. However, these recombinant P450s were neither characterized nor isolated.

CYP79D1 and CYP79D2 have been cloned from cassava (*Manihot esculenta* Crantz).[15] Both P450s catalyze the conversion of valine and isoleucine to the corresponding oximes, thus catalyzing the first committed step in the biosynthesis of the two cyanogenic glucosides linamarin and lotaustralin in cassava.[15] Attempts to express CYP79D1 and CYP79D2 in *Escherichia coli* using constructs and conditions similar to those tested to achieve expression of the homologous sorghum CYP79A1[16,17] were unsuccessful. This article describes the cloning strategies and culture conditions for the functional expression of CYP79D1 and CYP79D2 in *P. pastoris* and discusses the properties of the isolated recombinant CYP79D1.

Construction of Plasmids for P450 Expression in *P. pastoris*

Principles

The vector pPICZc (Invitrogen) containing the methanol-inducible *AOX1* promoter for the control of gene expression and the *Sh ble* bleomycin gene encoding resistance against zeocin is used to achieve intracellular expression in *P. pastoris* wild-type strain X-33. To optimize expression, 5′- and 3′-untranslated regions are removed from the *CYP79D1* and *CYP79D2* cDNA clones before insertion into

[9] J. L. Cereghino and J. M. Cregg, *FEMS Microbiol. Rev.* **24**, 45 (2000).
[10] K. Sreekrishna, R. G. Brankamp, K. E. Kropp, D. T. Blankenship, J. T. Tsay, P. L. Smith, J. D. Wierschke, A. Subramaniam, and L. A. Birkenberger, *Gene* **190**, 55 (1997).
[11] M. A. Romanos, J. J. Clare, K. M. Beesley, F. B. Rayment, S. P. Ballantine, A. J. Makoff, G. Dougan, N. F. Fairweather, and I. G. Charles, *Vaccine* **9**, 901 (1991).
[12] J. J. Clare, F. B. Rayment, S. P. Ballantine, K. Sreekrishna, and M. A. Romanos, *Bio/Technology* **9**, 455 (1991).
[13] S. M. Boyle, M. P. Popp, W. C. Smith, R. M. Greenberg, and M. O. James, *Mar. Environ. Res.* **46**, 25 (1998).
[14] J. M. Trant, *Arch. Biochem. Biophys.* **326**, 8 (1996).
[15] M. D. Andersen, P. K. Busk, I. Svendsen, and B. L. Møller, *J. Biol. Chem.* **275**, 1966 (2000).
[16] B. A. Halkier, O. Sibbesen, and B. L. Møller, *Methods Enzymol.* **272**, 268 (1996).
[17] B. A. Halkier, H. L. Nielsen, B. Koch, and B. L. Møller, *Arch. Biochem. Biophys.* **322**, 369 (1995).

the expression vector. The vector contains an approximately 80-bp-long multiple cloning site. To obtain a potentially optimal translation initiation context, the start ATG of the P450s is positioned exactly as the start ATG of the highly expressed *P. pastoris AOX1* gene. No modifications are made in the amino acid sequence of the P450s. Silent mutations are introduced in codons two and six to comply with yeast codon usage.

Procedure

Escherichia coli strain TOP10F' is used for the transformation and propagation of recombinant plasmids. An *Xho*I site is introduced immediately downstream of the *CYP79D1* stop codon by polymerase chain reaction (PCR). The PCR product is restricted with *Xho*I and *Bsm*BI. The latter enzyme cuts 18 bp downstream of the start ATG codon. pPICZc is restricted with *Bst*BI and *Xho*I. The vector and PCR product are ligated together using an adapter made from annealed oligonucleotides: 5′-CGAAACGATGGCTATGAACGTCTCT-3′ (sense direction) and 5′-TGGTAGAGACGTTCATAGCCATCGTTT-3′. The adapter reconstructs the first 18 bp of *CYP79D1* (start codon underlined) introducing two silent mutations (double underlined) and reconstructs a short vector sequence removed by the *Bst*BI restriction, thereby positioning the *CYP79D1* start codon exactly as the start codon of *AOX1*. *CYP79D2* is cloned into pPICZc using the same adapter, as the first 24 bp of the coding sequence of *CYP79D1* and *CYP79D2* genes are identical.

Pichia pastoris is transformed by electroporation according to the manufacturer's manual (EasySelect *Pichia* expression Kit Version A, Invitrogen). The presence of *CYP79D1* or *CYP79D2* in zeocin-resistant colonies is confirmed by PCR analysis on the *P. pastoris* colonies.

Cell Culture Conditions

Principles

Both CYP79D1 and CYP79D2 are functionally expressed in *P. pastoris* as documented by the formation of *(E)*- and *(Z)*-2-methylpropanal oxime from valine in recombinant yeast cells (Fig. 1) and by the presence of an additional protein band with the expected mass as monitered by SDS–PAGE (Fig. 2A).[15] The formation of *(E)*- and *(Z)*-2-methylpropanal oxime is independent on the addition of NADPH-P450 oxidoreductase to the assay. This demonstrates that an endogenous *P. pastoris* reductase system is able to support electron donation to recombinant P450s.

The expression levels of CYP79D1 and CYP79D2 are examined in minimal media; in rich media induced at low, intermediate, or high OD_{600} values; and in rich media induced with a six times elevated level of methanol (3%) (Fig. 3). Expression levels are assessed as CYP79D1 or CYP79D2 activity per culture volume. Analysis of the expression level by enzyme activity measurements in intact *P. pastoris* cells serves as a very straightforward method, but the assay may have limited precision

 pPICZc   CYP79D1   CYP79D2

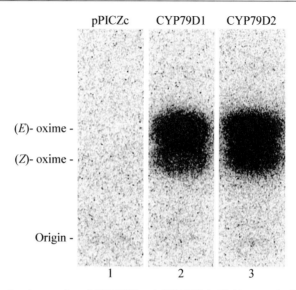

FIG. 1. Functional expression of CYP79D1 and CYP79D2 in *Pichia pastoris*. *P. pastoris* cells were incubated with L-[U-^{14}C]valine and extracted with ethyl acetate, and the product containing ethyl acetate extracts was analyzed by thin-layer chromatography. Lane 1, *P. pastoris* transformed with expression vector pPICZc (control); lane 2, pPICZc containing CYP79D1; and lane 3, pPICZc containing CYP79D2. *(E)*- and *(Z)*-oxime denote *(E)*- and *(Z)*-2-methylpropanal oxime, respectively. Adapted with permission from M. D. Andersen, P. K. Busk, I. Svendsen, and B. L. Møller, *J. Biol. Chem.* **275,** 1966 (2000).

due to possible interfering yeast components. Alternative methods include microsomal activity assays or carbon monoxide difference spectroscopy.[18] However, the activity and yield of protein in microsome preparations also vary considerably, and interfering yeast pigments may hamper precise CO difference spectroscopy. CYP79D1 and CYP79D2 show similar patterns of expression level at the conditions tested (Fig. 3). We conclude that optimal culture conditions are rich media induced at intermediate to high cell density for 24–30 hr. This time period coincides with the time required for the cultures to reach the stationary growth phase. *P. pastoris* cells cultured in minimal media never reach a high OD_{600} value, thereby preventing the formation of high amounts of P450 per liter of culture. However, the amount of P450 per *P. pastoris* cell could be high. Induction with high levels of methanol only seems to decrease the amount of recombinant enzyme activity, although the cultures grow to higher OD_{600} values. Induction at low OD_{600} values, although successful for expression of the homologous CYP79A1 in *E. coli*,[17] does not produce much recombinant P450 in *P. pastoris*.

At the transformation event, the expression cassette inserts into the *P. pastoris* genome at the *AOX1* locus via homologous recombination. Multiple gene insertion

[18] T. Omura and R. Sato, *J. Biol. Chem.* **239,** 2370 (1964).

A

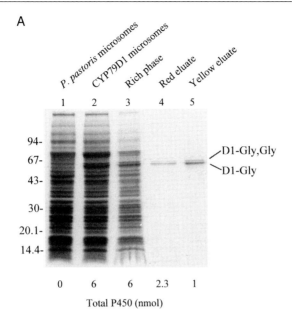

B

FIG. 2. Isolation of recombinant CYP79D1 and the glycosylation pattern of the N-terminal sequence. (A) Polypeptide profiles as analyzed by SDS–PAGE and staining with Coomassie Brilliant Blue. Lane 1, *P. pastoris* microsomes; lane 2, *P. pastoris* microsomes expressing CYP79D1; lane 3, Triton X-114 rich phase; lane 4, eluate from Reactive Red 120 agarose; and lane 5, eluate from Reactive Yellow 3A agarose. Approximately the same amount (2.8 pmol) of P450 was applied to lanes 2–5, and an assessment of total P450 content in the individual fractions is given below each lane. The values of lanes 2 and 3 are semiquantitative due to the presence of interfering cytochrome oxidase. The two protein bands observed in the red and yellow eluate fractions correspond to differential glycosylation of CYP79D1. (B) The glycosylation pattern of the N-terminal sequence of recombinant CYP79D1. N-Gly, glycosylated Asn. Adapted with permission from M. D. Andersen, P. K. Busk, I. Svendsen, and B. L. Møller, *J. Biol. Chem.* **275,** 1966 (2000).

events at the locus may provide higher levels of recombinant protein.[10,12] However, these spontaneous multiple insertions events are rare and occur in approximately 1–10% of the zeocin-resistant transformants. A total of 33 CYP79D1 or CYP79D2 *P. pastoris* transformants were examined with respect to expression level after 24 hr of induction. Of these, 3 *P. pastoris* transformants did not express detectable recombinant enzymatic activity, although the presence of the *CYP79D* gene in the

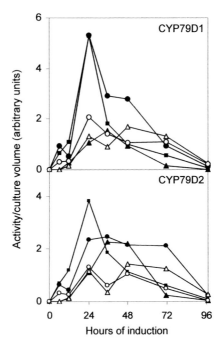

FIG. 3. Expression level of CYP79D1 and CYP79D2. *P. pastoris* cultures expressing CYP79D1 or CYP79D2 were followed during induction periods up to 96 hr. The activity per culture volume was evaluated by measuring the conversion of [^{14}C]valine to ^{14}C-labeled 2-methylpropanal oxime. Cells from overnight starter cultures were harvested and inoculated into inducing media at time 0 hr using different cell densities and media: (■) cultures induced in rich media at OD_{600} 7–8; (●) cultures induced in rich media at OD_{600} 0.5–1; (△) cultures induced in rich media by diluting starter cultures thousandfold; (○) cultures induced with elevated levels of methanol (3%) in rich media at OD_{600} 0.5–1; and (▲) cultures induced in minimal media. See text for further definitions of media.

P. pastoris genome has been confirmed with PCR. The expression level in the 30 other transformants varied within a factor of 4 as evaluated by the activity measured in intact cells. Differences in the expression level of the different transformants were consistent between experiments.

Procedure

To produce starter cultures, single colonies of *P. pastoris* are grown (28°, 220 rpm) for approximately 22 hr in rich media [BMGY: 1% yeast extract, 2% peptone, 0.1 M K$_2$HPO$_4$/KH$_2$PO$_4$ (KP$_i$, pH 6.0), 1.34% yeast nitrogen base, 4 × 10^{-5}% biotin, 1% glycerol, 100 μg/ml zeocin] or in minimal media (MGY: 1.34% yeast nitrogen base, 4 × 10^{-5}% biotin, 1% glycerol, 100 μg/ml zeocin). Cells from the starter cultures are pelleted (1500g, 10 min, room temperature), resuspended in a small volume of inducing media, and used for inoculation. Cells from starter cultures grown in BMGY media are used to inoculate rich media BMMY (BMGY with 1% methanol as inducer and devoid of glycerol) at either 1000-fold dilution or by diluting to OD_{600} 0.5–1 or to OD_{600} 7–8. Cells from starter cultures grown in minimal media (MGY) are inoculated in inducing minimal media MM (MGY with 1% methanol as inducer and devoid of glycerol). Cultures grown in rich media with elevated levels of the inducer methanol are also tested. To this end, BMMY is inoculated to OD_{600} 0.5–1 and methanol is added to 3%. All cultures are grown (28°, 300 rpm) for 96 hr in the inducing media with readdition of methanol to 0.5% (or 3%) each 24 hr.

For *in vivo* activity assays, *P. pastoris* cells (200 μl) are pelleted and resuspended in 100 μl 50 mM Tricine (pH 7.9) and incubated with 0.35 μCi L-[U-^{14}C]valine (246 mCi/mmol; Amersham). After incubation (30 min, 30°), the reaction is stopped by the addition of ethyl acetate. The product containing ethyl acetate phases is analyzed using thin-layer chromatography.[15]

Pichia pastoris Cell Lysis

Principle

Pichia pastoris cells are vortexed with glass beads in order to break the cell wall and lyse the cells. This is a fairly harsh procedure and may be the cause of low yields of CYP79D2 recovered in the resulting microsomes.[15] As an alternative, a more gentle lysis method based on incubation with lyticase (Zymolyase) was tested. Lyticase is expected to degrade yeast cell walls in a manner analogous to *E. coli* cell wall degradation using lysozyme. However, lysis of *P. pastoris* with lyticase provides lower yields of P450 in microsomes compared to the glass bead method. The reason for this is unknown, but the incubation period at 30° included in the procedure may cause denaturation or proteolytic degradation of sensitive proteins such as P450s. Alternatively, the lyticase activity may not be sufficiently effective in degrading the *P. pastoris* cell wall. Generally, the glass bead method provides CYP79D1 in microsomes in yields of up to 30 nmol per liter culture, whereas the lyticase method never gives more than 6 nmol CYP79D1 per liter culture. Cell lysis is performed immediately after harvest of the cultures, as storage of the cell pellet at −20° results in 50% lowered yields of CYP79D1.

Procedure

Pichia pastoris cells are harvested (3000g, 10 min, 4°) and washed once in buffer A [50 mM KP$_i$ (pH 7.9), 1 mM EDTA, 5% glycerol, 2 mM DTT, 1 mM phenylmethylsulfonyl fluoride] before being resuspended to OD_{600} 130 in buffer A. Typically, a 500-ml culture having OD_{600} 21 at harvest is resuspended to 80 ml. An equal volume of acid-washed glass beads (0.5 mm, Sigma) is added, and the cells are broken by vortexing (eight times for 30 sec, 4°, cooling on ice in between vortexing). The lysate is centrifuged (12,000g, 10 min, 4°) to remove cell debris, and the resulting supernatant is centrifuged (165,000g, 1 hr, 4°) to recover a microsomal pellet. Microsomes are resuspended in buffer A and stored at −80°.

Spectroscopy on *Pichia pastoris* System

Yeasts are known to contain endogenous P450 proteins[19] (drnelson.utmem.edu/ CytochromeP450.html). However, at the experimental conditions used, these are expressed in low amounts in *P. pastoris* because endogenous P450s are never detectable by CO difference spectroscopy. This property renders *P. pastoris* well suited for the detection of heterologous P450 expression by spectroscopy alone. However, the yeast cytochrome oxidase protein present in *P. pastoris* forms a spectrum in the presence of CO, which may interfere with P450 spectroscopy.[14] In the present study, the expression level of CYP79D1 is high, and the CO difference spectrum produced by the cytochrome oxidase (maximum at 430 nm, minimum at 445) is only visible as a shoulder on the 450-nm peak. In experiments where the expression level of the P450 is low, P450s can be measured in the absence of cytochrome oxidase by (1) culturing the cells for extended periods (90 hr), upon which the amount of cytochrome oxidase decreases to below the detection limit (data not shown), (2) retaining the *P. pastoris* cytochrome oxidase on a DEAE column, thereby separating it from P450,[15] and (3) removing cytochrome oxidase from the P450 by TritonX-114 phase partitioning performed in borate buffer,[20] whereupon the P450s partition to the TritonX-114 poor phase and the *P. pastoris* cytochrome oxidase partitions to the rich phase.[15,20]

Characterization of Recombinant CYP79D1

CYP79D1 is purified from *P. pastoris* microsomes by TritonX-114 phase partitioning, ion exchange, and dye column chromatography (Fig. 2A).[15] Recombinant CYP79D1 produces a typical CO difference spectrum and forms a type I substrate-binding spectrum in the presence of L-valine. CYP79D1 is further characterized with respect to substrate specificity. The spectroscopy and substrate specificity characterizations show the properties as expected for the native enzyme.[15] This strongly suggests that *P. pastoris* produces P450 proteins with native properties.

Samples that contain the CYP79D1 protein as monitored by CO difference spectroscopy and activity measurements always produce two distinct, closely migrating polypeptide bands on SDS–PAGE (Fig. 2A). N-terminal amino acid sequencing identifies both bands as derived from CYP79D1 (Fig. 2B). The initial

[19] D. R. Nelson, *Arch. Biochem. Biophys.* **369**, 1 (1999).
[20] M. D. Andersen and B. L. Møller, *Protein Expr. Purif.* **13**, 366 (1998).

methionine of CYP79D1 is removed by the yeast processing system. Sequencing of the first 15 residues demonstrates that the upper band (D1-Gly,Gly) is glycosylated at both asparagine positions possible, whereas the lower band (D1-Gly) only is glycosylated at the first asparagine (Fig. 2B). The two different glycosylation patterns explain the occurrence of two bands. Glycosylation at the N-terminal part of CYP79D1 is in agreement with the localization of the N-terminal end in the lumen of the endoplasmatic reticulum accessible for the glycosylation machinery. Reports of native P450 glycosylations are scarce[21,22] and hence CYP79D1 is not expected to be glycosylated *in planta*. Therefore, the observed glycosylation of recombinant CYP79D1 probably reflects that the expression has been obtained using a yeast system. The relative mobility of recombinant CYP79D1 on SDS–PAGE corresponds well to the mass deduced from the coding sequence. Accordingly, the CYP79D1 is expected to have a simple glycosylation structure. Hyperglycosylations are generally observed as a disadvantage of *Saccharomyces cerevisiae* expression systems. In contrast, *P. pastoris* adds smaller size oligosaccharides to the protein (8–14 mannose residues per side chain[23,24]).

Expression Level

Barnes *et al.*[25] have performed an interesting comparison of the expression levels of bovine CYP17 in mammalian cells, insect cells, *S. cerevisiae*, and *E. coli*. In that study, *E. coli* and insect cell expression systems gave the highest yields of P450 protein.[25] Expression in yeast, *S. cerevisiae*, compared poorly to the other systems, and the expression level was as low as 2.5 nmol P450 per liter culture before cell lysis. The level of expression in yeast is typically improved considerably by coexpression with a P450 reductase. *Arabidopsis thaliana* contains two P450 reductases (ATR1 and ATR2) of which ATR1 generally appears to be most efficient.[26,27] After cell lysis, CYP79D1 is recovered in *P. pastoris* microsomes in yields up to 30 nmol per liter culture. These levels correspond to 125–180 pmol P450/mg microsomal protein, thus in the middle to upper range of what has been obtained in *S. cerevisiae*. Whereas *P. pastoris* does not seem to provide as high yields as an *E. coli* expression system, it constitutes an attractive alternative to *S. cerevisiae* expression systems.

[21] O. Shimozawa, M. Sakaguchi, H. Ogawa, N. Harada, K. Mihara, and T. Omura, *J. Biol. Chem.* **268,** 21399 (1993).

[22] D. Werck-Reichhart, Y. Batard, G. Kochs, A. Lesot, and F. Durst, *Plant Physiol.* **102,** 1291 (1993).

[23] R. Montesino, R. Garcia, O. Quintero, and J. A. Cremata, *Protein Expr. Purif.* **14,** 197 (1998).

[24] L. S. Grinna and J. F. Tschopp, *Yeast* **5,** 107 (1989).

[25] H. J. Barnes, C. M. Jenkins, and M. R. Waterman, *Arch. Biochem. Biophys.* **315,** 489 (1994).

[26] F. Cabello-Hurtado, Y. Batard, J. P. Salaun, F. Durst, F. Pinot, and D. Werck-Reichhart, *J. Biol. Chem.* **273,** 7260 (1998).

[27] T. Robineau, Y. Batard, S. Nedelkina, F. Cabello-Hurtado, M. LeRet, O. Sorokine, L. Didierjean, and D. Werck-Reichhart, *Plant Physiol.* **118,** 1049 (1998).

The amount of CYP79D1 in 1 liter of *P. pastoris* culture corresponds roughly to the amount in 0.4 kg of immature folded leaves and petioles of cassava. Purification from plant material generally results in a very low yield due to pigments and co-occurring phenolic compounds. Thus, sorghum CYP79A1 was purified from *E. coli* in 15% yield compared to a 1% yield from plant material.[17,28] Considering the different yields and recognizing that one cassava plant provides approximately 0.3 g of the immature folded leaves and petioles, about 22,000 grown-up cassava plants would be necessary to provide the same amount of purified CYP79D1 as obtained from 1 liter of *P. pastoris* culture.

Conclusions and Final Comments

In conclusion, we find that *P. pastoris* is an excellent and very promising host for the expression of P450 enzymes. The current study is not comprehensive with respect to the full potential of *P. pastoris*, but we believe that a sound foundation has been laid for further use and optimization of *P. pastoris* for P450 expression. Future experimentation should address the effect of P450 gene dosage and of coexpression of P450 reductase on expression levels and the effects of using different host strains of *P. pastoris*. The expression level of the 5-HT_{5A} receptor in *P. pastoris* was increased two to eight times when using the protease-deficient strain SMD1163.[29] In contrast, the expression level of another membrane-bound protein, the peptide transporter PepT1, was not changed by shifting into the SMD1163 protease-deficient *P. pastoris* strain.[30]

Acknowledgments

We thank Dr. Barbara Halkier for fruitful discussions and Dr. Søren Bak for critical reading of the manuscript. This work was supported by the Danish National Research Foundation, Danish International Development Agency, and Danish Agricultural and Veterinary Research Council.

[28] O. Sibbesen, B. Koch, B. A. Halkier, and B. L. Møller, *Proc. Natl. Acad. Sci. U.S.A.* **91,** 9740 (1994).

[29] H. M. Weiss, W. Haase, H. Michel, and H. Reilander, *FEBS Lett.* **377,** 451 (1995).

[30] F. Doring, S. Theis, and H. Daniel, *Biochem. Biophys. Res. Commun.* **232,** 656 (1997).

[33] Partial Recoding of P450 and P450 Reductase cDNAs for Improved Expression in Yeast and Plants

By ALAIN HEHN, MARC MORANT, and DANIELE WERCK-REICHHART

Introduction

Heterologous expression is an unavoidable step for the functional characterization of P450 proteins, allowing unambiguous identification of natural, physiological substrates, and evaluation of the metabolism of xenobiotics by a protein isolated from its natural context and from other P450s present in the same organism. Achieving high levels of expression, however, becomes critical when recombinant enzymes are needed for industrial purposes or metabolic engineering in crop plants. Several heterologous expression systems have been used for functional screening of just confirmation of P450 activities, including *Escherichia coli*, yeasts, insect or mammalian cell cultures, and plants. The most frequently used, and one of the most amenable to industrial applications, is the yeast *Saccharomyces cerevisiae,* which combines the advantages of microorganisms for engineering, growth, and storage with those of an eukaryotic translation machinery and membrane environment. Various *S. cerevisiae* strains have been constructed that overexpress yeast,[1] mammalian,[2] or plant[3] P450 reductases, or combinations of P450 reductases and cytochrome b_5,[4] for enhanced P450 expression and activity. A large proportion of P450s from various organisms can be expressed successfully at high levels in such engineered yeasts, but some of them are expressed at extremely low levels or are totally recalcitrant to yeast expression. A similar problem is encountered when proteins are expressed in other organisms, e.g., in whole plants for the determination of the impact of a gene expression *in planta* or for the production of valuable metabolites.

Several mechanisms of impaired gene expression or protein synthesis have been described. A problem that is frequently encountered is a variation in codon usage. A change in codon usage implies a change in GC content,[5] in secondary structures, in translation initiation and codon context, and in potential splice sites in a given mRNA. Alone or in conjunction, these factors affect the efficiency of translation, often resulting in proteins being poorly expressed or mistranslated.[6,7]

---

[1] P. Urban, C. Cullin, and D. Pompon, *Biochimie* **72,** 463 (1996).
[2] J. C. Gautier, P. Urban, P. H. Beaune, and D. Pompon, *Eur. J. Biochem.* **211,** 63 (1993).
[3] P. Urban, C. Mignotte, M. Kazmaier, F. Delorme, and D. Pompon, *J. Biol. Chem.* **272,** 19176.
[4] D. Pompon, B. Louerat, A. Bronine, and P. Urban, *Methods Enzymol.* **272,** 51 (1996).
[5] R. D. Knight, S. J. Freeland, and L. F. Landweber, *Genome Biol.* **2**: research 0010.1 (2001).
[6] J. F. Kane, *Curr. Opin. Biotech.* **6,** 494 (1995).
[7] C. Kurland and J. Gallant, *Curr. Opin. Biotech.* **7,** 489 (1996).

                                  0076-6879/02 $35.00

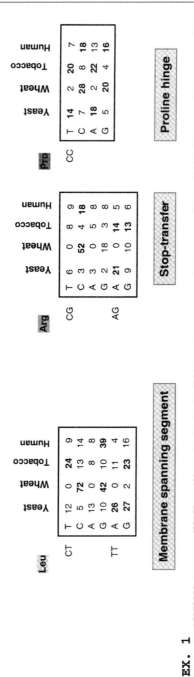

FIG. 1. Strongly biased codons found frequently at the 5' terminus of P450 cDNAs. The sequences of CYP73A17 from wheat (EX. 1) and of CYP73A1 from Jerusalem artichoke (EX. 2) are used as examples, with strongly biased codons highlighted. Average codon frequency (per thousand) in different organisms is shown in the boxes above the sequence.

The codon bias can be different between source and recombinant organisms (e.g., wheat and yeast). Codon bias also varies among genes for a same organism, becoming stronger for highly expressed genes, probably due to the need to ensure translation accuracy and efficiency, and to match abundance of the corresponding tRNAs.[8-13] Finally, variations in codon usage are observed within a single gene. Local clusters of rare codons or just codon choice was proposed to induce a pause protein synthesis or specific structures that favor the folding of protein subdomains.[14]

Some common structural features of membrane-anchored proteins, such as P450s and P450 reductases, imply a high frequency of strongly biased codons at the N terminus. Those encode hydrophobic residues forming the membrane-spanning segment (Leu, Ala, Val), positively charged amino acids found in the stop-transfer signal (Arg), and the proline cluster that forms the hinge between the membrane anchor and the globular part of the protein (Fig. 1). The presence of frequent low-usage codons was reported to cause severe translational inhibition or errors in E. coli[6,7] and yeast.[15] Clusters of rare codons, especially tandem repeats, are more likely to lead to translation failure when they were located close to the initiation site than when they were located farther in the coding sequence.[16-18] It was assumed that stalling of ribosomes at the beginning of a message interferes with the formation of new initiation complexes or that the stability of the translation complex, and the chances of the ribosomes not to fall off when pausing becomes too long at a minor codon, increases as a function of the distance already translated. In addition, the chances to obtain an aberrant protein increase as a mistake occurs upstream in the sequence.

General Strategy

Several strategies have been used to solve problems of codon bias. The first is a supplementation in rare tRNAs.[19-22] This first approach only addresses the

[8] T. J. Ikemura, J. Mol. Biol. **158**, 573 (1982).
[9] P. M. Sharp, T. M. F. Tuohy, and K. R. Mosurki, Nucleic Acids Res. **14**, 5125 (1986).
[10] P. M. Sharp and E. Cowe, Yeast **7**, 657 (1991).
[11] S. L. Fennoy and J. Bailey-Serres, Nucleic Acids Res. **21**, 5294 (1993).
[12] H. Chiapello, F. Lisacek, M. Caboche, and A. Hénaut, Gene **209**, 1 (1998).
[13] L. Duret, Trends Genet. **16**, 287 (2000).
[14] M. Oresic and D. Shalloway, J. Mol. Biol. **281**, 31 (1998).
[15] A. Hoekema, R. A. Kastelein, M. Varres, and H. A. De Boer, Mol. Cell. Biol. **7**, 2914 (1987).
[16] G. F. Chen and M. Inouye, Nucleic Acids Res. **18**, 1465 (1990).
[17] A. H. Rosenberg, E. Goldman, J. J. Dunn, F. W. Studier, and G. Zubay, J. Bact. **175**, 716 (1993).
[18] E. Goldman, A. H. Rosenberg, G. Zubay, and F. W. Studier, J. Mol. Biol. **245**, 467 (1995).
[19] U. Brinkmann, R. E. Mattes, and P. Buckel, Gene **85**, 109 (1989).
[20] R. A. Spanjaard, K. Chen, J. R. Walker, and J. van Duin, Nucleic Acids Res. **18**, 5031 (1990).
[21] B. J. Del Tito, J. M. Ward, J. Hodgson, C. J. L. Gershater, H. Edwards, L. A. Wysocki, F. A. Watson, G. Sathe, and J. F. Kane, J. Bact. **117**, 7086 (1995).
[22] T. Kleber-Janke and W. M. Becker, Protein Expr. Purif. **19**, 419 (2000).

problem of a single or of a few codons. The second is a complete codon reengineering of the sequence of interest. It has the advantage of help solving all the problems resulting from inappropriate codon usage and from other factors that could interfere with efficient expression.[23–25] Clustering of biased codons near the translation start of P450s or P450 reductases offers the possibility to selectively exchange unfavorable codons for those preferred by the host organism just at the 5' end of the coding sequences.

This strategy is often sufficient to obtain dramatic increases in the production of recombinant proteins, as (1) it suppresses problems of translation initiation and (2) it probably also helps decrease the total number of codons corresponding to rare tRNAs (the so-called hungry codons) below the threshhold that causes a complete depletion of the pool and early translation termination.[26]

Protocol

The length of the 5' coding sequence that needs to be recoded varies with the structure of individual genes and depends on the presence of tandem repeats of hungry codons at variable distances from the ATG. Optimizing the first 10 codons is the most critical, although total suppression of translation inhibition is usually not achieved until proteins are recoded beyond the Arg and Pro-rich region at the N terminus. Codons corresponding to rare tRNAs in the recombinant organisms www.kazusa.or.jp/codon) are replaced systematically with codons most encountered most frequently in these organisms on 100 to 120 bp from the ATG. Repetition of the same codon is avoided and the most frequent ones are used alternately. Recoding is performed by polymerase chain reaction (PCR) using *Hifi, Pfu,* or any high-fidelity DNA polymerase using a single megaprimer (120–150 mer) in 5' and a classical 20- to 25-mer in 3' of the amplified segment (Fig. 2). Suitable restriction sites are added at the 5' ends of each primer. PCR conditions are chosen so as to minimize errors, i.e., using high concentrations of template and small numbers of amplification cycles. Typically, the PCR mix contains 100 ng of template DNA (pBluescript + P450 coding sequence), 100 nM of each primer, 50 μM dNTPs, 1.5 units of *Hifi* (Roche, Basel, Switzerland) DNA polymerase, and the appropriate buffer in a total volume of 50 μl. The polymerase is added after 3 min heating at 94°. Amplification is performed using four cycles of 1 min 30 sec at 94°, 2 min 30 sec at 50°, and 3 min 30 sec at 72°, followed by nine cycles of 1 min 30 sec at 94°, 2 min 30 sec at 53°, 3 min 30 sec at 72°, and 10 min elongation at 72°. If *Pfu* (Stratagene,

[23] F. J. Perlak, R. L. Fuchs, D. A. Dean, S. L. McPherson, and D. A. Fischhoff, *Proc. Natl. Acad. Sci. U.S.A.* **88,** 3324 (1991).

[24] J. Haas, E. C. Park, and B. Seed, *Curr. Biol.* **6,** 315 (1996).

[25] G. J. A. Rouwendal, O. Mendes, E. J. H. Wolbert, and A. Douwe de Boer, *Plant Mol. Biol.* **33,** 989 (1997).

[26] S. Varenne and C. Ladzunski, *J. Theor. Biol.* **120,** 99 (1986).

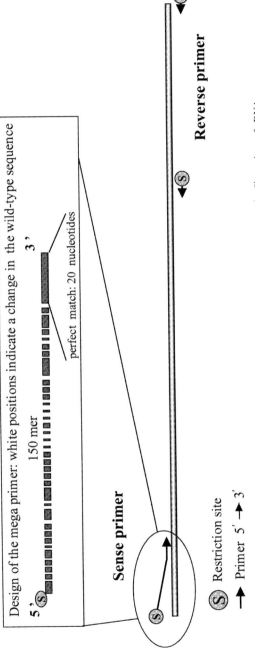

FIG. 2. PCR strategy for the synonymous replacement of nonoptimal codons at the 5′ terminus of cDNAs.

Gene	Segment recoded	Origin	% GC	Expression before recoding	Expression recoded
Example 1					
CYP73A17	5'-ter 18 bp	Wheat (**monocot**)	66	0	91
	5'-ter 39 bp			0	121
	5'-ter 111 bp			0	256
reductase TAR1	5'-ter 117 bp	Wheat	50	22	397
Example 2					
CYP73A1	5'-ter 99 bp	J. artichoke (**dicot**)	48	185	277
Example 3					
CYP86A5	5'-ter 120 bp	Wheat	70	0	136
	5'-ter 120 bp + inner 75 bp			0	161

La Jolla, CA) polymerase is used for the amplification, an additional step is needed for adding A overhangs. In this case, 5 units of *Taq* DNA polymerase (Invitrogen, Rockville, MD) is added to the medium, which is incubated for another 10 min at 72°. Amplified DNA fragments are purified on the gel, eluted, cloned into a T-overhang vector, and checked by sequencing before digestion and insertion into the expression vector.

Applications

The method is suitable for allowing or improving expression in any recombinant organism as soon as codon usage differs between the source of the cDNA and the system chosen for protein expression (e.g., for the expression of genes from monocots in yeasts or dicots). It is also useful for increasing protein expression when overall codon usage does not differ significantly between source and recombinant organisms, but local clusters, especially tandem repeats, of codons corresponding to rare tRNAs are present at the 5′ terminus of a coding sequence (see example 2 later). In this case, expression of a wild-type gene is usually not abolished completely in the recombinant organism, as the total number of nonoptimal codons does not lead to a complete depletion of the pool of rare tRNAs. Removal of low-usage codons 5′ of the cDNA, however, enhances translation initiation and reduces the risk of ribosomes to stall or fall off at the early stage of the protein synthesis. The result is a significant gain in yield of recombinant protein.

If elimination of clusters of nonoptimal codons at the 5′ end seems particularly critical, the synonymous replacement of tandem repeats of low-usage codons farther downstream in the coding sequence can also significantly enhance the expression of recombinant protein (see example 3). The gain in protein yield achieved via optimization of an internal sequence probably depends on the size and organization of the cluster of rare codons.

Examples of Recoding Impact on Expression in Yeast

Example 1

Among higher plants, the genome of grasses has a higher GC content than dicots. This implies a strong bias in codon usage compared to dicots or *S. cerevisiae*. The result is a low or insignificant expression of most P450 or P450 reductase genes from wheat or other monocots in classical yeast expression systems. Partial recoding of the 5′ end of these genes increases protein expression dramatically (Fig. 3). The increase in protein expression is proportional to the length of cDNA

FIG. 3. Examples of improved expression in *S. cerevisiae* resulting from codon optimization. The levels of P450 expression are given in $pmol \cdot mg^{-1}$ microsomal protein; P450-reductase expression is quantified measuring enzyme activity ($pmol \cdot min^{-1} \cdot mg^{-1}$ protein) in recombinant yeast microsomes.

recoded. Expression seems to reach a maximum when the first 100–110 bp have been optimized.[27]

Example 2

Codon usage in genes from dicotyledonous plants does not differ significantly from that of yeast genes. Clusters of low-usage codons are, however, found at the 5' end of some P450 sequences from dicots (Fig. 1). Codon optimization of such cDNAs leads to a significant increase in the yield of recombinant protein.[27]

Example 3

In this example, a cluster of low-usage codons is present in a more central segment of the coding sequence, corresponding to amino acids 158 to 182, in a P450 from a dicot (Fig. 3). A replacement of the nonoptimal codons with elimination of tandem repeats increases the protein expression by 25%.

Example of Recoding Impact on Expression in Plants

As mentioned earlier, there is a strong codon bias between monocots and dicots among higher plants. The most common model plant systems used for recombinant protein expression are the dicots tobacco and *Arabidopsis* because these plants are very easy to transform and routine procedures are available in most laboratories. Most attempts at expressing P450 genes from monocots in tobacco under the control of stong constitutive promoters (such as a double 35S promoter of the cauliflower mosaic virus) led to very deceiving results. Transcripts were present, sometimes at very high levels, but the protein was absent or barely detectable (Fig. 4). As in the case of yeast, codon optimization at the 5' terminus of the cDNA relieves translation inhibition and leads to a strong increase in protein expression. In the example illustrated in Fig. 4, the length of coding sequence that had to be optimized to relieve translation inhibition was longer in plants than in yeast.[27]

Limitations of the Method

The depletion of the pool of rare tRNAs will depend on the total number of cognate codons in the transcript population present at a given time. This implies that the risk for a depletion increases with the size and number of the protein(s) to be expressed and with the number of copies of the exogenous transcript(s) present in a recombinant cell. As a consequence, recoding of longer segments might be needed to obtain a good expression of very large proteins. For the same reasons,

---

[27] Y. Batard, A. Hehn, S. Nedelkina, M. Schalk, K. Pallett, H. Schaller, and D. Werck-Reichhart, *Arch. Biochem. Biophys.* **379**, 161 (2000).

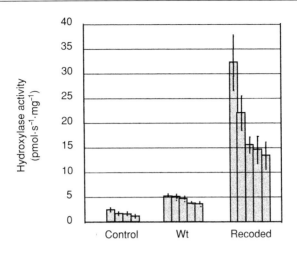

FIG. 4. Example of improved expression in tobacco resulting from codon optimization of a wheat P450 sequence. Each bar represents an individual T0 transformant. Levels of CYP73A17 expression are measured as cinnamate 4-hydroxylase activity in three microsomal membrane preparations from each plant. Control: plants transformed with a void plasmid. Wt: plants transformed with the wild-type CYP73A17 cDNA under the control of a double CaMV 35S promoter. Recoded: plants transformed with the same construction after optimization of the first 111 bp in 5′ of the cDNA.

simultaneous expression of P450 and reductase isolated from a same organism can impair each other, even when both protein are partially recoded. One possible way to circumvent this type of problems is to decrease the level of transcripts, either by using a weaker promoter or by decreasing the transcription rates. Reduced transcription rates can be, for example, achieved by decreasing the temperature during the induction of protein synthesis.

Acknowledgments

The support of Aventis Crops Science and of the Association Nationale de la Recherche Technique to A.H. and M.M. is gratefully acknowledged. The method described in this article is covered by a French and international patent.

[34] Selective Covalent Labeling with Radiolabeled Suicide Substrates for Isolating P450s

By CHRISTIAN HELVIG, NATHALIE TIJET, IRÈNE BENVENISTE, FRANCK PINOT, JEAN-PIERRE SALAÜN, and FRANCIS DURST

Introduction

Systematic sequencing of whole genomes, development of reverse genetics methods, and rapid progress in proteomics are providing tools of unprecedented power for the identification of gene functions. Despite these advances, equating the rapidly expanding number of P450 genes being deposited in the databases with a biological function remains a daunting task. This is particularly true for the plant P450 genes. Even the simple model plant *Arabidopsis* has five times the P450 number of humans, with 244 full-length genes forming 45 families (drnelson.utmem.edu/Arablinks.html). This diversity reflects the bewildering array of secondary metabolites synthesized by plants (*in toto* well over 100,000 distinct chemicals). A number of these chemicals have interesting organoleptic, pharmacological, or biocidal properties, or may be used as dyes or flavors. It is now well established that P450 plays a pivotal role by catalyzing the numerous hydroxylation, epoxidation, C–O and C–C coupling, C–O, C–N, and C–C cleavage, isomerization, and dehydration reactions that are essential for the diversification and complexification of the chemical structures of these plant secondary metabolites.[1–3] So, an alternative and complementary route to the identification of enzymes and the genes encoding them is to trace back from a known substrate or its metabolite(s) to the protein catalyzing the bioconversion, using a "reverse metabolomics" approach. This approach is becoming increasingly attractive with the recent development of sensitive and high-throughput methods for the analysis of metabolite dynamics (metabolomics), which allow to identify metabolites that are synthesized or broken down in response to developmental changes or external stimuli. However, classical protein purification methods will be at best very tedious and probably fail to relate thousands of potential substrates and hundreds of gene sequences. There are, for example, 66 members of the CYP71 family in *Arabidopsis thaliana,* too many probably to be resolved on the basis of their physicochemical properties only.

A promising route is offered by introducing a substrate-specific step into the separation procedure. This may be achieved by using so-called suicide substrates

[1] M. A. Schuler, *Crit. Rev. Plant Sci.* **15,** 235 (1996).
[2] C. Chapple, *Annu. Rev. Plant Physiol. Plant Mol. Biol.* **49,** 311 (1998).
[3] R. A. Kahn and F. Durst, *Recent Adv. Phytochem.* **34,** 151 (2000).

(really mechanism-based inhibitors) to introduce a covalent tag into the target enzyme and to allow tracking it during separation. As an example of the power of this approach and its shortcomings, this article describes the isolation of the first plant fatty acid ω-hydroxylase and its subsequent cloning.

Background

Mechanism-based enzyme inactivators are irreversible inactivators of specific target enzymes that catalyze their own destruction by unmasking a latent functional group during the catalytic cycle of the enzyme. The reactive intermediate thus formed may either evolve to a normal reaction product, which is then released from the enzyme, or become covalently attached. It is the ratio between these two competing reactions ($K_{dissociation}/K_{binding}$) that decides primarily the rate of enzyme inactivation. Even in the case of low $K_{binding}$, an excess of suicide substrates will entail complete inactivation, and therefore tagging of the enzyme molecules present. In principle, such inhibitors are very specific because they inactivate only those enzymes that recognize them as substrates. Inactivation of a target P450 may occur when the reactive species that is unmasked during an attempted oxidation step is attached covalently to the enzyme active site.[4] This strategy has been used to identify amino acids that occupy the active site of the enzyme.[5] This chapter describes the use of this covalent tagging technique to isolate a target P450 protein, the obtention of specific peptide sequences, and the subsequent cloning of the corresponding cDNA.

Target Enzyme

The target enzyme is a plant fatty acid hydroxylase. Fatty acid ω-hydroxylation is essential for the synthesis of the plant cuticle and is probably involved in signaling.[6] As shown previously, mechanism-based inhibitors containing a terminal acetylene are potent irreversible inhibitors of both mammalian[7,8] and plant[9,10]

[4] P. R. Ortiz de Montellano and M. A. Correia, in "Cytochrome P450: Structure, Mechanisms, and Biochemistry" (P. R. Ortiz de Montellano, ed.), 2nd Ed., p. 305. Plenum, New York, 1995.

[5] K. A. Regal, M. L. Schrag, U. M. Kent, L. C. Wienkers, and P. F. Hollenberg, Chem. Res. Toxicol. 13, 262 (2000).

[6] K. Wellesen, F. Durst, F. Pinot, I. Benveniste, K. Nettesheim, E. Wisman, S. Steiner-Lange, H. Saedler, and A. Yephremov, Proc. Natl. Acad. Sci. U.S.A. 98, 9694 (2001).

[7] C. A. Cajacob, W. K. Chan, E. Shephard, and P. R. Ortiz de Montellano, J. Biol. Chem. 263, 18640 (1986).

[8] S. Muerhoff, D. E. Williams, N. O. Reich, C. A. Cajacob, P. R. Ortiz de Montellano, and B. S. S. Masters, J. Biol. Chem. 264, 749 (1989).

[9] A. Simon, D. Reichhart, F. Durst, and P. R. Ortiz de Montellano, Xenobiotica 14s, 83 (1984).

[10] J.-P. Salaun, A. Simon, F. Durst, N. O. Reich, and P. R. Ortiz de Montellano, Arch. Biochem. Biophys. 260, 540 (1988).

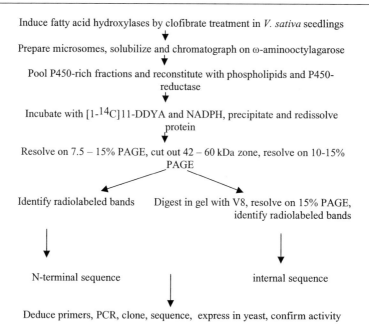

Induce fatty acid hydroxylases by clofibrate treatment in *V. sativa* seedlings

Prepare microsomes, solubilize and chromatograph on ω-aminooctylagarose

Pool P450-rich fractions and reconstitute with phospholipids and P450-reductase

Incubate with [1-^{14}C]11-DDYA and NADPH, precipitate and redissolve protein

Resolve on 7.5 – 15% PAGE, cut out 42 – 60 kDa zone, resolve on 10-15% PAGE

Identify radiolabeled bands   Digest in gel with V8, resolve on 15% PAGE, identify radiolabeled bands

N-terminal sequence   internal sequence

Deduce primers, PCR, clone, sequence, express in yeast, confirm activity

FIG. 1. Covalent labeling of target P450.

fatty acid ω-hydroxylases. In *Vicia sativa* microsomes incubated with terminal acetylene analogs and NADPH, a fraction of the acetylenic substrate, probably after activation to a ketene intermediate, forms a covalent link to the P450 apoprotein.[11]

Methodology

A flow chart of the main steps for covalent tagging, protein isolation and subsequent cloning, heterologous expression, and confirmation of catalytic activity is presented in Fig. 1.

Reactants

[1-^{14}C]Lauric acid (45 Ci/mol) and [1-^{14}C]linoleic acid (58 Ci/mol) are from NEN-DuPont (England). 11-[1-^{14}C]Dodecynoic acid (11-DDYA) (42.3 Ci/mol) is synthesized by established procedures.[11] Clofibrate and NADPH are from Sigma Chemie (La Verpillière, France). Dilauroylphosphatidylcholine is from Calbiochem (La Jolla, CA).

[11] C. Helvig, C. Alayrac, C. Mioskowski, D. Koop, D. Poullain, F. Durst, and J. P. Salaun, *J. Biol. Chem.* **272,** 414 (1997).

Preparation of Microsomes

Vicia sativa L. (var. Lolita) seeds are from S.A. Blondeau, BP 1, 59235 Bersée, France. Four-day-old etiolated seedlings are aged for 48 hr in a 1 mM clofibrate emulsion in distilled water. The extraction buffer contains 100 mM sodium phosphate buffer, pH 7.4, containing 1 mM EDTA and 10% glycerol (v/v); 15 mM 2-mercaptoethanol, 40 mM sodium ascorbate, 10% (w/w of fresh tissue) insoluble polyvinylpyrrolidone (PVP), and 1 mM phenylmethylsulfonyl fluoride (PMSF) (from a stock solution in methanol). The seedlings are homogenized in 1.5 volume chilled extraction buffer with a Waring blender or with an Ultra-Turrax at 8000 rpm. Homogenates are filtered through a 50-μm nylon mesh and centrifuged for 15 min at 10,000g. The supernatant is then centrifuged for 60 min at 100,000g at 4°. The microsomal pellets are resuspended in 50 mM sodium phosphate buffer, pH 7.4, 1.5 mM 2-mercaptoethanol, and 30% (v/v) glycerol. To prepare microsomes from yeast cells expressing CYP94A1, cells are grown for 3 days at 30° on SGI liquid minimum medium (1 g/liter lactocasamino acids, 7 g/liter yeast nitrogen base, 20 g/liter glucose, 20 mg/liter tryptophan). The culture is centrifuged at 7500g for 10 min at room temperature, and the pellet is resuspended in complete YPI medium (10 g/liter yeast extract, 10 g/liter Bacto-peptone, 20 g/liter galactose) and grown overnight at 30°. Cells are broken with glass beads (0.45 mm diameter), microsomes are prepared by differential centrifugation, and the pellet is resuspended in 50 mM Tris–HCl, pH 7.4, 1.5 mM 2-mercaptoethanol, and 20% (v/v) glycerol and is stored at −20°. Microsomal protein is quantified by a microassay procedure from Bio-Rad (Hercules, CA), and P450 is measured by the method of Omura and Sato.[12]

RNA Isolation and Analysis

Total RNA is isolated from clofibrate-treated *V. sativa* seedlings using the procedure of Lesot *et al.*[13] About 30 μg RNA per lane is denatured, separated on 1.2% agarose-formaldehyde gels, and transferred to Hybond N$^+$ membranes (Amersham). The blot is hybridized with the ^{32}P-labeled VAG11 fragment at 65° (see later). 18S Ribosomal DNA from radish is used as an internal control. After hybridization, the blot is washed twice with 2× SSC (1× SSC = 0.15 M NaCl/0.015 M sodium citrate), 0.1% sodium dodecyl sulfate (SDS) at room temperature for 15 min and twice with 0.2× SSC, 0.1% SDS at 55° for 30 min.

Chemical Tagging of ω-Hydroxylase

Microsomes from clofibrate-treated *V. sativa* seedlings are solubilized by 1.5 mg CHAPS/mg microsomal protein. CHAPS is decreased to 0.2% by dilution

---

[12] T. Omura and R. Sato, *J. Biol. Chem.* **239**, 1097 (1964).
[13] A. Lesot, I. Benveniste, M. P. Hasenfratz, and F. Durst, *Plant Cell Physiol.* **31**, 1177 (1990).

of the solubilizate, and P450 is partially purified and concentrated by chromatography on ω-aminooctylagarose equilibrated with 100 mM potassium phosphate, pH 7.2, containing 10% glycerol and 1.5 mM 2-mercaptoethanol. Bound protein is eluted with a 0 to 0.3% (w/v) linear gradient of Emulgen 911. P450-enriched fractions are concentrated on Centricon 30, and monooxygenase activity is reconstituted with concentrated NADPH P450 reductase from Jerusalem artichoke.[14] Excess detergent is removed with Biobeads SM2, and the enzyme preparation is incubated with [1-^{14}C]11-DDYA at room temperature for 30 min. Dilauroylphosphatidylcholine (DPLC) homogenized by sonication is added, and incubation is prolonged for 30 min at 4°. Labeling of the P450 is started by the addition of 1 mM NADPH with a regeneration system. Typical incubations contain 50 pmol P450, 50 pmol NADPH P450 reductase, 15 μg DLPC, and 100 μM [1-^{14}C]11-DDYA. After 30–60 min of incubation at 27°, the reaction is stopped by the addition of 1 ml cold acetone to precipitate proteins and to remove unreacted substrate and metabolites.

Isolation of ω-Hydroxylase

Precipitated proteins are collected by centrifugation (15 min at 3000g), dissolved in 50 μl of 180 mM Tris–HCl, pH 6.8, containing 5% SDS (w/v), 5% 2-mercaptoethanol (v/v), 30% glycerol (v/v), and 0.025% bromphenol blue (w/v), and analyzed by SDS–PAGE using 7.5 to 15% acrylamide. After Coomassie blue staining, gels are soaked for 20 min in Amplify and dried. Radioactivity is detected after 8–15 days of exposure using a Bio Imaging analyzer (Fujix Bas 2000, Fuji, Inc., Japan). To improve resolution, the portion of the gel comprising peptides between 48 and 58 kDa is cut out and chromatographed on a 10–15% acrylamide SDS gel.

Coomassie staining reveals that five peptides, labeled p1 to p5, are resolved in this manner. Two peptides, p3 and p4, are found to be radiolabeled after autoradiography. The two peptides are transferred on Problot membranes and sequenced. The sequence of peptide p4 shows only weak homology to a P450 N terminus, but the sequence for peptide p3, MFQFLLEVLLPYLLPLLLYILPF (the three residues underlined were later found to be H, L, and T, respectively), shows the hydrophobic stretch, which is typical for the N-terminal membrane anchor domain of microsomal P450 enzymes. In a parallel experiment, p3 is hydrolyzed directly in a 15% polyacrylamide gel with V8 protease. The gel is migrated twice, resolving five peptides: p3a to p3e, which are sequenced. Peptides p3a and p3b contain the N terminus of p3, p3c and p3d prove to be contaminants, but the sequence for p3e, LMNLYP-PVPMMNAKE, contains the LYPPVP motif. This motif is found in several P450s of the CYP4 family, approximately 70 residues N-terminal to the heme-binding

---

[14] I. Benveniste, A. Lesot, M.-P. Hasenfratz, G. Kochs, and F. Durst, *Biochem. Biophys. Res. Commun.* **177**, 105 (1991).

domain. The polymerase chain reaction (PCR) strategy to clone the ω-hydroxylase is based on the assumption that peptide p3e is part of the enzyme protein.

Cloning of ω-Hydroxylase cDNA

The sense primer 5′-TA(T/C)CCICCIGTICCIATG-3′ corresponding to six amino acids of internal peptide p3e is synthesized and used with antisense oligo(dT)$_{18}$ NotI for reverse transcription (RT)-PCR amplification of specific cDNAs employing RNA from clofibrate-treated *V. sativa* seedlings. After filling in the 3′ overhangs, the PCR amplification product is digested with NotI, ligated into NotI/EcoRV-cut pBluescript SK(+), and sequenced. The excised RT-PCR NotI/EcoRV fragment, named VAG11, is labeled with [^{32}P]dCTP by random priming and is used to screen at high stringency 300,000 recombinant clones from a λZAP cDNA library from clofibrate-treated *V. sativa* seedlings. From 18 positive clones, 8, which have insert sizes above 1800 bp, are further PCR analyzed using internal primers from VAG11 and appear identical. Clone VAGH111 is sequenced and found to code for a full-length cytochrome P450, which is classified as CYP94A1 by the P450 nomenclature committee.[15]

Heterologous Expression of CYP94A1 in Yeast

To confirm that VAGH111 codes for the target P450, *Saccharomyces cerevisiae* strain WAT11, which overexpresses a plant P450 reductase, is transformed with VAGH111 as described by Pompon *et al.*[16] Reformatting and cloning of VAGH111 into the vector pYeDP60 are performed by PCR amplification using specific primers to delete the 5′ and 3′ noncoding regions of VAGH111 and introduce a *Bam*HI site immediately upstream of the initiation codon and an *Eco*RI site immediately downstream of the stop codon. The nucleotide sequence of both strands is determined, and yeast is transformed using the lithium acetate procedure of Schiestl and Gietz.[17]

Enzyme Activity

The standard assay contains in a final volume of 0.2 ml, 0.19–0.43 mg of microsomal protein, 20 mM phosphate buffer, pH 7.4, 1 mM NADPH plus a regenerating system (6.7 mM glucose 6-phosphate and 0.4 IU of glucose-6-phosphate dehydrogenase), 375 μM 2-mercaptoethanol, and radiolabeled substrate. The reaction is initiated with NADPH at 27°, stopped with 0.2 ml acetonitrile/acetic acid

[15] N. Tijet, C. Helvig, F. Pinot, R. Lebouquin, A. Lesot, F. Durst, J. P. Salaun, and I. Benveniste, *Biochem. J.* **332**, 583 (1998).

[16] D. Pompon, B. Louerat, A. Bronine, and P. Urban, *Methods Enzymol.* **272**, 51 (1996).

[17] R. H. Schiestl and R. D. Gietz, *Curr. Genet.* **16**, 339 (1989).

(99.8/0.2, v/v), and mixed, and 100-μl aliquots are spotted directly onto silica plates developed with a mixture of diethyl ether/light petroleum (bp 40–60°)/formic acid (70/30/1, v/v/v) for lauric acid and (50/50/1) for oleic acid. Areas corresponding to polar metabolites are scrapped directly into counting vials or eluted with ether and subjected to RP-HPLC analysis using a mixture of water/acetonitrile/acetic acid (25/75/0.2) and (45/55/0.2) as described previously.[18] The radioactivity of RP-HPLC effluents is monitored with a computerized on-line solid scintillation counter (Ramona-D RAYTEST, Germany). For gas chromatography-mass spectrometry (GC-MS) analysis, metabolites are methylated and silylated as described and subjected to GC-MS analysis. The fragmentation pattern of MeTMS derivatives of ω-hydroxylated FA is similar to that already described.[19] Lauric acid, which is a model substrate for plant fatty acid hydroxylases, and oleic acid, which is ω-hydroxylated to form the main precursor of C_{18} cutin monomers,[18] are hydroxylated with a turnover number of 20 ($V/K = 1.4$) and 14.3 ($V/K = 0.4$). The reaction products are identified as the corresponding ω-hydroxy acids by cochromatography with synthetic reference compounds and by GC-MS analysis. The enzyme appears specific for fatty acids, as physiological substrates to other plant P450 enzymes, such as cinnamic acid or kaurene, or different herbicides known to be metabolized by plant P450 are not oxidized.

Comments

Since the cloning of CYP94A1, more plant fatty acid hydroxylases have been isolated using 94A1 as a probe. They have been expressed in yeast, their kinetic parameters determined, and the induction patterns have been studied. In retrospect, we can now analyze the different parameters that have lead to success and to the isolation of this particular P450. Success depends on several factors.

1. Affinity of the enzyme for the substrate. In principle, the extend of suicide inactivation is determined by the partition ratio of the suicide substrate and the relative amounts of enzyme and inhibitor, rather than by affinity. While this holds true for a pure enzyme in solution, it may not be the case with microsomes, which may contain other P450 forms having an affinity for this substrate. We know at present that at least two other fatty acid ω-hydroxylases, CYP94A2[20] and CYP94A3, are expressed in *V. sativa* microsomes. Kinetic studies have shown that they all accept laurate as substrate but that only 94A1 has high activity with

[18] F. Pinot, J.-P. Salaun, H. Bosch, A. Lesot, C. Mioskowski, and F. Durst, *Biochem. Biophys. Res. Commun.* **184**, 183 (1992).

[19] F. Pinot, H. Bosch, C. Alayrac, C. Mioskowski, A. Vendais, F. Durst, and J. P. Salaun, *Plant Physiol.* **102**, 1313 (1993).

[20] R. A. Kahn, R. Le Bouquin, F. Pinot, I. Benveniste, and F. Durst, *Arch. Biochem. Biophys.* **391**, 180 (2001).

oleic acid (C18:1). A posteriori, this suggests that much better discrimination among the three isoforms could be achieved using 17-ODNYA (Z)9-octadecen-17-ynoic acid, the terminal acetylene analog of oleic acid or the corresponding 9,10-epoxide.[19]

2. Relative amount of the target enzyme in microsomes. That 94A1 was tagged predominantly, not 94A2 and 94A3, also results from the clofibrate treatment of the seedlings prior to microsome preparation. Subsequent induction studies with the three isoforms have shown that only 94A1 is induced.[15] Under these conditions, the ratio of 94A1/A2 + A3 mRNAs reaches well over 100, ensuring optimal conditions for selective labeling of 94A1.

3. Elimination of nonspecific binding. Ideally, the reactive ketene species is generated in the immediate vicinity of a strong nucleophile, leading to a high partition ratio and a very low escape of active compound from the catalytic site. The presence of 1.5 mM 2-mercaptoethanol (or other -SH-containing compounds such as glutathione) during enzyme inactivation was essential. In the absence of these compounds, several other microsomal proteins were labeled,[11] indicating that the ketene intermediate formed when the enzyme oxidized 11-DDYA was longer lived than generally assumed and that a fraction of it could exit from the active site and label neighbor proteins.

4. The choice of the type of suicide substrate used. Obviously, inhibitors leading mainly to covalent tagging of the protein will be prefered over those labeling the heme, which might be lost during the subsequent purification procedure. Acetylenic analogs of substrates appear well suited for this purpose.[4] A report on the autocatalytic inactivation of the 8'-hydroxylase of abscisic acid, a plant hormone, by 8'-propargylabscisic acid[21] shows that this technique may be used to target other putative P450 enzymes.

[21] J. Cutler, P. A. Rose, T. M. Squires, M. K. Loewen, A. C. Shaw, J. W. Quail, J. E. Krochko, and S. R. Abrams, *Biochemistry* **39**, 13614 (2000).

[35] Cloning of cDNAs Encoding P450s in Flavonoid/Isoflavonoid Pathway from Elicited Leguminous Cell Cultures

By SHIN-ICHI AYABE, TOMOYOSHI AKASHI, and TOSHIO AOKI

Introduction

The first cloning of a plant cytochrome P450 (P450) cDNA (*CYP71A1*) was reported in 1990.[1] Cinnamic acid 4-hydroxylase (CYP73 family) cDNAs[2–4] and flavonoid 3′,5′-hydroxylase (CYP75 family) genes[5] were also functionally characterized very early in the history of plant P450 research. Both enzymes are involved in flavonoid biosynthesis. Then, however, progress in the characterization of P450 cDNAs of the flavonoid pathway slowed until 1998, when the cDNAs encoding (2*S*)-flavanone 2-hydroxylase (CYP93B1),[6] pterocarpan 6a-hydroxylase (CYP93A1),[7] and isoflavone 2′-hydroxylase (CYP81E1)[8] were identified. Further P450 cDNAs of the flavonoid pathway, i.e., flavone synthase II (CYP93B subfamily),[9,10] 2-hydroxyisoflavanone synthase (CYP93C subfamily),[11–14] flavonoid 3′-hydroxylase (CYP75B2),[15] and flavanone 6-hydroxylase (CYP71D9),[16] have now been characterized.

---

[1] K. R. Bozak, H. Yu, R. Sirevåg, and R. E. Christoffersen, *Proc. Natl. Acad. Sci. U.S.A.* **87**, 3904 (1990).

[2] T. Fahrendorf and R. A. Dixon, *Arch. Biochem. Biophys.* **305**, 509 (1993).

[3] M. Mizutani, E. Ward, J. DiMaio, D. Ohta, J. Ryals, and R. Sato, *Biochem. Biophys. Res. Commun.* **190**, 875 (1993).

[4] H. G. Teutsch, M. P. Hasenfratz, A. Lesot, C. Stoltz, J. M. Garnier, J. M. Jeltsch, F. Durst, and D. Werck-Reichhart, *Proc. Natl. Acad. Sci. U.S.A.* **90**, 4102 (1993).

[5] T. A. Holton, F. Brugliera, D. R. Lester, Y. Tanaka, C. D. Hyland, J. G. T. Menting, C.-Y. Lu, E. Farcy, T. W. Stevenson, and E. C. Cornish, *Nature* **366**, 276 (1993).

[6] T. Akashi, T. Aoki, and S. Ayabe, *FEBS Lett.* **431**, 287 (1998).

[7] C. R. Schopfer, G. Kochs, F. Lottspeich, and J. Ebel, *FEBS Lett.* **432**, 182 (1998).

[8] T. Akashi, T. Aoki, and S. Ayabe, *Biochem. Biophys. Res. Commun.* **251**, 67 (1998).

[9] T. Akashi, M. Fukuchi-Mizutani, T. Aoki, Y. Ueyama, K. Yonekura-Sakakibara, Y. Tanaka, T. Kusumi, and S. Ayabe, *Plant Cell Physiol.* **40**, 1182 (1999).

[10] S. Martens and G. Forkmann, *Plant J.* **20**, 611 (1999).

[11] T. Akashi, T. Aoki, and S. Ayabe, *Plant Physiol.* **121**, 821 (1999).

[12] W. Jung, O. Yu, S. M. Lau, D. P. O'Keefe, J. Odell, G. Fader, and B. McGonigle, *Nature Biotechnol.* **18**, 208 (2000).

[13] C. L. Steele, M. Gijzen, D. Qutob, and R. A. Dixon, *Arch. Biochem. Biophys.* **367**, 146 (1999).

[14] N. Shimada, T. Akashi, T. Aoki, and S. Ayabe, *Plant Sci.* **160**, 37 (2000).

[15] F. Brugliera, G. Barri-Rewell, T. A. Holton, and J. G. Mason, *Plant J.* **19**, 441 (1999).

[16] A. O. Latunde-Dada, F. Cabello-Hurtado, N. Czittrich, L. Didierjean, C. Schopfer, N. Hertkorn, D. Werck-Reichhart, and J. Ebel, *J. Biol. Chem.* **276**, 1688 (2001).

P450s of flavonoid biosynthesis can be classified into two categories: aromatic carbon hydroxylases (flavonoid 3′,5′- and 3′-hydroxylases, isoflavone 2′-hydroxylase, and flavanone 6-hydroxylase) and oxygenases/oxidases of nonaromatic carbons, including those involved in the construction of flavone and isoflavone skeletons [(2S)-flavanone 2-hydroxylase, pterocarpan 6a-hydroxylase, flavone synthase II, and 2-hydroxyisoflavanone synthase]. All P450s of the latter category now known belong to the CYP93 family and, except for pterocarpan 6a-hydroxylase, use (2S)-flavanones as the substrate. The stereochemical coincidence of the configuration of the ligands (p-hydroxyphenyl, methylene, and hydrogen) around C-2 and C-6a of (2S)-flavanones and (6aR)-pterocarpan, the substrates for CYP93s, has been recognized to explain the apparent difference in the substrate specificities among pterocarpan 6a-hydroxylase and other CYP93s.[11]

2-Hydroxyisoflavanone synthase has been of particular interest from the viewpoint of the mechanism of both its catalysis and its physiological functions.[17,18] It catalyzes the hydroxylation of a nonaromatic carbon (C-2) and, at the same time, constructs the isoflavonoid skeleton through 1,2-aryl migration. It is the most important enzyme in the biosynthesis of legume-specific isoflavonoid phytoalexins, as well as of daidzein and genistein, the constitutive ingredients of soybean, which possess health-promoting phytoestrogenic effects.

The three P450 cDNAs encoding 2-hydroxyisoflavanone synthase, (2S)-flavanone 2-hydroxylase, and isoflavone 2′-hydroxylase, which are involved in the biosynthesis of phytoalexins and a unique retrochalcone, have been characterized from suspension-cultured cells of the leguminous plant Glycyrrhiza echinata (licorice) by polymerase chain reaction (PCR) amplification of P450 fragments.[6,8,11,19] The use of elicitor-treated cells as the source of a cDNA library was the reason for the successful cloning. In these cells, medicarpin, a pterocarpan phytoalexin, and retrochalcone are both induced to accumulate to the maximum concentration in about 24 hr after elicitation.[20,21] The P450s involved in the pathways shown in Fig. 1, none of which had been identified at the molecular level at the time, were expected to be induced on elicitation. The full-length cDNAs of the inducible P450s were then expressed heterologously in insect or yeast cells, and the recombinant microsomes were used for enzyme assays. This article describes the methods of cloning these P450 cDNAs from elicited cultures of G. echinata and the procedures for the functional characterization with heterologous eukaryotic

[17] R. A. Dixon and C. L. Steele, Trends Plant Sci. 4, 394 (1999).
[18] R. A. Dixon, in "Comprehensive Natural Products Chemistry" (U. Sankawa, ed.), Vol. 1, p. 773. Elsevier, Oxford, 1999.
[19] T. Akashi, T. Aoki, T. Takahashi, N. Kameya, I. Nakamura, and S. Ayabe, Plant Sci. 126, 39 (1997).
[20] S. Ayabe, K. Iida, and T. Furuya, Phytochemistry 25, 2803 (1986).
[21] K. Nakamura, T. Akashi, T. Aoki, K. Kawaguchi, and S. Ayabe, Biosci. Biotechnol. Biochem. 63, 1618 (1999).

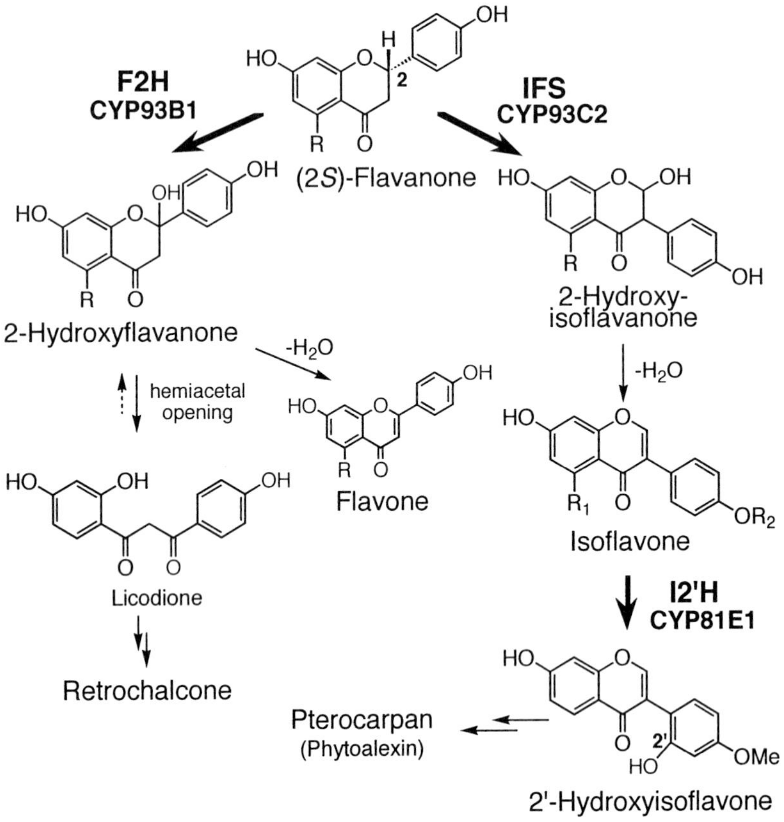

FIG. 1. Biosynthesis of isoflavonoids, retrochalcone, and flavones. F2H, (2S)-flavanone 2-hydroxylase; I2'H, isoflavone 2'-hydroxylase; IFS, 2-hydroxyisoflavanone synthase.

cells. Detailed methods of assaying the enzyme activities and the preparation of the 2-hydroxyisoflavanone sample by a large-scale incubation are presented.

Some notes on the enzyme reactions and the products are relevant here. First, licodione, the direct precursor of retrochalcone, is synthesized from (2S)-5-deoxyflavanone[22] and is the spontaneous chain tautomer of 2-hydroxyflavanone, a putative hemiacetal intermediate in the flavone synthase II reaction.[23] Therefore, the P450 licodione synthase is identical to (2S)-flavanone 2-hydroxylase (CYP93B1) and is closely related to flavone synthase II. Actually, the later cloning of genuine flavone synthase II cDNAs (CYP93B3, CYP93B4) from torenia and snapdragon was performed successfully using G. echinata CYP93B1 cDNA as a

---

[22] K. Otani, T. Takahashi, T. Furuya, and S. Ayabe, Plant Physiol. 105, 1427 (1994).
[23] G. Stotz and G. Forkmann, Z. Naturforsch. 36c, 737 (1981).

probe for screening.[9] Licodione can be identified by thin-layer chromatography (TLC), but detection of it by high-performance liquid chromatography (HPLC) is difficult because it exists in a keto–enol tautomeric mixture.[24] Licodione easily loses water on acid treatment and is converted to the stable 7,4'-dihydroxyflavone. 2-Hydroxynaringenin, the product of the same reaction from 5-hydroxyflavanone, is also acid labile and is converted into the flavone apigenin. Therefore, (2S)-flavanone 2-hydroxylase can be best assayed by measuring flavone formation after the direct reaction product is treated with acid. Second, 2-hydroxyisoflavanone, the first P450 (CYP93C) reaction product in the phytoalexin pathway with an isoflavonoid skeleton, is also very labile to acid and loses water to give an isoflavone.[25] In the routine assay of 2-hydroxyisoflavanone synthase (CYP93C2), the formation of the isoflavones daidzein and genistein from 5-deoxyflavanone and 5-hydroxyflavanone, respectively, after the acid treatment of the reaction mixture is determined by HPLC. However, 2-hydroxyisoflavanone can be isolated from a large-scale incubation of the yeast microsome expressing CYP93C2 with the 5-deoxyflavanone substrate. We have been successful in establishing a new scheme of biosynthesis of a methylated isoflavone, formononetin, using in vitro-synthesized 2-hydroxyisoflavanone as the real substrate for methyltransferase.[26] Finally, isoflavone 2'-hydroxylase, an aromatic carbon hydroxylase of the isoflavonoid pathway, is essential in the biosynthesis of phytoalexins because most, if not all, isoflavonoid phytoalexins have an oxygen function at the corresponding position. The detection and separation of 2'-hydroxyisoflavone products from isoflavone substrates are performed easily using HPLC. Chromatographic data on the substrates and the products of these enzyme reactions are summarized in Table I.

Cultured Plant Cells, Elicitation, and Preparation of cDNA Library

Two callus cell lines of G. echinata, one of which was established in the late 1960s (Ge-1 line)[27] and another which was newly derived from young leaves and petioles in 1998 (AK-1 line),[21] are maintained on Murashige–Skoog's agar [0.9% (w/v)] medium supplemented with either β-indoleacetic acid (1 mg/liter) and kinetin (0.1 mg/liter) (for the Ge-1 line) or 2,4-dichlorophenoxyacetic acid (0.1 mg/liter) and kinetin (1 mg/liter) (for the AK-1 line). Suspension cultures are initiated by inoculating the callus cells into liquid media of the same composition but without agar, and elicitation is performed by the aseptic addition of 0.1–0.2% (w/v) yeast extract (Difco Laboratories, Detroit, MI) to the cultures. Poly(A)$^+$

[24] Y. Kirikae, M. Sakurai, T. Furuno, T. Takahashi, and S. Ayabe, Biosci. Biotechnol. Biochem. 57, 1353 (1993).
[25] M. F. Hashim, T. Hakamatsuka, Y. Ebizuka, and U. Sankawa, FEBS Lett. 271, 219 (1990).
[26] T. Akashi, Y. Sawada, T. Aoki, and S. Ayabe, Biosci. Biotechnol. Biochem. 64, 2276 (2000).
[27] T. Furuya, K. Matsumoto, and M. Hikichi, Tetrahedron Lett. 27, 2567 (1971).

TABLE I
TLC AND HPLC DATA OF FLAVONOIDS[a]

Substituent	Flavonoid subclass and compound[a]	λ_{max} (nm)	Solvent						
			R_f in TLC				R_t (min) in HPLC		
			A[b]	B[c]	C[c]	D[c]	I	II	III
	Flavanone								
R=OH	Naringenin	292	0.46	0.35	0.75			8.9	
R=H	Liquiritigenin	277	0.38	0.53		0.75	18.5	6.5	
	2-Hydroxyisoflavanone								
R=OH	2,5,7,4'-Tetrahydroxyisoflavanone			0.71					
R=H	2,7,4'-Trihydroxyisoflavanone	285	0.20	0.76			5.6		
	2-Hydroxyflavanone and tautomer								
R=OH	2,5,7,4'-Tetrahydroxyflavanone				0.38				
	Licodione					0.59			
	Isoflavone								
R₁=OH, R₂=H	Genistein	262		0.30					24.6
R₁=R₂=H	Daidzein	248		0.38			20.7		13.9
R₁=H, R₂=CH₃	Formononetin	248						16.6	
	Flavone								
R=OH	Apigenin	340			0.65			12.4	
R=H	7,4'-Dihydroxyflavone	330				0.41		8.5	
	3-Hydroxyflavanone								
	3,5,7,4'-Tetrahydroxyflavanone			0.63					
	3,7,4'-Trihydroxyflavanone	231		0.64			8.0		
	2'-Hydroxyisoflavone								
R₁=OH, R₂=H	2'-Hydroxygenistein	262							13.0
R₁=R₂=H	2'-Hydroxydaidzein	248							9.0
R₁=H, R₂=CH₃	2'-Hydroxyformononetin	248						10.6	

[a] See Fig. 1 for the structures in each flavonoid subclass.
[b] Solvent for silica gel TLC: A, toluene : ethyl acetate : methanol : light petroleum = 6 : 4 : 1 : 3 (v/v).
[c] Solvent for cellulose TLC: B, 15% acetic acid; C, chloroform : acetic acid : water = 10 : 9 : 1 (v/v); D, 30% acetic acid. I, 40% methanol; II, 50% methanol and 3% acetic acid; III, 40% methanol and 3% acetic acid.

RNA is isolated from 1 g of the cells 6 to 12 hr after elicitation using the Quick Prep mRNA purification kit (Amersham Pharmacia Biotech, Buckinghamshire, England) or the Straight A's mRNA isolation system (Novagen, Madison, WI). The cDNA library is constructed using the ZAP-cDNA synthesis kit (Stratagene, La Jolla, CA).

Amplification of P450 Fragments and Screening of Full-Length cDNAs

Inosine-containing 5'-degenerate primers are constructed from the conserved sequence of plant P450s. Primer I [5'-(T/C/A)TI(C/G)CITT(T/C)(G/A)GIIIIGGI (A/C)(G/C)I(A/C)G-3'] is based on the heme-binding domain (FxxGxxxCxG), but the sequence corresponding to the cysteine that binds to heme is not included. Primer II [5′-(A/G)(A/C)IT(T/A)(T/C)IIICCI(G/A)(A/T)(G/A)AG(G/A)TT-3′] is based on about 30 amino acids upstream from this domain. PCR is carried out using 5′-degenerate primers and the 3′-nonspecific antisense primer (T7 primer, 5′-AATACGACTCACTATAG-3′) complementary to the λZap II vector sequence. Both PCR products amplified from primer I containing the cysteine-binding CxG motif and primer II with the FxxGxxxCxG motif are assumed to be fragments of P450 cDNA.

Eight P450 fragments are obtained from the cDNA library of the *G. echinata* Ge-1 line treated with 0.2% yeast extract for 8 hr. Screening the Ge-1 cDNA library, as well as the library of the elicited AK-1 line using the fragments, yields four full-length sequences. While CYP73A14 is identified as cinnamic acid 4-hydroxylase, the other three display little sequence homology (<40%) with functionally identified plant P450s at the amino acid level and are therefore assigned to new subfamilies, i.e., CYP81E1, CYP93B1, and CYP93C2, by Dr. D. Nelson of the P450 Nomenclature Committee.

Heterologous Expression

P450s function in an electron transfer system with cytochrome P450 reductase as the electron donor. To test the enzymatic function of P450s with heterologous expression systems, cDNAs are generally expressed in eukaryotes such as insect and yeast cells.[28,29] A baculovirus–insect expression system allows the production of a high level of recombinant protein, but the construction of a recombinant baculovirus and maintenance of the cell line require time-consuming procedures and expensive reagents. However, handling yeast cells is relatively easy, although the expression level is sometimes low. Previous studies employed recombinant yeast cells transformed with a cDNA encoding plant P450 reductase for the expression of plant P450s,[30] but heterologous P450 reductase seemed unnecessary in the case of functional expression of 2-hydroxyisoflavanone synthase, (2S)-flavanone 2-hydroxylase, and isoflavone 2′-hydroxylase. While insect cells were used for the initial identification of (2S)-flavanone 2-hydroxylase cDNA, more convenient yeast cells have become the standard assay material for P450s of the (iso)flavonoid pathway in our laboratory.

Expression of P450s in Insect Cells

The expression vector is constructed by subcloning the coding region with modified end sequences amplified by PCR with KOD polymerase (Toyobo, Tokyo, Japan) into the pFASTBAC1 donor plasmid (Invitrogen, Carlsbad, CA). The

[28] F. P. Guengerich, W. R. Brain, M.-A. Sari, and J. T. Ross, *Methods Enzymol.* **206,** 130 (1991).
[29] C. A. Lee, T. A. Kost, and C. J. Serabjit-Singh, *Methods Enzymol.* **272,** 86 (1996).
[30] D. Pompon, B. Louerat, A. Bronine, and P. Urban, *Methods Enzymol.* **272,** 51 (1996).

recombinant baculovirus is constructed using the Bac-to-Bac Baculovirus Expression system (Invitrogen). For the expression of P450 proteins, *Spodoptera frugiperda* (Sf9) cells (Invitrogen) maintained in TMN-FH medium[31] are infected with recombinant virus at a rate of 5 to 10 plaque-forming units per cell and incubated at 25°. Hemin (2 μg/ml) is added to the culture medium to supplement the low endogenous levels of heme in insect cells. Cells are harvested 110 hr after infection, washed with phosphate-buffered saline, and stored at −80°.

Expression of P450s in Yeast Cells

For the expression of plant P450s in yeast cells, we use the expression vector pYES2 (Invitrogen) and the protease-deficient *Saccharomyces cerevisiae* strain BJ2168 (a; *prc1407, prb1-1122, pep4-3, leu2, trp1, ura3-52;* Nippon Gene, Tokyo, Japan). The coding regions with tailored end sequences possessing restriction enzyme sites are obtained by PCR. The product is digested with the relevant restriction enzymes, and the fragment is cloned into the corresponding site of pYES2 downstream from the *GAL1* promoter in sense orientation. The plasmids are transferred into BJ2168 cells, and the transformants are selected on a medium composed of 6.7 mg/ml yeast nitrogen base without amino acids, 20 mg/ml glucose, 30 μg/ml leucine, 20 μg/ml tryptophan, and 5 mg/ml casamino acid. For the induction of the protein, cells grown in the medium for 10 hr are transferred to 40 volumes of induction medium, i.e., YPGE medium[30] supplemented with 20 mg/ml galactose and 2 μg/ml hemin but without glucose. After 24 hr, the cells are harvested, frozen in liquid nitrogen, and stored as described earlier.

Preparation of Microsomes

All procedures are carried out at 4°. The insect and yeast cells are disrupted with glass beads (0.35–0.60 mm diameter) in 3 volumes of 100 mM potassium phosphate (pH 7.5) containing 300–400 mM sucrose and 3.5 mM (for cells expressing CYP81E1 protein) or 14 mM (for cells expressing CYP93 proteins) 2-mercaptoethanol. The lysate is centrifuged successively at 10,000g and 15,000g for 10 min each, and the resultant supernatant is ultracentrifuged at 160,000g for 90 min. The precipitates are suspended homogeneously in the disruption buffer described earlier and used as the enzyme preparation (1.0–3.0 mg/ml protein).

The expression of the heterologous plant P450 protein is estimated by the CO difference spectrum. To 2 ml microsome suspension, 1 ml of 100 mg/ml sodium dithionite is added. An aliquot is bubbled with carbon monoxide for 30 sec in a 50-ml conical polypropylene tube (Becton Dickinson Labware, Bedford, MA), and A_{450} and A_{490} are measured with the nonbubbled aliquot as the blank. The

---

[31] D. R. O'Reilly, L. K. Miller, and V. A. Luckow, *in* "Baculovirus Expression Vectors: A Laboratory Manual," p. 109. Oxford Univ. Press, New York, 1994.

molarity of P450 (mM) is determined according to the formula $(A_{450}-A_{490})$/91.[32] Usually, 50–100 pmol/mg protein of P450 is observed in the microsome of recombinant yeast in our experiments.

Enzyme Assay with Recombinant Microsomes

The radiochemical assay is sharp and clear, with the substrates being the natural enantiomer and the products easily traced from the label, but the cost is high and the experimental conditions are strictly restricted. Nonradiochemical methods are easy, but sometimes may not give unambiguous results due to endogenous contamination in the enzyme reaction products. Accordingly, each method should be properly suited to the purpose of the study; e.g., the first functional characterization of 2-hydroxyisoflavanone synthase and (2S)-flavanone 2-hydroxylase was performed by radiochemical assay, and subsequent determinations of substrate specificities were done by nonradiochemical assay.

Enzymatic Synthesis of ^{14}C-Labeled (2S)-Flavanones

(2S)-Flavanones (liquiritigenin and naringenin) are the common substrates for 2-hydroxyisoflavanone synthase and (2S)-flavanone 2-hydroxylase. They are produced in plant cells from 4-coumaroyl-CoA and malonyl-CoA catalyzed by chalcone synthase, chalcone isomerase, and, in the 5-deoxyflavonoid pathway, polyketide reductase.[33,34] For the radiochemical assays of P450s with (2S)-flavanone substrates, ^{14}C-labeled (2S)-liquiritigenin and (2S)-naringenin are synthesized enzymatically with plant cell-free extracts. G. echinata cells (Ge-1 or AK-1 line, 10 g) treated with 0.2% yeast extract for 12 hr are homogenized in 10 ml of 100 mM potassium phosphate buffer (pH 6.0) containing 20 mM sodium ascorbate and 25% glycerol using a chilled mortar and sea sand. Homogenates are filtered through cheesecloth and centrifuged at 10,000g for 10 min at 4°. The supernatant is then mixed with 2.5 g of Dowex 1-X2 (equilibrated with 100 mM potassium phosphate buffer, pH 6.0) and left standing for 20 min. The solution obtained by filtration is incubated with 130 nmol of 4-coumaroyl-CoA, synthesized by the method of Stöckigt and Zenk,[35] and 85 nmol of [2-^{14}C]malonyl-CoA (185 kBq) in the presence or absence of 0.5 mM NADPH at 30° for 2 hr. The ethyl acetate extracts of the reaction mixture are subjected to silica gel TLC [Kieselgel F$_{254}$ (Merck, Darmstadt, Germany)] with solvent A (see Table I), and the radioactive product, either

[32] T. Omura and R. Sato, *J. Biol. Chem.* **239**, 2379 (1964).

[33] W. Heller and G. Forkmann, *in* "The Flavonoids: Advances in Research since 1986" (J. B. Harborne, ed.), p. 499. Chapman and Hall, London, 1994.

[34] G. Forkmann and W. Heller, *in* "Comprehensive Natural Products Chemistry" (U. Sankawa, ed.), Vol. 1, p. 713. Elsevier, Oxford, 1999.

[35] J. Stöckigt and M. H. Zenk, *Z. Naturforsch.* **30c**, 352 (1975).

$(2S)$-[^{14}C]liquiritigenin from the incubation with NADPH or $(2S)$-[^{14}C]naringenin from the incubation without NADPH, is separated and eluted with ethyl acetate.

Radiochemical Assay of 2-Hydroxyisoflavanone Synthase and (2S)-Flavanone 2-Hydroxylase

For the assay of 2-hydroxyisoflavanone synthase, $(2S)$-[^{14}C]liquiritigenin or $(2S)$-[^{14}C]naringenin (0.08 nmol, 6.4 kBq/nmol) in 30 μl 2-methoxyethanol is incubated with 1 ml yeast microsomes expressing CYP93C2 in the presence of 1 mM NADPH in the total volume of 1.05 ml. After incubation for 2 hr at 30°, the reaction is terminated by adding 30 μl acetic acid. The simultaneous addition of 1 μg each of the substrate and the presumed isoflavone product (daidzein or genistein) without the ^{14}C label as carriers improves the identification of the reaction products on chromatograms. The ethyl acetate extract of the mixture is subjected to cellulose TLC [Funacel SF (Funakoshi, Tokyo, Japan)] and is analyzed by TLC radiochromatography/autoradiography. The following radioactive compounds are observed on TLC (see Table I): 2-hydroxyisoflavanones, isoflavones formed spontaneously during the reaction/work-up and 3-hydroxyflavanone by-products, in addition to unreacted flavanone substrates. The identity of 2-hydroxyisoflavanones is then demonstrated by the formation of isoflavones by acid treatment. The concentrated ethyl acetate extracts of the reaction mixture are dissolved and stirred in 10% HCl in methanol, typically at 50° for 10 min, followed by ethyl acetate extraction and cellulose TLC analysis. This procedure abolishes the radioactivity of the 2-hydroxyisoflavanone spots and enhances that of the isoflavone spots greatly.

For the $(2S)$-flavanone 2-hydroxylase assay, the same substrates are incubated with the insect microsome harboring CYP93B1 as described earlier, and the ethyl acetate extracts of the mixture are analyzed by cellulose TLC. A much simpler distribution pattern of radioactivity among the spots on TLC is observed: licodione and putative 2-hydroxynaringenin are the only radioactive products (see Table I) from $(2S)$-[^{14}C]liquiritigenin and $(2S)$-[^{14}C]naringenin, respectively. The acid-catalyzed conversion of licodione and 2-hydroxynaringenin into 7,4'-dihydroxyflavone and apigenin is also performed by the same method.

Nonradiochemical Assay of 2-Hydroxyisoflavanone Synthase, (2S)-Flavanone 2-Hydroxylase and Isoflavone 2'-Hydroxylase

The incubation mixture and the assay conditions are the same as that for the radiochemical assay, except that the mixture contains either 10 μg each of cold racemic flavanones (liquiritigenin or naringenin) instead of radiolabeled ones for 2-hydroxyisoflavanone synthase and $(2S)$-flavanone 2-hydroxylase or 1 μg each of isoflavones (daidzein, genistein, or formononetin) for isoflavone 2'-hydroxylase. Products are analyzed by reversed-phase HPLC. We use a Shim-pack CLC-ODS column (6.0 × 150 mm; Shimadzu, Kyoto, Japan) with various proportions of methanol/water/acetic acid as the solvent at a flow rate of 1 ml/min at 40°. Table I shows the retention times (R_t) of the substrates and products observed with each solvent used. The eluate is monitored with an UV detector at the wavelength of the λ_{max} of the presumed product, also shown in Table I. In the assay of 2-hydroxyisoflavanone synthase and (2S)-flavanone 2-hydroxylase, it is important to compare the HPLC chromatograms of the direct reaction products and those after acid treatment to ensure that acid-labile 2-hydroxy(iso)flavanone products are converted into stable (iso)flavones. In contrast, in the assay of isoflavone 2'-hydroxylase, the products, i.e., 2'-hydroxyformononetin, 2'-hydoxydaidzein, and 2'-hydroxygenistein, give isolated peaks on the chromatograms and are clearly separated from the peaks of each substrate (see Table I).

Large-Scale Preparation of 2,7,4'-Trihydroxyisoflavanone

Yeast cells expressing CYP93C2 are harvested from 6 liters of the induction medium described earlier. Microsomes (ca. 30 mg protein) are prepared and incubated with 3 mg liquiritigenin in the presence of 0.5 mM NADPH in the total volume of 15 ml. After incubation at 25° for 2 hr, the ethyl acetate extract of the reaction mixture is subjected to silica gel TLC with solvent A, and 2,7,4'-trihydroxyisoflavanone is recovered (see Table I). Cultivation of yeast cells, assay, and separation by TLC are performed 10 times, and partially purified 2,7,4'-trihydroxyisoflavanone is collected from a total of 60 liters of yeast culture. The resultant product is purified further by reversed-phase HPLC with 35% methanol in water (R_t, 10 min). By this method, about 2 mg of 2,7,4'-trihydroxyisoflavanone is obtained. The ^{1}H nuclear magnetic resonance spectrum of the final product in acetone-d_6 is identical with the reported spectrum of 2,7,4'-trihydroxyisoflavanone.[25]

[36] Selected Cell Cultures and Induction Methods for Cloning and Assaying Cytochromes P450 in Alkaloid Pathways

By TONI M. KUTCHAN and JOACHIM SCHRÖDER

Introduction

Multiple cytochrome P450-dependent oxidases and monooxygenases have been found to be involved in plant alkaloid biosynthesis. The biosynthesis of selected members of alkaloid classes for which the enzymatic steps have been elucidated, such as the monoterpenoid indole alkaloid vindoline and the isoquinoline alkaloids morphine, macarpine, and berberine, involves highly substrate-specific cytochromes P450.[1] Although *in vitro* enzyme assays have been successfully developed for most of these cytochromes P450, purification to apparent homogeneity of cytochromes P450 from plant tissues has, in many cases, proved elusive due to the low abundance and instability of these proteins. A direct molecular genetic approach circumvents the difficulties associated with cytochrome P450 purification. Because many alkaloid biosynthetic pathways can be induced in plant cell culture, the differential expression of cytochrome P450 encoding genes involved in these pathways can be exploited in the cloning of these genes. The protocols described herein have been used successfully in the laboratories of the authors to clone and functionally express cytochrome P450-encoding cDNAs of alkaloid biosynthesis.

Selected Cell Cultures

Plant cell suspension cultures that have been optimized for the production or for the induction of secondary metabolites, such as alkaloids, have been in use since the 1970s. Notable examples of high-production strains are *Coptis japonica* (berberine, 12% dry weight), *Berberis wilsoniae* (jatrorrhizine, 12% dry weight), and *Rauwolfia serpentina* (raucaffricine, 3% dry weight). These systems have been very useful for studies on the enzymology of alkaloid biosynthesis. Cell suspension cultures that can be induced to biosynthesize alkaloids serve as experimental systems for the differential identification of cDNAs that encode enzymes involved in the biosynthesis of these alkaloids. Two notable examples are the California poppy *Eschscholzia californica,* which accumulates benzo[*c*]phenanthridine alkaloids in response to addition of an inducer chemical to the culture, and the Madagascar

---

[1] T. M. Kutchan, *in* "The Alkaloids" (G. Cordell, ed.), Vol. 50, p. 257. Academic Press, San Diego, 1998.

periwinkle *Catharanthus roseus,* in which selected genes of monoterpenoid indole alkaloid biosynthesis are induced in response to light.[2,3]

Eschscholzia californica

Cell suspension cultures are grown routinely in 1-liter conical glass flasks containing 400 ml of Linsmaier–Skoog medium over 7 days at 23° on a gyratory shaker (100 rev min^{-1}) in diffuse light (750 lux). Two hundred milliliters (67 g fresh weight, 5.4 g dry weight) of 7-day-old suspension cultures are transferred aseptically to 200 ml fresh Linsmaier–Skoog medium.[4] For smaller scale experiments, 150 ml of cell suspension can be maintained in 300-ml conical glass flasks. Seventy-five milliliters (25 g fresh weight, 2 g dry weight) of 7-day-old suspension culture are transferred aseptically to 75 ml fresh Linsmaier–Skoog medium.

Catharanthus roseus

Cell suspension cultures (*C. roseus* L.G. Don, line CP3a)[5] are grown routinely in 200-ml Erlenmeyer flasks containing MX medium.[6] After growth in darkness for 7 days at 25° on a gyratory shaker (110 rev min^{-1}), 20 ml of the culture (20 g) is added aseptically to 40 ml fresh medium to start a new growth period.

Inducers

A large number of biotic and abiotic chemicals will induce *E. californica* cell cultures to accumulate secondary metabolites. These include protein extracts of the yeast *Saccharomyces cerevisiae* and the bacterium *Bacillus subtilis,* membrane-derived octadecanoid compounds such as 12-oxophytodienoic acid and methyl jasmonate, a variety of barbiturates such as barbituric acid and phenobarbital, and inorganic chemicals such as NaF. The octadecanoids have been shown to induce secondary metabolite accumulation in a very broad range of plant species in cell culture. One of the most effective members of this class is methyl jasmonate. Because of its commercial availability (Serva, Sigma), it is also one of the best studied.

Light has not been studied systematically as an inducer of alkaloid biosynthetic pathways. The inducibility of tabersonine 16-hydroxylase (T16H, CYP71D12)

[2] H. Gundlach, M. J. Müller, T. M. Kutchan, and M. H. Zenk, *Proc. Natl. Acad. Sci. U.S.A.* **89,** 2389 (1992).

[3] G. Schröder, E. Unterbusch, M. Kaltenbach, J. Schmidt, D. Strack, V. De Luca, and J. Schröder, *FEBS Lett.* **458,** 97 (1999).

[4] E. M. Linsmaier and F. Skoog, *Physiol. Plant.* **18,** 100 (1965).

[5] H.-P. Vetter, U. Mangold, G. Schröder, F.-J. Marner, D. Werck-Reichhart, and J. Schröder, *Plant Physiol.* **100,** 998 (1992).

[6] T. Murashige and F. Skoog, *Physiol. Plant.* **15,** 473 (1962).

activity by light was detected in experiments investigating whether the regulation observed in developing seedlings[7] might have been retained in undifferentiated cell suspension cultures.

Methyl Jasmonate Induction of Alkaloid Accumulation in E. californica

To induce secondary metabolite accumulation, 7-day-old cells are transferred to fresh Linsmaier–Skoog medium. Growth is allowed to proceed at 23° on a gyratory shaker (100 rev min^{-1}) in diffuse light (750 lux) for 3 days. On day 3, methyl jasmonate diluted in ethanol is added to a final concentration of 100 μM (5 μl of dilution ml of cell suspension culture). Control cells are treated in an identical manner except that 5 μl/ml of suspension culture of ethanol is added instead of methyl jasmonate. Aliquots of cells (50 ml) are removed aseptically at 0, 3, 6, 9, 12, and 24 hr after the addition of methyl jasmonate or ethanol. The cells are collected by vacuum filtration, shock frozen in liquid nitrogen, and stored at −80°. These samples are later used for RNA isolation using standard protocols for plant cells.[8]

Light Induction of C. roseus Suspension Cultures

Induction is performed with a light field from six white light lamps (Philips 40W/18) and four UV-A fluorescent lights (Osram, 40W/73). The light field (8 W m^{-2}) is positioned 45 cm above the culture flasks, which are on a small gyratory shaker. Although irradiation is sufficient, the best induction of T16H is obtained if the cells are subjected simultaneously to a nutritional downshift. This is carried out by collecting the cells on a G2 glass sinter filter and transfer of 5.3 g fresh weight into 200-ml quartz glass Erlenmeyer flasks containing 40 ml 8% sucrose in water. The cells are harvested after 1.5 and 3 hr by vacuum filtration, shock frozen in liquid nitrogen, and stored at −70°. The choice of the time points is based on the expectation that the highest mRNA concentration precedes the maximum enzyme activity.[3]

Homology-Based Cloning of Alkaloid Biosynthetic Cytochrome P450 cDNAs

The first cDNAs encoding plant cytochromes P450 of known function were isolated using either polyclonal antibodies or internal amino acid sequences obtained from purified protein. These include allene oxide synthase from flax,[9] cinnamate

[7] B. St-Pierre and V. De Luca, Plant Physiol. 109, 131 (1995).

[8] H. P. Pauli and T. M. Kutchan, Plant J. 13, 793 (1998).

[9] W. C. Song, C. D. Funk, and A. R. Brash, Proc. Natl. Acad. Sci. U.S.A. 90, 8519 (1993).

hydroxylase from Jerusalem artichoke,[10] P450tyr from sorghum,[11] and berbamunine synthase from barberry.[12] As primary protein sequences for plant cytochromes P450 accumulated, conserved domains could be exploited for direct cloning of cDNAs, which eliminated the necessity to first purify the proteins.[13,14] The following are protocols with which to isolate cDNAs encoding selected cytochromes P450 of alkaloid biosynthesis.

(S)-N-Methylcoclaurine 3'-Hydroxylase (cyp80b1)

RNA is isolated from 3 to 5 g fresh weight methyl jasmonate-treated cell suspension cultures of *E. californica* 6 hr after addition of the inducer. The choice of this time point is based on previous results obtained with another inducible enzyme of benzo[*c*]phenanthridine alkaloid biosynthesis in *E. californica,* the berberine bridge enzyme.[15] RNA is transcribed into cDNA with M-MLV (murine moloney leukemia virus) reverse transcriptase and is used as a template for polymerase chain reaction (PCR).

DNA amplification is performed using *Pfu* DNA polymerase under the following conditions. In the presence of primer 1 : 5 min at 94°, 2 cycles of 94° for 30 sec, 40° for 3 min, and 72° for 30 sec. After addition of primer 2 : 3 cycles of 94° for 30 sec, 40° for 3 min, and 72° for 30 sec and then 30 cycles of 94° for 30 sec, 50° for 60 sec, and 72° for 30 sec. After the final cycle, the mixture is cooled to 4°. Amplification primers: primer 1, 5'-TTG GAT CCI GA[A,G] III TT[C,T] III GA[A,G] IGI TT-3'; primer 2, 5'-AAG GGC CCI II[G,A] CAI III [C,G]TI III CCI III CC[G,A] AA-3'.

The amplified DNA is resolved by agarose gel electrophoresis, and the DNA band of approximately the expected size (130 bp) is isolated and subcloned into *Sma*I restriction-digested pUC18. If *Taq* DNA polymerase is used instead of *Pfu* DNA polymerase, the fragment can be subcloned into pGEM-T (Promega, Madison, WI). The nucleotide sequence of individual clones is determined until no new sequences are identified. For *E. californica,* the nucleotide sequence was determined for 74 individual clones, at which point new sequences were no longer found. Nineteen individual cytochrome P450 encoding sequences were identified.

Partial cDNAs encoding cytochromes P450 are tested for differential expression by RNA gel blot analysis. cDNAs that hybridize to the RNA transcript that

[10] H. G. Teutsch, M. P. Hasenfratz, A. Lesot, C. Stoltz, J. M. Garnier, J. M. Jeltsch, F. Durst, and D. Werck-Reichhart, *Proc. Natl. Acad. Sci. U.S.A.* **90,** 4102 (1993).

[11] B. M. Koch, O. Sibbesen, B. A. Halkier, I. Svendsen, and B. L. Møller, *Arch. Biochem. Biophys.* **323,** 177 (1995).

[12] P. F. X. Kraus and T. M. Kutchan, *Proc. Natl. Acad. Sci. U.S.A.* **92,** 2071 (1995).

[13] A. H. Meijer, E. Souer, R. Verpoorte, and J. H. C. Hoge, *Plant Mol. Biol.* **22,** 379 (1993).

[14] M. K. Udvardi, J. D. Metzger, V. Krishnapillai, W. J. Peacock, and E. S. Dennis, *Plant Physiol.* **105,** 755 (1994).

[15] H. Dittrich and T. M. Kutchan, *Proc. Natl. Acad. Sci. U.S.A.* **88,** 9969 (1991).

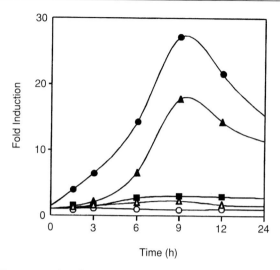

FIG. 1. Graphic representation of an RNA gel blot prepared from a methyl jasmonate-induced cell suspension culture of *E. californica*. RNA was isolated at 0, 1.5, 3, 6, 9, 12, and 24 hr after the addition of elicitor to the culture medium. The blot was hybridized to *cyp80b1* cDNA (▲), *cyp82b1* cDNA (■), berberine bridge enzyme cDNA (●), cytochrome P450 reductase cDNA (△), and a noninducible cytochrome P450 cDNA (○). All cDNAs are isolated from *E. californica*. From H. P. Pauli and T. M. Kutchan, *Plant J.* **13,** 793 (1998), with permission of Blackwell Science Ltd.

accumulates in response to the addition of methyl jasmonate to the cell culture medium are potentially encoding cytochromes P450 of alkaloid biosynthesis. Full-length clones are isolated either by screening a cDNA library using the partial clone as hybridization probe or by rapid amplification of CDNA ends-polymerase chain reaction (RACE-PCR). An example of phosphorimager-quantitated RNA gel blot analysis using the full-length cDNAs obtained in this manner as hybridization probes is given in Fig. 1.

Tabersonine 16-Hydroxylase (cyp71d12)

Standard techniques are used to establish a cDNA library with RNA isolated from cells induced for T16H activity. Cytochrome P450-specific sequences are amplified with established strategies,[13] i.e., a degenerate primer (5'-GG[T,G] [G,T][G,T][G,T] GG[T,G] III [C,A]GI ATT TGT CCI GG-3') for conserved sequences around the heme-binding site close to the C terminus of the protein and a primer from the phage vector. Amplification products of the expected size (350–450 bp) are cloned and sequenced, and putative cytochromes P450 are identified by similarity of the deduced C-terminal protein sequences to known cytochromes P450. Candidates predicting more than 90% identity to functionally identified cytochromes P450 are ignored. The others are hybridized in Northern blots to

RNA from cell cultures subjected to various induction conditions, and candidates revealing a correlation with the induction kinetics of T16H activity are selected. Full-length clones are obtained either by screening a cDNA library or by 5′-RACE PCR.

Functional Expression of Alkaloid Biosynthetic Cytochrome P450 cDNAs

Bacterial, yeast, and insect cell culture have been used as heterologous hosts for the functional expression of cDNAs encoding plant cytochromes P450 of alkaloid biosynthesis. Each system requires special consideration with respect to the construction of expression vectors and to the source of cytochrome P450 reductase.

Expression of cyp80b1 in Insect Cell Culture and Assays

The functional expression of alkaloid biosynthetic cDNAs in insect cell culture using a baculovirus-based expression vector has proven quite successful. The vector system has been improved over the years, such that isolation of pure, recombinant virus is routine and reliable.[16,17]

The open reading frame of the *cyp80b1* cDNA free of 5′- and 3′-flanking regions is generated by PCR using *Pfu* DNA polymerase and is subcloned into *Sma*I-digested pUC18. The 1504-bp *Bam*HI/*Not*I restriction fragment from pUC18/ *cyp80b1* is ligated into *Bam*HI/*Not*I-digested pFastBac1 (Life Technologies, Karlsruhe, Germany). The expression levels observed are highest when the start codon of the cDNA is placed as close as possible to the 5′ end of the polylinker of the vector pFastBac1. pFastBac1/*cyp80b1* is then transposed into baculovirus DNA in the *Escherichia coli* strain DH10Bac. Competent DH10Bac cells are available commercially from Life Technologies, and the procedure is carried through as outlined by the manufacturer. The recombinant baculovirus is then transfected into *Spodoptera frugiperda* Sf9 cells according to the Life Technologies protocol. After amplification of the recombinant baculovirus, the infected insect cells are used for enzyme assays and CO difference spectra measurements as follows.

Cells from 50 ml of a *S. frugiperda* Sf9 cell suspension culture (containing 2×10^6 cells/ml) are collected by centrifugation (900g, 10 min, room temperature), resuspended gently in 8 ml of TC-100 medium (Sigma) containing 10% fetal calf serum, and mixed gently with various amounts of recombinant baculovirus. After incubation for 1 hr at 28° with agitation at 80 rpm, 40 ml of TC-100 medium containing 10% fetal calf serum and 100 μl of hemin (1 mg/ml) (Sigma) are added. Incubation is continued for 3 days at 28° and 140 rpm. The optimal amount of

---

[16] T. M. Kutchan, A. Bock, and H. Dittrich, *Phytochemistry* **35**, 353 (1994).

[17] F.-C. Huang and T. M. Kutchan, *Phytochemistry* **53**, 555 (2000).

recombinant baculovirus results in enlarged cells with not more than 30% of the cells disrupted.

The infected cells are collected by centrifugation (4000g, 5 min, room temperature) and are washed twice with PBS buffer (130 mM NaCl, 7 mM Na$_2$HPO$_4$, 3 mM NaH$_2$PO$_4$, pH 7.4). The washed cells are resuspended in 3.5 ml of 100 mM Tricine (pH 7.5) containing 5 mM thioglycolic acid and are used directly for enzyme activity measurements.

The enzyme assay for (S)-N-methylcoclaurine 3′-hydroxylase activity (reaction shown in Fig. 2A) contains the following components: 200 mM Tricine/NaOH, 400 μM NADPH, 100 μM (S)-N-methylcoclaurine, and 70–100 μl of the insect cell culture suspension (obtained as described earlier and containing 300 μg protein

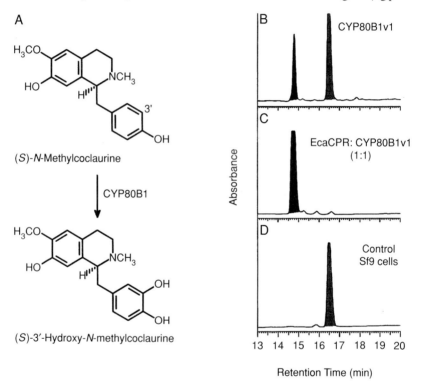

FIG. 2. (S)-N-Methylcoclaurine 3′-hydroxylase (CYP80B1). (A) Reaction catalyzed by CYP80B1. HPLC chromatograms of an aliquot of an enzyme assay containing (S)-N-methylcoclaurine, NADPH and (B) S. frugiperda Sf9 cells expressing CYP80B1v1, (C) S. frugiperda Sf9 cells expressing both CYP80B1v1 and the E. californica cytochrome P450 reductase (EcaCPR), and (D) uninfected S. frugiperda Sf9 cells. Retention times: (S)-N-methylcoclaurine, 16.5 min; (S)-3′-hydroxy-N-methylcoclaurine, 14.7 min. cyp80b1v1 is the first of two alleles isolated that encode CYP80B1 in E. californica. From H. P. Pauli and T. M. Kutchan, Plant J. **13**, 793 (1998), with permission of Blackwell Science Ltd.

or 0.5 pmol cytochrome P450) in a total volume of 200 μl. Incubation is allowed to proceed for 1 hr at 30°, after which 20 μl of a 20% trichloroacetic acid solution is added to stop the reaction. Insoluble debris is removed by centrifugation (15,000g, 2 min, room temperature). One hundred microliters of the supernatant is transferred to an HPLC glass ampule, and 90 μl of that solution is resolved by chromatography. The system used to resolve *(S)*-*N*-methylcoclaurine from the enzymatic product *(S)*-3′-hydroxy-*N*-methylcoclaurine is a Merck (Darmstadt, Germany) Lichrospher 60 RP-Select B column (5 μm; 4 × 250 mm) and the following solvent system: (A) 97.99% (v/v) H_2O, 2% CH_3CN, 0.01% (v/v) H_3PO_4; (B) 1.99% (v/v) H_2O, 98% CH_3CN, 0.01% H_3PO_4; gradient 0–25 min 0–46% B; 25–26 min 46–100% B; 26–30 min 100–0% B; 30–32 min 0% B; flow 1 ml min^{-1} with detection at 282 nm. Examples of HPLC chromatograms of *(S)*-*N*-methylcoclaurine 3′-hydroxylase enzyme tests are given in Fig. 2. Enzyme assays carried out in the absence (Fig. 2B) and in the presence (Fig. 2C) of the *E. californica* cytochrome P450 reductase indicate that the cytochrome P450 reductase level in *S. frugiperda* Sf9 cells may be limiting in the enzyme assay.

Microsomes for CO difference spectrum measurements are isolated from infected insect cells in the following manner. Seven hundred fifty milliliters of the recombinant baculovirus-infected insect cell culture is centrifuged (4000g, 5 min, room temperature) to remove medium 3 days after infection. The cells are washed twice with PBS buffer (130 mM NaCl, 7 mM Na_2HPO_4, 3 mM NaH_2PO_4, pH 7.4). The cell pellet is resuspended gently in 30 ml of 100 mM Tricine/NaOH (pH 7.5) containing 5 mM thioglycolic acid. Fifteen milliliters of the cell suspension is frozen in liquid nitrogen and stored at −80°. If stored in small aliquots, this material can be used for up to 1 year for enzyme tests. The remaining 15 ml of cell suspension is sonicated four times (10 pulses at output 5, 50% duty cycle) with a Branson sonifier cell disruptor. After this step, the microsomes have to be protected from light. Cell debris is removed by centrifugation (8000g, 20 min, 4°). Microsomes in the supernatant are sedimented by centrifugation at 105,000g for 65 min at 4° in an ultracentrifuge. The firmly packed microsomal pellet is resuspended in 1.5 ml of 100 ml Tricine/NaOH (pH 7.5) containing 5 mM thioglycolic acid at 4° and is homogenized in a chilled glass homogenizer with a Teflon pestle. The microsomal suspension is used immediately for measurement of the CO difference spectrum (Fig. 3).

The difference spectrum of the microsomal preparation is measured on a Cary Model 3Bio UV-visible spectrophotomer with cuvettes of 1 cm optical path at room temperature (20–25°). Six hundred microliters of the microsomal suspension is homogenized with 400 μl of 100 mM Tricine/NaOH (pH 7.5) containing 5 mM thioglycolic acid and 0.5% (v/v) Emulgen 913 in a chilled glass homogenizer. Insoluble debris is removed by centrifugation (15,000g, 15 min, 4°). The supernatant (250 μl) is transferred into each of two Eppendorf tubes containing 750 μl of 100 mM Tricine/NaOH (pH 7.5) (4°) containing 5 mM thioglycolic

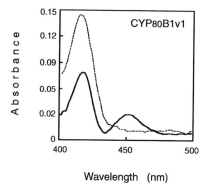

Wavelength (nm)

FIG. 3. CO-difference spectra of CYP80B1v1 expressed functionally in *S. frugiperda* Sf9 cells. Solid lines are spectra obtained with microsomes isolated from recombinant baculovirus-infected cells, and dashed lines are spectra obtained with microsomes isolated from uninfected cells. From H. P. Pauli and T. M. Kutchan, *Plant J.* **13,** 793 (1998), with permission of Blackwell Science Ltd.

acid and 20 mg of solid sodium dithionite and is mixed by vortexing. Microsomal preparations are placed in both the sample and reference cells of the spectrophotometer. After recording the baseline, the contents of the sample cell are gassed carefully with CO for 2 min, and the spectral shift from 420 to 450 nm is measured immediately.

Expression of cyp71d12 in Escherichia coli and Enzyme Assays

The procedures are optimized for expression with the plasmid construct described here. The principles have been applied successfully to other plant cytochromes P450, but it should be emphasized that each construct requires independent optimization of some parameters (e.g., bacterial strain, length and temperature of protein expression).[18–20]

The CYP71D12 protein is expressed as a translational fusion with the cytochrome P450 reductase. The technique was first described in 1987 by Murakami *et al.*[21] for the functional expression of animal cytochromes P450 in yeast. Detailed discussions of the fusion protein strategies and the expression of animal cytochromes P450 in *E. coli* have been published.[22,23] The *cyp71d12* cDNA is modified in two steps. First, the 5' end is modified to create a suitable membrane

[18] M. Hotze, G. Schröder, and J. Schröder, *FEBS Lett.* **374,** 345 (1995).

[19] M. Kaltenbach, G. Schröder, E. Schmelzer, V. Lutz, and J. Schröder, *Plant J.* **19,** 183 (1999).

[20] S. Irmler, G. Schröder, B. St-Pierre, N. P. Crouch, M. Hotze, J. Schmidt, D. Strack, U. Matern, and J. Schröder, *Plant J.* **24,** 797 (2000).

[21] H. Murakami, Y. Yabusaki, T. Sakaki, M. Shibata, and H. Ohkawa, *DNA* **6,** 189 (1987).

[22] H. J. Barnes, *Methods Enzymol.* **272,** 3 (1996).

[23] C. W. Fisher, M. S. Shet, and R. W. Estabrook, *Methods Enzymol.* **272,** 15 (1996).

anchor, and this step is also used to introduce a *Nco*I cloning site at the start AUG (CC**ATG**G). The membrane anchor of CYP71D12 in the construct contains MALLLAVF, encoded by CC **ATG** GCT TTA CTA TTA GCA GTT TTT, and both the peptide and the DNA sequence have been shown to be useful for expression in the bacteria.[22,23] The second modification is the replacement of the stop codon by a *Sal*I site encoding a SerThr linker for the translational fusion to the cytochrome P450 reductase. The cDNA *Nco*I/*Sal*I fragment is inserted into a cassette plasmid that provides the promoter regulating the protein expression and allows the C-terminal fusion of the cytochrome P450 with the cytochrome P450 reductase from *C. roseus*.[18] The fusion protein (129 kDa) is expressed in *E. coli* strain RM82 containing plasmid pUBS520.[24] Expression of a protein of the correct size is tested in immunoblots with antiserum against the cytochrome P450 reductase.[18]

Precultures in 2 ml LB medium with the appropriate antibiotics (0.1 mg ml^{-1} carbenicillin, 0.025 mg ml^{-1} kanamycin) are grown overnight at 37° to $A_{600} \leq 3$. They are then diluted to $A_{600} = 0.06$ into 250 ml induction medium [225 ml TB medium,[25] antibiotics as in the preculture, 25 ml 0.70 M potassium phosphate (pH 7.5), and 0.0625 ml trace elements[26]]. Growth is allowed to proceed at 37° to $A_{600} = 0.3$ (about 1.5 hr). Then 28 mg δ-aminolevulinic acid is added to supply sufficient precursors for hemin biosynthesis, and growth is continued until $A_{600} = 0.6$ (about 0.5 hr). The culture is cooled rapidly on ice to 25°, and protein expression is induced by the addition of isopropyl-β-thiogalactoside (IPTG) to 1 mM final concentration. The cultures are harvested after 4 hr at 25° by centrifugation for 15 min at 4° and 5000g. The cells are washed by suspension (10 ml 10 mM Tris–HCl, pH 7.5, 0.15 M NaCl) and centrifugation. They are then shock frozen in liquid nitrogen and stored at −70°.

Membranes are prepared as follows. The cell pellets are resuspended in 2 ml TSE [75 mM Tris–HCl, pH 7.5, 0.25 M sucrose, 0.25 mM EDTA, 0.34 mg phenylmethylsulfonyl fluoride (PMSF), 1 mg DNase I, 0.15 mg RNase A], and the cells are broken by passing them twice through a French pressure cell. The extract is centrifuged for 15 min at 7800g at 4° to remove insoluble material, and the supernatant fluid is used to obtain the membranes by centrifugation for 15 min at 350,000g and 4° in the TLA rotor of the Beckman centrifuge TL-100. The pelleted membranes are resuspended in TGE (75 mM Tris–HCl, pH 7.5, 10% glycerol, 0.25 mM EDTA) to a protein concentration of 20 mg/ml, and they are homogenized in the cold in a Dounce homogenizer. The samples are frozen in liquid nitrogen and stored at −70°. The storage of small portions is recommended because repeated thawing and freezing leads to large losses of enzyme activity.

[24] P. M. Schenk, S. Baumann, R. Mattes, and H.-H. Steinbiss, *BioTechniques* **19**, 196 (1995).
[25] J. Sambrook, E. F. Fritsch, and T. Maniatis, "Molecular Cloning: A Laboratory Manual." Cold Spring Harbor Laboratory Press, Cold Spring Harbor, NY, 1989.
[26] P. Sandhu, Z. Guo, T. Baba, M. V. Martin, R. H. Tukey, and F. P. Guengerich, *Arch. Biochem. Biophys.* **309**, 168 (1994).

A   B

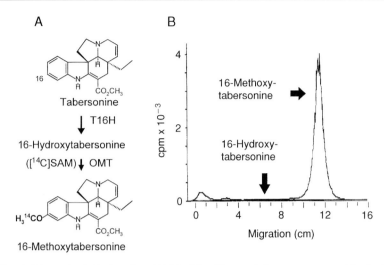

FIG. 4. Tabersonine 16-hydroxylase. (A) Principle of the coupled enzyme assay used for the determination of T16H activity. (B) Example for a radioactivity scan of a TLC plate analyzing the formation
of radioactive 16-methoxytabersonine. 16-Hydroxytabersonine, the product of the T16H reaction, is
detected under UV light.

The standard T16H assay couples the hydroxylation of tabersonine with the
next step in the pathway to vindoline, an O-methyltransferase (OMT) reaction
(Fig. 4A), and the product analysis identifies the labeled 16-methoxytabersonine
obtained in the presence of radiolabeled S-adenosyl-L-methionine as methyl group
donor. The assay (0.1 ml) contains 30 μM tabersonine, 0.1 M Tris–HCl (pH 8.0),
a NADPH regenerating system (10 mM glucose 6-phosphate, 17.5 mU glucose-
6-phosphate dehydrogenase, Boehringer Mannheim, Germany), 4 mM dithiothreitol, 40–80 μg $E.$ $coli$ membranes containing the T16H fusion protein, 20 μg
leaf extract as O-methyltransferase source,[7] and 18 μM S-adenosyl-L-[methyl-
^{14}C]methionine (2 GBq/mmol, 220,000 dpm, Amersham Life Science, England).
The incubations are started by the addition of 5 μl freshly prepared 10 mM NADPH.
The reactions are terminated after 1 hr at 30° by adding 42 μl 0.1 M NaOH
(pH increase to about 10), and the incubations are extracted twice with 0.2 ml ethyl
acetate. The organic solvent is removed in $vacuo$ (Speed-Vac), and the residue is
dissolved in 10 μl methanol. The radioactive 16-methoxytabersonine is quantified[5]
after thin-layer chromatography (TLC) on a silica gel with a fluorescence indicator (60F$_{254}$, Merck, Germany). The solvent is ethyl acetate/hexane (1 : 1, v/v). The
R_f values for tabersonine, 16-methoxytabersonine, and 16-hydroxytabersonine are
0.79, 0.72, and 0.51, respectively. Figure 4B shows an example of the radioactivity
scan of a TLC plate. SAM is poorly soluble in ethyl acetate, and therefore usually
only the methylated radioactive product is detectable. However, the leaf extract

used as the OMT source also contains OMTs with other substrate specificities, and the methylation of endogenous substrates can obscure the results at very low T16H activities. Tabersonine, 16-hydroxytabersonine, and 16-methoxytabersonine are all detectable under UV light (312 nm). Inspection of TLC plates reveals the presence of 16-hydroxytabersonine when highly active T16H preparations are used, indicating that OMT activity may become a limiting factor in the coupled assay. The small differences in spectra of tabersonine and 16-hydroxytabersonine preclude a reliable optical assay quantifying the hydroxylation reaction directly. HPLC conditions for the separation of the compounds have been described.[3]

Acknowledgments

Research in the laboratories of the authors was supported by grants from the Deutsche Forschungsgemeinschaft Schr 21/5 (J.S.), SFB 369 (T.M.K.), and Fonds der Chemischen Industrie (J.S. and T.M.K.).

[37] Isolation and Functional Characterization of Cytochrome P450s in Gibberellin Biosynthesis Pathway

By Chris A. Helliwell, W. James Peacock, and Elizabeth S. Dennis

Introduction

Gibberellins (GA) are an important class of plant growth regulators and in recent years, many of the genes encoding the enzymes responsible for GA biosynthesis have been isolated.[1] Like many other pathways of secondary metabolism in plants, GA biosynthesis includes steps catalyzed by cytochrome P450s (Fig. 1); these P450 encoding genes have proved to be difficult to isolate.

Gibberellin biosynthesis starts from geranylgeranyl diphosphate. Two terpene cyclases, copalyl diphosphate synthase and kaurene synthase, catalyze cyclization reactions to produce ent-kaurene. Cytochrome P450s then catalyze the oxidation of ent-kaurene to GA_{12}. The steps of the pathway from GA_{12} to the bioactive GAs are catalyzed by soluble dioxygenases.

We isolated cDNA clones encoding members of the CYP701A and CYP88A subfamilies of P450 proteins from Arabidopsis and pumpkin and have shown

---

[1] P. Hedden and A. L. Phillips, Trends Plant Sci. **5,** 523 (2000).

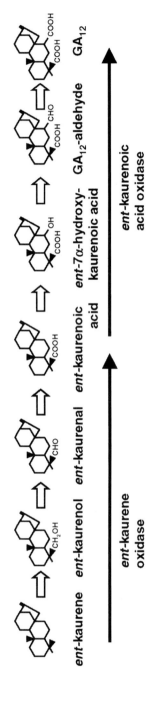

Fig. 1. Cytochrome P450-mediated steps of the GA biosynthesis pathway. *ent*-Kaurene oxidase catalyzes the three sequential oxidations at C-19 of the GA skeleton, and *ent*-kaurenoic acid oxidase catalyzes the three sequential oxidations at C-7.

that these subfamilies each catalyze three steps of the GA biosynthesis pathway.[2-5]

One method to isolate genes encoding P450s in the GA biosynthesis pathway is to take advantage of a mutation in a gene encoding one of the pathway enzymes. We used the *Arabidopsis ga3* mutant, which has a characteristic GA responsive dwarf phenotype and was thought to be impaired in *ent*-kaurene oxidation, to isolate the gene encoding CYP701A3, *ent*-kaurene oxidase.[2] The genome resources available for *Arabidopsis,* including a large collection of single nucleotide polymorphisms between the ecotypes Lansdberg *erecta* and Columbia, allow the use of map-based cloning strategies to assist in the isolation of mutated genes. The mutant approach depends on being able to isolate and maintain a mutant. This is not possible where a mutation in a gene is lethal or where functionally redundant genes are present. In *Arabidopsis,* large portions of the genome are duplicated, making functional redundancy a major factor.[6] For example, there was no *Arabidopsis* mutant with reduced *ent*-kaurenoic acid oxidation, the next step in the pathway after *ent*-kaurene oxidase. In this case, we used a method based on transcript abundance to identify candidate genes.[5] This article describes the polymerase chain reaction (PCR) strategy used to isolate cDNAs encoding *ent*-kaurene oxidase and a putative *ent*-kaurenoic acid oxidase from developing pumpkin seeds. It also describes the yeast assay used to demonstrate that CYP701A subfamily members are *ent*-kaurene oxidases and that CYP88A subfamily members are *ent*-kaurenoic acid oxidases.

Preparation of Pumpkin Seed cDNA

The developing pumpkin seed is a rich source of GA biosynthetic enzyme activity, with the peak of GA biosynthesis activity occurring during the expansion of cotyledons.[7] Immature pumpkins are obtained and the seeds dissected to identify pumpkins in which the cotyledons are between 20 and 80% of seed length. Cotyledon and liquid endosperm are then harvested from the developing seeds of these pumpkins. Total RNA is extracted by a standard method,[8] and

[2] C. A. Helliwell, C. C. Sheldon, M. R. Olive, A. R. Walker, J. A. D. Zeevaart, W. J. Peacock, and E. S. Dennis, *Proc. Natl. Acad. Sci. U.S.A.* **95,** 9019 (1998).

[3] C. A. Helliwell, A. Poole, W. J. Peacock, and E. S. Dennis, *Plant Physiol.* **119,** 507 (1999).

[4] C. A. Helliwell, P. M. Chandler, A. M. Poole, E. S. Dennis, and W. J. Peacock, *Proc. Natl. Acad. Sci. U.S.A.* **98,** 2065 (2001).

[5] C. A. Helliwell, M. R. Olive, L. Gebbie, R. Forster, W. J. Peacock, and E. S. Dennis, *Aust. J. Plant Physiol.* **27,** 1141 (2000).

[6] The Arabidopsis Genome Initiative, *Nature* **408,** 796 (2000).

[7] J. E. Graebe, *in* "Proceedings of the Seventh International Conference on Plant Growth Substances" (D. J. Carr, ed.), p. 151. Springer-Verlag, Berlin, 1970.

[8] J. Longemann, J. Schell, and L. Willmitzer, *Anal. Biochem.* **163,** 16 (1987).

poly(A)$^+$ mRNA is extracted using a Stratagene poly(A) Quik mRNA isolation kit (Stratagene, La Jolla, CA).

Polymerase Chain Reaction Amplification Strategy

Consensus sequences of the PERF and FXXGXXXCXG motifs from the five plant P450 sequences available at the time are used to design degenerate oligonucleotides (Fig. 2). The primers are used in a nested PCR strategy; a first amplification is carried out using a PERF primer and an oligo(dT) primer followed by a second reaction using an FXXGXXXCXG primer with an oligo(dT) primer. PCR reactions are carried out with Amplitaq DNA polymerase (PerkinElmer, Foster City, CA) using the manufacturer's buffer with 1.5 mM MgCl$_2$, 250 μM dNTPs,

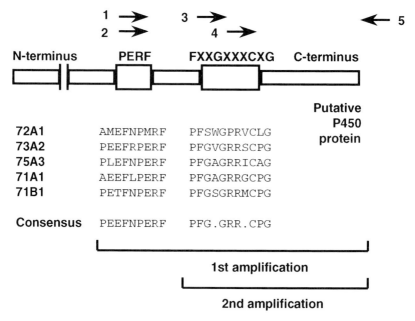

```
1 AGCTGGATCCNGA(A/G)(C/A)GNTTC
2 AGCTGGATCCNGA(A/G)(C/A)GNTTT
3 AGCTAAGCTT(T/C)ICCITT(T/C)(G/A/T)(G/C)I(G/T/A)(G/T/C)NGG
4 AGCTGGATCCGGI(A/C)(G/C)I(A/C)GI(A/G)IITG(C/T)(G/C)(C/T)NGG
5 oligo dT
```

Fig. 2. Identification of conserved P450 motifs and design of primers used to amplify P450 sequences from developing pumpkin seeds. A nested PCR approach was used with a first round amplification using primer 1 or 2 with the oligo(dT) primer (5) followed by a second round with either primer 3 or 4 with oligo(dT).

and 0.1 μg cotyledon cDNA per 50-μl reaction. The nested PCR strategy is used to increase the proportion of P450-specific PCR products produced.

PCR products are obtained with either primer 1 or 2 followed by secondary PCR with primer 3 but not with primer 4. The deduced amino acid sequences from the PCR products come from four different P450 subfamilies, two being from the CYP88A and CYP701A subfamilies. RNA gel blots show that both CYP701A1 and CYP88A2 are highly expressed in developing pumpkin seeds. We subsequently showed that CYP70A1 and CYP88A2 cDNAs encode GA biosynthetic P450s. These subfamilies are not closely related in sequence, suggesting that this method is a powerful way of identifying highly expressed P450s in cDNA from a particular source tissue independent of sequence.

Using Yeast Expression to Define Function of GA Biosynthetic P450s

To demonstrate conclusively whether a P450 catalyzes a step of the GA biosynthetic pathway, we have developed a simple yeast assay followed by Gas Chromatography-Mass Spectroscopy (GC-MS) analysis. We have successfully used two types of yeast strain, a basic strain G1315, which expresses the endogenous yeast cytochrome P450 reductase, and the engineered strains WAT11 and WAT21 (a gift from D. Pompon),[9] each of which is engineered to express one of two *Arabidopsis* cytochrome P450 reductases. Constructs to express the P450 cDNA are made by PCR amplifying the P450 sequences, removing 5' and 3' UTR sequences, and cloning the resulting P450 cDNA PCR product into the expression vector.

A straightforward LiCl yeast transformation procedure gives sufficient transformants.[10] Yeast strains expressing the P450 are selected by RNA gel blot analysis. We found that this is the simplest and most reliable method of identifying the yeast transformants to be assayed. RNA gel blots clearly identify yeast expressing the P450 mRNA, whereas it is often difficult to see the additional P450 protein band on gels of microsomal protein isolated from yeast (Fig. 3). RNA is prepared from overnight cultures using a hot acidic phenol method.[11] We typically use 10 μg of RNA for gel blot analysis.

To test the ability of the P450-expressing yeast to carry out a step of the GA biosynthesis pathway, an overnight culture of the yeast is grown in a rich media (YPG) or YPL [YPG with 2% (w/v) galactose replacing the glucose] when the P450 cDNA is expressed using a galactose-inducible promoter. The test is carried out on cells harvested from 1 ml of the overnight culture and is resuspended in

[9] D. Pompon, B. Lauerat, A. Bronine, and P. Urban, *Methods Enzymol.* **272,** 51 (1996).
[10] C. Cullin and D. Pompon, *Gene* **65,** 203 (1998).
[11] F. M. Ausubel, R. Brent, R. E. Kingston, D. D. Moore, J. G. Seidman, J. A. Smith, and K. Struhl, *in* "Current Protocols in Molecular Biology," p. 13.12.1. Wiley, New York, 1993.

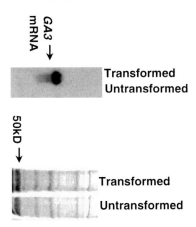

FIG. 3. Analysis of P450 expression in transformed yeast. (Left) RNA gel blot of yeast transformed with a GA3 expression construct and untransformed yeast. The GA3 mRNA is indicated. (Right) Coomassie-stained SDS–PAGE gel of microsomal protein from the same yeast strains as shown in GA3 mRNA. The arrow at 50 kDa indicates the predicted position of the GA3 protein.

1 ml of 100 mM Tris–HCl, pH 7.5, 0.5 mM NADPH, 0.5 mM FAD, and substrate dissolved in methanol. The final methanol concentration in the assay mixture is kept to 5% or less. The amount of substrate added to the assay mix is determined by the amount available, we typically use between 5 and 50 μg per assay of *ent*-kaurene, *ent*-kaurenol, *ent*-kaurenal, *ent*-kaurenoic acid, *ent*-7α-hydroxykaurenoic acid, or GA$_{12}$-aldehyde. The assay mix is transferred to an open 25-ml scintillation vial and incubated for 3 hr at 30°, shaking at 150 rpm. Substrates and products are then extracted once with 1 ml hexane and twice with 1 ml ethyl acetate; the three organic extracts are then pooled into 6-ml glass culture tubes and dried in a Speed-Vac. Care is taken not to overdry samples when potential products include *ent*-kaurenol or *ent*-kaurenal as these compounds tend to be lost with prolonged drying.

Products of the yeast incubation are identified using GC-MS. In many cases, it is possible to identify products directly in the hexane/ethyl acetate extract. Samples are derivatized by methylation and trimethysilation. Methylation is carried out by resuspending the dried extract in 50 μl methanol and 200 μl ethereal diazomethane and leaving at room temperature for 15 min. The sample is then redried in a Speed-Vac and transferred to a 1-ml reactivial in 3 × 200 μl volumes of methanol. The sample is dried down again and then is resuspended in 5 μl pyridine and 5 μl *N,O*-bis(TMS)trifluoroacetamide with 1% trimethylchlorosilane. The sample is derivatized for 30 min at 90° and is centrifuged to collect liquid in the bottom of the vial prior to injection into the GC. GC-MS analysis is carried out on 1 μl of the derivatized product. We use two GC columns: a BPX-5 [25 m × 0.22 mm internal diameter with a 0.25-μm-thick 5% phenyl (equivalent) polysilphenylene-siloxane stationary phase (SGE, Austin, TX)] and an HP-1 [25 m × 0.2 mm internal diameter with a 0.33-μm dimethyl polysiloxane stationary phase (Hewlett-Packard)]. For both columns the temperature profile is 60° for 1.5 min, 60–150° at 25° min^{-1}, 150–230° at 2° min^{-1}, and 230–300° at 70° min^{-1}. The HP-1 column achieves better separation of the GA pathway intermediates, particularly *ent*-kaurenoic acid and *ent*-kaurenol. We typically analyze the samples in a selected ion-monitoring (SIM) mode to putatively identify products by comparing spectra obtained by monitoring characteristic ions for the GA pathway intermediates to a library containing spectra for GA biosynthetic intermediates.[12] Once a product is identified, we carry out repeat injections in full scan mode and with parafilm standards to confirm the identity of the product and to determine the Kovat's retention index. The spectra and Kovat's retention index are then compared to those for known standards to confirm the authenticity of the compound detected. It is generally possible to authenticate the identity of the reaction products in this way. In cases where a putative product could be identified but final authentication is not possible due to the presence of contaminating compounds, the reaction products are purified further by reversed-phase C_{18} HPLC of the methyl ester followed by GC-MS analysis.

For HPLC purification, the derivatized reaction products are dried and resuspended in 80% methanol. The sample is injected onto a C_{18} reversed-phase column and eluted with an 80–100% gradient of solvent B (100% methanol) in solvent A (10% methanol in 2 mM acetic acid). The flow rate is 1 ml · min^{-1} collecting 1-ml fractions for up to 30 min. The fraction containing the desired product is determined by running derivatized ^{14}C-labeled compounds under the same conditions. We use this method to purify both *ent*-kaurenoic acid-methyl ester and GA_{12}-methyl ester from contaminating compounds.

Comments

Assigning function to the cytochrome P450s of plants will be a major challenge over the coming years. The identification of ~286 putative P450 encoding genes in the *Arabidopsis* genome indicates the size of the task.[6] Some method is needed to identify candidate genes from this large gene family. A good basis for the selection of candidate genes is to identify P450s with high expression levels in a tissue where the enzyme activity under investigation is known to be high. Assignment of function depends largely on assaying enzyme activities of heterologously expressed P450s. Yeast is the system of choice, particularly for plant P450s. Our experience suggests that strains such as WAT11 and WAT21, which express plant cytochrome P450 reductases, can give higher activities compared

---

[12] P. Gaskin and J. MacMillan, "GC-MS Analysis of the Gibberellins and Related Compounds: Methodology and a Library of Spectra." Cantock's Enterprises, Bristol, 1992.

to strains expressing the yeast reductase. To illustrate this point, CYP88A3 and CYP701A1 did not give detectable activities in the G1315 strain, but we were able to demonstrate the activity of both P450s in WAT11 and WAT21. Another crucial factor is the ability to detect small quantities of reaction products. In this respect the sensitivity of the GC-MS methodology used to detect intermediates of the GA biosynthesis pathway has been crucial to our success in demonstrating the activities of CYP70A1 and CYP88A family members as encoding *ent*-kaurene oxidase and *ent*-kaurenoic acid oxidase.

Author Index

Numbers in parentheses are footnote reference numbers and indicate that an author's work is referred to although the name is not cited in the text.

A

Abe, K., 123
Abramczyk, H., 111
Abrams, S. R., 359
Ackermann, J. M., 278, 279
Ackland, M. J., 138
Adachi, S., 16, 80, 85(19)
Adam, B.-L., 257
Adams, M. D., 249
Adams, P. D., 82
Adamson, D. J. A., 188, 194
Aebersold, R., 256
Afzelius, L., 138, 142(13)
Agundez, J. A., 61, 63
Akaike, T., 123
Akarsu, A. N., 54
Akashi, T., 360, 361, 361(6; 8; 9; 11), 363, 363(21)
Akhrem, A. A., 95
Akong, M., 333
Aktan, S. G., 54
Akwa, Y., 244
Alayrac, C., 354, 358, 359(11; 19)
Aldridge, T. C., 237
Alex, A. A., 138
Alexandersson, M., 255
Allen, A. O., 110
Alozie, I., 54
Alpen, E. L., 105(15), 106
Altaman, C. J., 261
Altschul, S. F., 173
Amano, M., 104, 115(8)
Amarneh, B., 123
Ames, B. N., 301, 304
Andersen, M. D., 333, 334, 335(15), 336, 337, 339(15), 340, 340(15)
Anderson, K. L., 54
Anderson, N. L., 250
Anderson, S. P., 237

Andersson, L. A., 129, 130(32)
Andersson, T. B., 133, 138, 142(13)
Antonarakis, S. E., 29
Anzenbacher, P., 145, 152, 153, 154, 156
Anzenbacherová, E., 152, 153, 154
Aoki, T., 360, 361, 361(6; 8; 11), 363, 363(9; 21)
Aoyama, T., 45, 46, 214
Appel, D., 123
Appel, R. D., 250
Arabidopsis Genome Initiative, 383, 387(6)
Archakov, A. I., 94, 95, 99, 101, 150, 152
Ariyoshi, N., 47, 59, 61, 63(7)
Arlotto, M. F., 301
Arold, N., 252
Arutyunyan, A. M., 107
Asada, A., 46
Asman, M., 162
Assmann, G., 255
Astle, W. F., 54
Ausubel, F. M., 385
Auwerx, J., 250, 255
Awni, W. M., 321
Axel, R., 84, 90(34)
Ayabe, S., 360, 361, 361(6; 8; 11), 362, 363, 363(9; 21)

B

Baba, T., 261, 306, 307, 309(24), 379
Back, D. J., 179
Bailey, J. E., 256
Bailey-Serres, J., 345
Baillie, T. A., 263, 264(24), 268(27), 269, 269(32), 270(32), 272(33; 39), 283
Bak, S., 3, 333
Baker, A. R., 114
Ballantine, S. P., 334
Balny, C., 145, 146, 147
Bamberg, K., 189

Subject Index

A

Adenovirus transduction, *see* CYP3A4
Adverse drug reactions
  cytochrome P450 alleles, 29
  economic impact, 28–29
  mortality, 28
AFM, *see* Atomic force microscopy
Alkaloid biosynthesis, P450s
  cell culture for induction
    Catharanthus roseus, 371–372
    Eschscholzia californica, 371–372
    overview, 370–371
  CYP71D12
    assay, 380–381
    expression in *Escherichia coli,* 378–379
    homology-based cloning, 374–375
  CYP80B1
    assay, 376–377
    expression in insect cells, 375–376
    homology-based cloning, 373–374
    microsome preparation and difference spectroscopy, 377–378
  inducers
    light, 371–372
    methyl jasmonate, 371–372
ALMOND
  CYP2C inhibitors, three-dimensional quantitative structure–activity relationship model, 142
  molecular interaction field calculation, 139, 142
Ames test
  Salmonella typhimurium tester strains, *see Salmonella typhimurium* P450 mutagenesis test
  species specificity, 301
AMoRe, molecular replacement, 83, 89–90
Antibody chip, proteomics, 257–258
Atomic force microscopy
  cytochrome P450 interaction analysis
    materials, 94–95, 101

protein preparation, 101–102
    redox partner complex imaging, 102
  principles, 94

B

BEAST, molecular replacement, 84, 89–91
BLAST
  CYP4A gene cluster mapping, 37, 39, 42–43
  GenBank searching, 3, 6–7, 9
  PSI-BLAST program for protein sequence alignment, 17–18
Branched DNA signal amplification assay
  applications, 170–171, 179
  chemiluminescence detection, 177
  kits and protocols, 175–177
  linearity, 179
  normalization of CYP gene expression levels, 177, 179
  principles, 170–172
  probe sets for human P450 genes
    capture extenders and specificity, 172–173
    CYP2D6 specificity for major alleles, 174–175
    synthesis, 172
    species-specific probe sets, 174

C

CAR, *see* Constitutive androstane receptor
CNS, molecular replacement, 82–83
Conservation index, amino acids in P450s, 25–28
Constitutive androstane receptor
  green fluorescent protein fusion protein
    advantages and limitations for liver targeting, 213
    chlorpromazine induction of nuclear translocation, 209
    construction, 209
    expression plasmid delivery to liver, 207–208
    fluorescence microscopy, 208, 210

ISBN 0-12-182260-5

90051
9 780121 822606